Aquafeed Formulation

Chapters

General Formulation Details

Key Points to Remember

Aquafeed Formulation

Edited by

Sergio F. Nates

AMSTERDAM • BOSTON • HEIDELBERG • LONDON
NEW YORK • OXFORD • PARIS • SAN DIEGO
SAN FRANCISCO • SINGAPORE • SYDNEY • TOKYO

Academic Press is an imprint of Elsevier

Academic Press is an imprint of Elsevier
The Boulevard, Langford Lane, Kidlington, Oxford OX5 1GB
225 Wyman Street, Waltham MA 02451

Notices
Knowledge and best practice in this field are constantly changing. As new research and experience broaden our understanding, changes in research methods, professional practices, or medical treatment may become necessary.

Practitioners and researchers may always rely on their own experience and knowledge in evaluating and using any information, methods, compounds, or experiments described herein. In using such information or methods they should be mindful of their own safety and the safety of others, including parties for whom they have a professional responsibility.

To the fullest extent of the law, neither the Publisher nor the authors, contributors, or editors, assume any liability for any injury and/or damage to persons or property as a matter of products liability, negligence or otherwise, or from any use or operation of any methods, products, instructions, or ideas contained in the material herein.

Library of Congress Cataloging-in-Publication Data
A catalog record for this book is available from the Library of Congress

British Library Cataloguing-in-Publication Data
A catalogue record for this book is available from the British Library

ISBN: 978-0-12-800873-7

For information on all Academic Press publications
visit our website at http://store.elsevier.com

  Working together
to grow libraries in
ELSEVIER Book Aid International developing countries

www.elsevier.com • www.bookaid.org

Publisher: Nikky Levy
Acquisition Editor: Patricia Osborn
Editorial Project Manager: Jaclyn Truesdell
Production Project Manager: Melissa Read
Designer: Greg Harris

Contents

List of contributors

Girish Channarayapatna Evonik (SEA) Pte. Ltd, Singapore
Girish Channarayapatna holds a Bachelor degree in Veterinary Science and Master degree in Poultry Science from University of Agricultural Sciences, College of Veterinary Medicine, Bangalore, India. He completed his PhD program in 2009 from University of Guelph, Animal and Poultry Science Department, Canada. His main research focus during his Master's and PhD program was on characterization and prevention of mycotoxicoses in poultry. He joined Evonik Industries in March 2010 and worked as the Technical Sales Manager for Asia South region for almost 4 years. He took over the new position as Director, Nutrition and Technical Sales for Asia South region from Jan 2014. He has published 6 scientific articles in peer-reviewed journals, 2 book chapters, 1 review article and several proceeding papers, abstracts and popular press articles.

Pedro Encarnação Biomin Singapore Pte Ltd, Singapore
Dr. Pedro Encarnação has an extensive background in aquaculture and nutrition. He has been involved in several research projects focusing on the improvement of feed formulations for aquaculture species and improving animal performance by the use of feed additives. He has an Honors Degree in Marine Biology and Fisheries and an MSc in Aquaculture from the University of Algarve (Portugal), and obtained is PhD in Animal Nutrition from the University of Guelph (Canada).
   He has been in Asia for more than 9 years as Biomin Director of Business Development for the Aquaculture industry.

Brett Glencross Ridley Agriproducts, Narangba, Queensland, Australia
Dr. Brett Glencross is the Technical Manager with the Aqua-Feed Division of Ridley, Australia's largest provider of animal nutrition solutions and products. He joined Ridley in March 2015 after 20 years in academia as a researcher. Prior to joining Ridley he was the Senior Principal Research Scientist for Aquaculture Feed Technologies research within the Aquaculture Research Program of the Commonwealth Scientific and Industrial Research Organisation (CSIRO) in Australia from 2009 to 2015. He has Honours and Masters Degrees in Biochemistry from the University of Western Australia and a PhD in Animal Nutrition from the University of Queensland.

Karthik Masagounder Evonik Industries AG, Hanau-Wolfgang, Germany
Karthik Masagounder has obtained his Bachelors in Fisheries Sciences and Masters
in Aquaculture from India and PhD from the University of Missouri-US focusing on
fish nutrition. After PhD, he continued his postdoctoral research in the US for another
2 years. He has published more than 15 papers including 8 peer-reviewed journal
papers. Karthik has been working with Evonik Industries in the Asia South region
as a Regional Technical Sales Manager from Jan 2012 until May 2015. Recently, he
moved to Germany and started as Research Manager in Aqua Nutrition with Evonik
Industries.

Zuridah Merican Aqua Research PLC, Singapore
Zuridah is the editor and publisher of *Aqua Culture Asia Pacific*, a magazine that
strives to be the beacon for the regional aquaculture industry. She is well known and
highly recognised by the public sector, academia and the industry in Asia, primarily and the rest of the world. She has been providing information to the Asia Pacific
industry for the past 11 years. She is based in Kuala Lumpur. She began her career in
aquaculture and fisheries with the Department of Fisheries, Malaysia in 1977. While
in Singapore, she carried out assessments on the mariculture industry and a review
on the feed industry in Asia. This was followed with consultancies on the industry
in Asia for European companies. When she was based in Europe, Zuridah was the
editor of UK based *International Aquafeed Magazine* for four years and then *Asian
Aquaculture* for three years when she returned to Asia.

César Molina-Poveda GISIS, Guayaquil, Ecuador
César Molina-Poveda has a B.Sc. with honors in Chemist, a M.Sc. in Shellfish
Biology, Fisheries and Culture from Bangor University (UK), and Ph.D. in Science
and Technology of Animal Production from Polytechnic University of Valencia
(Spain). He has also completed training courses at Mie University and the National
Research Institute of Aquaculture (both in Japan), and Artemia Referent Center at
Ghent University (Belgium). His expertise in shrimp and tilapia aquaculture has
built from 24 years' experience of applied research in both academic and industrial
background, which includes nutrition studies and grow-out production management.

Sergio F. Nates
Dr. Nates is currently the President of the Latin American Rendering Association and
the former President & CEO of the Fats and Proteins Research Foundation. He is also
the former Vice-Chairman of the Animal Co-Products Research and Education Center
(ACREC) at Clemson University. Dr. Nates is a member of the Board of Directors
of the Global Aquaculture Alliance and the Chairman of the Feed mill Certification
Committee – GAA.

Over the last twenty years of his career, Sergio has specialized in assisting the development of responsible fishing and aquaculture practices. He has also developed and
implemented comprehensive management and research programs worldwide; product
development, ingredient value models, formulation standards and quality assurance
programs. He is the author of several book chapters and over 100 publications.

Sheila Ramos Evonik (SEA) Pte. Ltd, Singapore
Sheila Ramos works for Evonik Industries as a Technical Sales Manager for the Asia South region. Sheila received her B.S. in Agriculture, majoring in Animal Science, and M.S. in Animal Nutrition from the University of the Philippines at Los Baños. She has been with Evonik for almost 11 years, and is popular in the region in helping the feed industries to improve their feed quality with Evonik's analytical services. Sheila has been a speaker in various conferences and has published more than 15 papers related to raw material and feed quality as well as advanced amino acid nutrition concepts.

Ingolf Reimann Evonik Industries AG, Hanau-Wolfgang, Germany
Dr. Ingolf Reimann works for Evonik Industries as Head of Analytical Services for the Animal Nutrition business. He received his degree in Analytical Chemistry from University of Duesseldorf, Germany. Ingolf has been with Evonik since 2002. His focus is to provide the feed industry globally with accurate and fast analytical services including AMINOLab and AMINONIR, the handling of analytical data and the extraction of value out of analytics.

Dagoberto Sanchez AppliedAquaNutrition Consulting, Lima, Peru
Dr. Dagoberto Sanchez is a Global Aqua Nutrition Consultant, located in Lima, Peru. He has 25 years of career in the aquaculture industry, feed formulation and manufacturing, with multifunctional experience in aquatic nutrition, R&D and aqua-farming. Prior to being a global consultant he supported Skretting as Latin America Business Development Manager and worked for 15 years as the Nutrition and R&D Director in Alicorp (Nicovita). He has an MBA in Peru and Masters Degrees in Mariculture from Texas A&M University and a PhD in Animal Nutrition from Texas A&M University in the USA.

A. Victor Suresh United Research (Singapore) Pte. Ltd, Singapore
Victor Suresh learned feed formulation at his first job with the Ralston Purina Company that he joined after completing his PhD in aquaculture at Southern Illinois University, Carbondale. In the course of his 20 year old professional career he has formulated feeds for aquatic species farmed in Asia and the Americas and has used four different software packages for feed formulation. He presently heads United Research, a firm specializing in formulation and R&D services in the feed sector, and lives in Singapore.

Acknowledgments

I would like to express my sincere gratitude to the authors. Their efforts to produce exceptional manuscripts and exciting contributions lie within the cover of this book. To Dr Victor Suresh for always listening and giving me words of encouragement. Thanks are due to all members of the staff of Elsevier involved in the preparation of this book.

Finally, no acknowledgments by me would be complete without a thank you to Dr Tom Zeigler, who introduced me to the field of aquaculture nutrition, and whose passion and leadership had lasting effects.

Sergio F. Nates

Introduction

Sergio F. Nates

I.1 Introduction

Rapid growth of aquaculture worldwide has become increasingly dependent upon the use of external feed inputs, and in particular upon the use of compound aquafeeds. In addition, changes in production technology and marketing, and changes in feed ingredients are key structural changes necessary for the aquaculture sector to grow (Tacon et al., 2011).

The aquaculture feed industry is responsible for converting raw materials of agricultural origin into feeds. These feeds are not only important in terms of cost but also in terms of nutrition, as some of these feeds are the primary source of animal and plant protein required by cultivated aquaculture species for normal development. In addition, this is a broad industry employing people with a variety of skills, including process engineers, economists, marketing experts, shellfish and fish scientists, regulatory experts, quality control technicians, and transportation and distribution specialists.

Feed is the largest single cost item in aquaculture production and since it accounts for 50–60% of the total cost, any saving on feed, though small, may greatly reduce the total cost and increase returns. Formulating aquaculture feeds requires the use of combinations of several ingredients since most feedstuffs have been shown to have significant nutrient and functional limitations and cannot be used individually at very high levels in the diets of most aquaculture species. Adopting local ingredient alternatives for the formulation of an aquafeed mix is a logical step for aquaculture producers to remain profitable. Challenges and obstacles include material availability, farming, initial cost competitiveness, and handling and processing. Consequently, feed formulation is an important aspect of the aquaculture industry and accurate formulation must be overcome before alternative aquaculture feed formulas can be fully developed successfully.

I.2 Feed ingredients

Use of locally available raw materials as ingredients in aquaculture feed contributes to a sustainable utilization of resources as well as potential growth in aquaculture production with less environmental impact. In addition, the evaluation of feed ingredients is critical to feed development. It is vital to discriminate the effects on feed intake from the effects on the utilization of nutrients from ingredients for growth and other metabolic processes (Glencross et al., 2007).

Ingredients used in practical aquaculture diets can be classified as protein (amino acid) sources; energy sources; essential lipid sources; vitamin supplements; mineral supplements; and special ingredients to enhance growth, pigmentation, or sexual development in the species, or to enhance physical properties, palatability, or preservation of the feed (Hardy, 2000).

The number of ingredients used to feed aquaculture species in different countries is very high. Moreover, there is considerable variability for each ingredient considering both its chemical composition and its nutritional value, as a result of factors associated with its production or processing. This situation justifies the development of national reference tables adapted to the specificities of each production system.

The ingredients studied are organized into the following groups:

1. Cereal grains and by-products
2. Fruits and tubers
3. Molasses
4. Vegetable protein concentrates
5. Fibrous foods
6. Concentrated animal protein
7. Fats, oils, and glycerin
8. Minerals and micro-ingredients

Likewise, all ingredients have advantages and disadvantages within a formulation. The advantages are normally associated with nutritional contributions, availability of essential nutrients (amino acids and lipids), and the disadvantages are related to antinutritional factors, the presence of contaminants, the presence of molds and the possible production of mycotoxins, low or variable quality, poor digestibility, susceptibility to oxidation, costs, availability, and sustainability.

With the significant exception of soybean meal, plant protein sources are generally nutritionally imbalanced in terms of essential amino acids, particularly lysine, and the first limiting amino acid in cereals. Unless supplemented with animal protein sources and crystalline amino acids, plant-based diets may not meet the requirements for critical amino acids for most commercial aquaculture species, especially carnivorous ones.

On the other hand, animal protein ingredients are normally used to balance the amino acid contents of diets and to increase protein content of the final feed. In many countries, feed manufacturers ensure that animal protein ingredients do not fall below minimum levels in fish and shrimp diets, especially for the larvae and juvenile stages whose amino acid requirements are high. The requirements for essential amino acids are progressively reduced as animals grow older, and it is possible to meet the needs of adults with diets containing lower levels of animal protein and relatively higher levels of plant protein. Fish meal, poultry meal, feather meal, and blood meal are the animal protein sources most widely used in aquaculture diets.

There is also a vital need to seek effective ingredients that can either partially or totally replace fish meal and other ingredients as protein sources in aquafeeds. Algal products can be used to enhance the nutritional value of food and animal feed owing to their chemical composition (Dewi et al., 2014); they play a crucial role in aquaculture (Jamali and Ahmadifard, 2015).

I.3 Nutritional requirements

The nutrient balance of feed ingredients influences feed utilization and growth of aquaculture species, and there is generous information on the nutrient content of feed ingredients produced by different industries worldwide. However, many requirements are, at best, only rough approximations of the optimum amounts of nutrients for practical diets to grow aquaculture species to harvestable size. Management, environmental factors, and size can have an effect on dietary nutrient levels for optimum performance. Nevertheless, nutrient requirement data that are available serve relatively well as a basis to formulate highly productive, economical diets for commercial aquaculture. In formulating a diet for a species where nutrient requirements are not known, the requirements for a related species whose nutrient requirements are known can be used. Generally, most variation of nutrients required among classes should be expected between warm- and coldwater species, fresh- and saltwater species, and finfishes and crustaceans.

It is essential to know the nutritional requirements, particularly for protein, lipid, and energy, for optimum growth of the species as well as in formulating a balanced diet. Improper protein and energy levels in feed increases production cost and deteriorates water quality. Insufficient energy in diets causes protein waste due to the increased proportion of dietary protein used for energy and the produced ammonia can reduce the water quality. Moreover, feed ingredients should deliver the necessary nutrients in amounts to meet the requirements of aquaculture species (Tacon, 1987). However, the amounts of total amino acids contained in feed ingredients are often much higher than the amounts that are digestible. Feed formulations based on digestible amino acids have been shown to increase body weight gain and feed intake and can improve body composition.

According to a study published by FAO (2003) protein is the key building block for feed formulation systems and the main and most expensive component of feeds (Shiau, 1998). Besides, the concept of "an ideal protein" used as a method of determining the essential amino acid requirements of fish species was first suggested by Wilson (1989). The dietary essential amino acid requirements determined using the ideal protein concept (based on the whole-body essential amino acid pattern) can serve as a valuable index for formulating the diets of cultivated aquaculture species until their dietary essential amino acid requirements are empirically established using amino acid test diets (Wilson, 1991).

Qualitative amino acid requirements appear identical for all fish species examined; arginine, histidine, isoleucine, leucine, lysine, methionine, phenylalanine, threonine, tryptophan, and valine are all required for normal growth and metabolism (Wilson and Halver, 1986). On the other hand, several studies have shown that shrimp require 10 essential amino acids: arginine, methionine, valine, threonine, isoleucine, leucine, lysine, histidine, phenylalanine, and tryptophan (Lim and Akiyama, 1995; Lovell, 2002; Fox et al., 2006).

Lipids are a group of natural organic compounds comprising fats, oils, phospholipids, and sterols. Dietary lipids are utilized in fish as a major energy source to spare proteins, provide essential fatty acids needed for proper functioning of many physiological processes and maintenance of membrane fluidity and permeability as well as for growth and survival. Dietary lipids also influence the flavor and texture of prepared feeds and flesh quality (Stickney and Hardy, 1989). In addition, dietary lipids are a highly digestible and concentrated source of energy that supplies 8–9 kcal/g, about double that supplied by either carbohydrate or protein (Chuang, 1990).

Carbohydrates are a source of cheap energy, but the ability of aquatic organisms, including shrimp and fish, to utilize them is limited (Shiau, 1997). This is due to a lack of ability to digest and regulate plasma glucose concentrations (Zainuddin and Aslamyah, 2014). The use of carbohydrates by fish and shrimp is less efficient than by land animals (Mohapatra et al., 2003). In addition, fish are known to have a limited ability for digestion and metabolism of carbohydrates and hence, excessive intake of this nutrient may result in nutritional problems (Hemre et al., 2002). Excess carbohydrates reduce the growth rate and are often accompanied by poor feed utilization (Zhou et al., 2015).

I.4 Feed ingredient testing

Proximate analysis is usually the first step in the chemical evaluation of a feed ingredient, where the material is subjected to a series of relatively simple chemical tests so as to determine the content of moisture, crude protein, lipid, crude fiber, ash, and digestible carbohydrate.

In vivo methods for feed ingredient testing require time-consuming trials with many live animals. In vitro methods, however, allow the quick assessment of nutritional value as well as the potential negative effects of any antinutritional compounds in the test material. In vitro pH-stat determination of digestibility is a promising new test technique. The new in vitro pH-stat assay simulates the digestion of a protein source by the enzymes of the target animal. Digestible energy cost per unit energy is the dominant cost pressure in the formulation of aquaculture diets, so it is pertinent to focus on the development of an assay for the assessment of available energy. As digestible energy is the easiest available energy parameter to measure in feed ingredients for shrimp and fish, it is a logical parameter on which to focus for this type of analysis.

On the other hand, many in vitro assays have been used with varying degrees of success to evaluate protein and ingredient quality, including the potassium hydroxide solubility test, nitrogen water solubility test, urease assay, and pepsin digestibility assay. Of these, probably the most common and rapid in vitro digestibility test for measuring protein quality is the pepsin digestibility assay that dates back to the early 1950s. A considerable body of research data relating protein to secondary productivity return in livestock has amassed since the mid-1940s. Yet, as good a predictor of productivity as the pepsin digestibility test is, when protein levels are constant but animal protein sources vary, productivity differences are seen.

Some methods are better at predicting digestible amino acids in vegetable and animal proteins. Protein analyses of animal excreta have shown that less-productive animals excrete higher levels of protein than more-productive animals. The quantity, not the digestibility, of the protein is expressed in such tests. Indeed, digestibility not only varies by source of protein but also within a category. For example, one fish meal can be more digestible than another. As a result, the original "0.2%" pepsin method came about. However, studies have shown that pepsin digestibility analysis turns out higher digestibility rates than metabolic studies suggest, although more dilute concentrations have been used in an attempt to correct for this difference. In essence, there is no reliable mathematical relationship between the digestibility of one pepsin dilution and another.

The evaluation of aquafeed ingredients may benefit from recent advances in methodologies applied to the in vivo and in vitro measurement of digestibility in feeds for terrestrial animals (Cruz-Suárez et al., 2007; Lemos et al., 2009). One promising technique is the pH-stat in vitro determination of digestibility of feeds and feed ingredients. The assay simulates digestion of a protein source by the enzymes of the target animal. A significant correlation between the pH-stat in the in vivo and in vitro digestibility values exists when proteins from the same animal or plant origin are compared (Lemos and Nunes, 2007). The relatively low-cost method provides accurate results, is not environmentally affected, and enables a higher number of analyses than live animal experiments for a given time (Lemos et al., 2000).

Important to note is that some ingredients have a unique set of amino acids, along with other unwanted compounds, that contributes to the ration. The unwanted compounds, called antinutritive substances, can interfere with the digestion of the amino acids and therefore reduce the value of the ingredients. Moreover, predictions of the digestibility of ingredients can be inaccurate because the relationship between in vivo and in vitro digestibility can be different. The physical structure of ingredients can be partially inaccessible to enzymatic action or the presence of antinutritive substances. Enzymes split protein at specific junctions, but some antinutritive substances block these junctions, preventing proper digestion and reducing the value of the ingredient in a feedstuff. Cross-linkage formation reduces the rate of protein digestion, possibly by preventing enzyme penetration or blocking the sites of enzyme attack. Another interaction that can affect the results of digestibility analyses is the formation of complexes between starches and lipids. Such formations, which can occur in situ in the digestive tracts of several aquatic species, are thought to decrease digestibility and response to ingested carbohydrates.

1.5 Feed additives

Many feed additives are available that can improve fish and shrimp growth performance. Products that improve feed efficiency are particularly important since feed costs are a major expense in aquaculture production. Proper use of these products can improve aquaculture profits. Feed additives may be both nutritive and nonnutritive and work by either direct or indirect methods on the animal's system. Many of the products influence different systems and, therefore, the effects of one can be additive to another.

Attempts to use natural materials such as medicinal plants are widely accepted as feed additives to enhance the efficiency of feed utilization and aquaculture productive performance. Recently, medicinal plants and probiotics have been reported as potential alternatives, among other feed additives, to antibiotics in aquaculture diets (Dada and Olugbemi, 2013).

Phytogenics comprises a wide range of substances and thus has been further classified according to botanical origin, processing, and composition. Phytogenic feed additives include herbs, which are nonwoody flowering plants known to have medicinal properties; spices, which are herbs with an intensive smell or taste, commonly added to human food; essential oils, which are aromatic oily liquids derived from plant materials such as flowers, leaves, fruits, and roots; and oleoresins, which are extracts derived by nonaqueous solvents from plant material (Jacela et al., 2010).

Some additives, amino acids and their metabolites, and vitamins are important regulators of key metabolic pathways necessary for feed intake, nutrient utilization, maintenance, growth, immunity, behavior, larval metamorphosis, reproduction, and resistance to environmental stressors and pathogenic organisms in various aquaculture species (Tincy et al., 2014).

Nutraceuticals can act as buffering agents in biological systems by reducing the deleterious effects of stressors and by improving growth. High dietary protein supplementation has an enhancing effect against different stressors. Dietary supplementation of different vitamins (e.g., vitamins C and E) can mitigate stress in shellfish and finfish. Supplementation of nutraceuticals can also help in mitigating multiple stressors, including temperature, salinity, and exposure to pesticides, and augmented growth and modulated nonspecific immune functions (Manush et al., 2005).

Organic acids and their salts are generally regarded as safe and have been approved to be used as feed additives in animal production. The use of organic acids has been reported to protect shrimp and fish by competitive exclusion, enhancement of nutrient utilization, and growth and feed conversion efficiency (Romano et al., 2015). The organic acids in no dissociated form can penetrate the bacteria cell wall and disrupt the normal physiology of certain types of bacteria.

I.6   Feed formulation

Ingredients could be chosen from a well-known list when preparing a supplemental feed, so that a feed mixture having the desired crude protein content is obtained.

In addition, with increased computer capabilities and improved software, feed rations can be calculated in almost unlimited numbers. During the past few years, computer use has progressively increased to minimize the time needed for calculation of ration formulation (Al-Desseit, 2009). The use of linear programming for determining the least-cost formulation of feed based on current market prices and small changes in relative prices can cause significant changes in demand for available feed ingredients. The application of the programming techniques for feed formulation parallel to the introduction of intensive systems of animal production in many countries is essential. Developing extensive nutrient databases (based on wide-ranging research) can give a competitive advantage to a feed company (Gosh et al., 2011).

Linear programming to minimize feed cost with respect to a set of restrictions (requirements and ingredient minima and maxima) was developed in the 1950s (Baum et al., 1953). The techniques presently applied to feed formulation are essentially unchanged. In addition to finding the ingredient mixture to meet diet specifications at minimal cost, the sensitivity analysis option of Excel's Solver routine can be used to determine the shadow prices of ingredients (Waldroup (mimeo, n.d.)). Shadow prices are the highest prices at which the ingredient would be included in the solution. Shadow prices can also be the low cost required for ingredients that are too expensive to come into the formula. Shadow prices can be used to develop usage curves relating ingredient prices to the amount of ingredient that would be used in a particular feed formula (Udo et al., 2011).

I.7   Feed production and quality

Balanced aquaculture feed production is a process where multiple variables are involved: raw materials, nutritional formulations, transportation, market performance, and we might even include climate, which undoubtedly also regulates the agroindustrial activities that delineate the supply and prices of agricultural by-products. However, when put in place, the key factor is the feed mill operation.

Feed mill operations are dynamic systems. Each element of that system and all their functions are interrelated to achieve one objective: a safe and well-balanced finished feed. Some of the factors that affect this dynamic could be unmanageable as market policies, and others as simple as adjusting screws. Nonetheless, as the aquaculture industry becomes increasingly conscious of costs and benefits, it is searching for more "functional" feeds, many of which are augmented with key ingredients and compounds that promote animal growth and survival. In addition, plant proteins are increasingly used as alternatives to proteins from animal sources. The optimum production of feedstuffs with optimum dry matter conversion of feed to weight depends largely on ingredient quality and nutrient availability for the species in question. The determination of digestibility of major nutrients is one of the main steps in the evaluation of their bioavailability for a given species. Measuring digestibility protein is the most important feed nutrient for aquaculture of high-value animals.

I.8   Best practices in formulation

Associated with the rapid growth of aquaculture, new intensive cultivation techniques have been used which generally have greater environmental impact than traditional culture techniques. However, this rapid increase occurred at a time in which the general public has an increased level of concern about the environmental consequences of human activity. In this regard, the production of aquaculture food products continues to be positive but it should consider the possible side effects of their increased activity on animal welfare and their impact on the environment. Thus the four pillars that will hold the aquaculture production venture are: food safety and quality, health and animal welfare, environmental integrity, and social responsibility (Tucker and Hargreaves, 2008).

In addition, several management techniques can be used in feed preparation, handling, and delivery that can affect animal performance and nutrient excretion, consequently affecting the surrounding environment. For instance, pelleting and reducing the particle size (grinding) of a ration increases the digestibility of the ration for aquaculture species, improves N and P utilization and reduces excretion by 5–15% each (Turcios and Papenbrock, 2014). With respect to the pollution generated by aquaculture, nitrogen and phosphorus are considered as waste components of fish farming, causing serious environmental problems. In addition, several fish excrete nitrogenous waste products by diffusion and ion exchange through the gills, urine, and feces. Decomposition and reuse of these nitrogenous compounds is especially important in aquaculture using recirculation systems due to the toxicity of ammonia and nitrite and the chance of hypertrophication of the environment by nitrate (Brown et al., 1999).

Several new technologies are being developed and tested to enhance the nutrient content or utilization of feed ingredients, or to alter the availability of nutrients in current commercial feeds. This includes enzymes, genetically modified feed ingredients, and feed processing technologies to enhance the availability of nutrients to meet the needs of specific animals and reduce excretion of nutrients. These specialty feeds and new technologies will provide nutrients in a proper balance that will allow "precision-feeding" of aquaculture species.

References

Al-Desseit, B., 2009. Least-cost broiler ration formulating using linear programming technique. J. Anim. Vet. Adv. 8 (7), 1274–1278.

Baum, E.L., Fletcher, H.B., Standelman, W.J., 1953. An application of profit maximizing techniques to experimental input–out data. Poult. Sci. 32, 378–381.

Brown, J.J., Glenn, E.P., Fitzsimmons, K.M., Smith, S.E., 1999. Halophytes for the treatment of saline aquaculture effluent. Aquaculture 175, 255–268.

Chuang, J.L., 1990. Nutrient requirements, feeding and culturing practices of *Penaeus monodon*: a review. F. Hoffmann-La Roche Ltd, Basel, 62 pp.

Cruz-Suárez, L.E., Nieto-Lopez, M., Guajardo-Barbosa, C., Tapia-Salazar, M., Scholz, U., Ricque-Marie, D., 2007. Replacement of fish meal with poultry by-product meal in practical diets for *Litopenaeus vannamei*, and digestibility of the tested ingredients. Aquaculure 272 (1–4), 466–476.

Dada, A.A., Olugbemi, B.D., 2013. Dietary effects of two commercial feed additives on growth performance and body composition of African catfish *Clarias gariepinus* fingerlings. Acad. J. Food Sci. 7 (9), 325–328.

Dewi, A.P.W.K., Nursyam, H., Hariati, A.M., 2014. Response of fermented *Cladophora* containing diet on growth performances and feed efficiency of Tilapia (*Oreochromis* sp.). Int. J. Agron. Agric. Res. 5 (6), 78–85.

FAO, 2003. Health management and biosecurity maintenance in white shrimp (*Penaeus vannamei*) hatcheries in Latin America. FAO Fisheries Technical Paper No. 450. Rome, FAO, 70 pp. Available at: <http://www.fao.org/docrep/007/y5040e/y5040e00.htm>.

Fox, J.M., Davis, D.D., Wilson, M., Lawrence, A.L., 2006. Current status of amino acid requirement research with marine penaeid shrimp. In: Cruz Suárez, L.E., Ricque Marie, D., Salazar, M.T., Nieto López, M.G., Villarreal Cavazos, D.A., Puello Cruz, A.C., et al. (Eds.), Avances en Nutrición Acuícola VIII. VIII Simposium Internacional de Nutrición Acuícola. 15–17 Noviembre Universidad Autónoma de Nuevo León, Monterrey, Nuevo León, México, pp. 182–196.

Gosh, D., Sathianandan, T.V., Vijayagopal, P., 2011. Feed formulation using linear programming for fry of catfish, milkfish, tilapia, Asian sea bass, and grouper in India. J. Appl. Aquacult. 23, 85–101.

Glencross, B.D., Booth, M., Allan, G.L., 2007. A feed is only as good as its ingredients – a review of ingredient evaluation strategies for aquaculture feeds. Aquacult. Nutr. 13, 17–34.

Hardy, R.W., 2000. New developments in aquatic feed ingredients and potential of enzyme supplements. In: Crus-Suárez, L.E., Ricque-MArie, D., Tapia-Salazar, M., Olvera-Novoa, M.A.Y., Civera-Cerecedo, R. (Eds.), Avances en Nutrición Acuícola V Memorias del Simposium Internacional de Nutrición Acuícola, Mérida, Yucatán, Mexico, pp. 216–226.

Hemre, G.I., Mommsen, T.P., Krogdahl, Å., 2002. Carbohydrates in fish nutrition: effects on growth, glucose metabolism and hepatic enzymes. Aquacult. Nutr. 8, 175–194.

Jacela, J.Y., DeRouchey, J.M., Tokach, M.D., Goodband, R.D., Nelssen, J.L., Renter, D.G., et al., 2010. Feed additives for swine: fact sheets – prebiotics and probiotics, and phytogenics. J. Swine Health Prod. 18 (3), 132–136.

Jamali, H., Ahmadifard, N., 2015. Evaluation of growth, survival and body composition of larval white shrimp (*Litopenaeus vannamei*) fed the combination of three types of algae. Int. Aquat. Res. 7, 115–122.

Lemos, D., Nunes, A.J.P., 2007. Prediction of culture performance of juvenile *Litopenaeus vannamei in vitro* (pH-stat) degree of feed protein hydrolysis with species specific enzymes. Aquacult. Nutr. 14 (2), 181–191.

Lemos, D., Ezguerra, J.M., Garcia-Carreño, F.L., 2000. Protein digestion in penaeid shrimp: digestive proteinases, proteinase inhibitors and feed digestibility. Aquaqculture 186 (1), 89–105.

Lemos, D., Lawrence, A.L., Siccardi, A.J., 2009. Prediction of apparent protein digestibility of ingredients and diets by *in vitro* pH-stat degree of protein hydrolysis with species-specific enzymes for juvenile Pacific white shrimp *Litopenaeus vannamei*. Aquaculture 295 (1), 89–98.

Lim, C., Akiyama, D.M., 1995. Nutrient requirements of penaeid shrimp. In: Lim, C., Sessa, D.J. (Eds.), Nutrition and Utilization Technology in Aquaculture AOCS Press, Champaign, IL.

Lovell, R.T., 2002. Diet and fish husbandry. In: Halver, J.E., Hardy, R.W. (Eds.), Fish Nutrition Academic Press, San Diego, CA, pp. 500.

Manush, S.M., Pal, A.K., Das, T., Mukherjee, S.C., 2005. Dietary high protein and vitamin C mitigate stress due to chelate claw ablation in *Macrobrachium rosenbergii* males. Comp. Biochem. Physiol. A 142, 10–18.

Mohapatra, M., Sahu, N.P., Chaudhari, A., 2003. Utilization of gelatinized carbohydrates in diets of *Labeoro hita* fry. Aquacult. Nutr. 9, 189–196.

Romano, N., Chick-Boon, K., Wing-Keong, N., 2015. Dietary microencapsulated organic acids blend enhances growth, phosphorus utilization, immune response, hepatopancreatic integrity and resistance against *Vibrio harveyi* in white shrimp, *Litopenaeus vannamei*. Aquaculure 4 (35), 228–236.

Shiau, S.Y., 1997. Utilization of carbohydrates in warmwater fish – with reference to tilapia, *Oreochromis niloticus* × *O. aureus*. Aquaculture 151, 79–96.

Shiau, S.Y., 1998. Nutrient requirements of penaeid shrimps. Aquaculture 164 (1), 77–93.

Stickney, R.R., Hardy, R.W., 1989. Lipid requirements of some warm water species. Aquaculture 79, 145–156.

Tacon, A.G.J., 1987. The nutrition and feeding of farmed fish and shrimp – a training manual. 1. The essential nutrients. GCP/RLA/075/ITA, Field Document 2. FAO, Rome. 117 pp.

Tacon, A.G.J., Hasan, M.R., Metian, M., 2011. Demand and supply of feed ingredients for farmed fish and crustaceans: trends and prospects. FAO Fisheries and Aquaculture Technical Paper No. 564. FAO, 87 pp.

Tincy, V., Mishal, P., Akhtar, M.S., Pal, A.K., 2014. Aquaculture nutrition: turning challenges into opportunities. World Aquacult. 45 (2), 67–69.

Tucker, C.S., Hargreaves, J.A., 2008. Environmental Best Management Practices for Aquaculture. Blackwell Publishing, Ames, IA, p. 592.

Turcios, A.E., Papenbrock, J., 2014. Sustainable treatment of aquaculture effluents – What can we learn from the past for the future? Sustainability 6, 836–856.

Udo, I.U., Ndome, C.B., Asuquo, P.E., 2011. Use of stochastic programming in least-cost feed formulation for African catfish (*Clarias gariepinus*) in semi-intensive culture system in Nigeria. J. Fish. Aquatic Sci. 6 (4), 447.

Waldroup, P.W., n.d. Using the MPS/360 linear programming system for feed formulation. Department of Animal Science, University of Arkansas, Fayetteville.

Wilson, R.P., 1989. Protein and amino acid requirements of fishes. In: Shiau, S.Y. (Ed.), Progress in Fish Nutrition. Marine Food Science Series No. 9 National Taiwan Ocean University, Keelung, Taiwan, pp. 51–76.

Wilson, R.P., 1991. Amino acid nutrition of fish: a new method of estimating requirement values. In: Collie, M.R., McVey, J.P. (Eds.), Proceedings of the US-Japan Aquaculture Nutrition Symposium. Newport, OR, October 28–29, 1991, pp. 49–54.

Wilson, R.P., Halver, J.E., 1986. Protein and amino acid requirements of fishes. Annu. Rev. Nutr. 6, 225–244.

Zainuddin, H., Aslamyah, S., 2014. Effect of dietary carbohydrate levels and feeding frequencies on growth and carbohydrate digestibility by white shrimp *Litopenaeus vannamei* under laboratory conditions. J. Aquacult. Res. Dev. 5 (6), 1000274.

Zhou, C., Xianping, G., Bo, L., Jun, X., Ruli, C., Mingchun, R., 2015. Effect of high dietary carbohydrate on the growth performance, blood chemistry, hepatic enzyme activities and growth hormone gene expression of Wuchang bream (*Megalobrama amblycephala*) at two temperatures. Asian-Australas. J Anim Sci. 28 (2).

Overview of the aquaculture feed industry

Zuridah Merican[1] and Dagoberto Sanchez[2]
[1]Aqua Research PLC, Singapore [2]AppliedAquaNutrition Consulting, Lima, Peru

1. Aquafeed in Asia

1.1 Introduction

In general, aquafeed producers in Asia have important supporting roles in the development of aquaculture in Asia. The feed industry was a major driver in commercial shrimp farming in Asia in the 1980s, which was introduced into Asia by Taiwanese feed companies. Aside from feed supply, feed producers determine how aquaculture will develop in a particular country through the development of culture technology, provision of technical services and in disease management. Some feed companies support grow-out farmers with seedstock, either from their own or associated hatcheries, and assist farmers in selling their harvests. Often the role of aquafeed producers is not limited to small-scale farms. Inevitably, the industry has been affected by the ups and downs within the production chain, such as the drop in shrimp production due to early mortality syndrome (EMS) in the last 4 years.

The industry in Asia is highly competitive, with a large number of players producing feeds for several species. Aside from pelleted feeds for the two species of marine shrimp, Indonesian feed companies produce extruded (floating, slow sinking, and sinking) and pelleted feeds for the common carp, tilapia, pangasius catfish, Clarias catfish, gouramy, milkfish, seabass, grouper, and a general freshwater fish feed. In China, feeds for marine fish include those for the yellow croaker, cobia, Asian seabass, seabream, red drum, flounder, and groupers. As the industry consolidates, the rate of growth is slowing. Success depends on volume production but this can be limited as the industry is intertwined with changes in the production chain and threats of diseases.

The opportunity for future expansion is country-specific. In China, producers are optimistic on a 10% growth rate to 2017, with demand for aquaculture products linked with population increases (Dong et al., 2013). In several countries, demand for pelleted and extruded fish feeds will rise as fish farmers continue to change from using farm-made or mash feeds as well as trash feeds to commercially produced pelleted or extruded feeds for freshwater and marine fish. Feeds for the marine fish sector are an opportunity for feed producers in most countries. However, the sector also needs to move up from being small scale to industrialized farming before the benefits of economies of scale for feed producers can be realized.

S. Nates (Ed): Aquafeed Formulation. DOI: http://dx.doi.org/10.1016/B978-0-12-800873-7.00001-4

This section covers the changes in the aquafeed industry in Asia in recent years. Official data on feed production and actual feed sales are not available for most countries and in this material, the authors relied on information from industry sources. The section describes the dynamics in the industry in recent times, followed by some industry developments in the leading producing countries.

1.2   A shift in equilibrium

In shrimp feeds, the onset of EMS, which was first reported in China in 2009, Vietnam in 2010, Malaysia in 2011, and Thailand in 2012 (Flegel et al., 2014), changed the equilibrium in the feed sector in these countries. Prior to 2010, shrimp feed volumes were growing at a CAGR of 9.6% in Thailand (calculated from data in Thongrod, 2013), 7.5% in Vietnam, and 11.8% Malaysia (AAP, 2008, 2009, 2010). This was the result of the surge in feed demand when farmers shifted to farming *Penaeus vannamei* shrimp at higher densities (from 80 post larvae $(PL)/m^2$) in comparison to that for the *Penaeus monodon* which average at $30 PL/m^2$. This led to an expansion of feed capacity to match the increase in shrimp production.

In Thailand, vannamei shrimp farming started in 2004. Aramsiriwat (2012) showed that the rate of growth in shrimp feed production averaged 9.7% per year from 2004 to 2012. According to Thongrod (2013) shrimp production rose from 530,000 tonnes in 2007 to 600,000 tonnes in 2011 and in parallel with this growth, feed volumes rose from 901,000 tonnes in 2007 to 1.02 million tonnes in 2011.

1.3   Lower demand with EMS

With EMS, feed markets shrank in affected countries as farmers failed in crossing the 40 days of culture. While some farmers stopped culturing, others consolidated their feed purchases, giving them more negotiating power over the feed mills. With increased competition, feed mills were desperate to maintain sales to reach their breakeven volumes. At the same time, costs were rising as feed mills had little negotiating power over commodity supplies, dictated by market prices. The results of EMS were unpredictable feed demand and sales; more frequent sales of starter feeds as farmers lost crops; and restarted culture cycles. This is accompanied by generally lower sales volumes of grower and finisher feeds. For each cycle, feed volumes were significantly reduced. Although there are no published data on feed sales, it is presumed that emergency harvests of shrimp size 100/kg (100 g/shrimp) also brought down shrimp production volumes and thus overall feed usage. To secure markets, some feed mills try to integrate by farming themselves while others have built alliances with farmers and developed contract farming. Wilson (2013) reported a 10% lower demand in 2012 as compared to 2011 at 740,000 tonnes. However, anecdotal reports from Thailand showed many feed mills running at only 40% of capacity in 2013. Malaysia had a small shrimp feed market at 125,000 tonnes in 2010 but this declined to 60,000 tonnes in 2013.

A major consequence of EMS is the decoupling of feed and post larvae. In the early 2000s, Charoen Pokphand Thailand introduced the purchase of feed with post larvae, otherwise known as bundling of feeds. This was popular especially when the bundling was of post larvae, genetically selected for faster growth and specific feeds, prerequisite for fast growth. During this time, post larvae were also in short supply and other feed companies modified this marketing strategy supplying post larvae from their own or affiliated hatcheries. Among the several causes of EMS is post larvae quality, and therefore farmers began to move away from this bundling concept.

1.4 Horizontal expansion and integration

After repeated crop failures with EMS, there has been a shift to farming fish, such as the Asian seabass and tilapia, in shrimp ponds. Feed mills encouraged this and even helped in marketing the fish. In this way, the feed producer continues the relationship with feed clients and has feed sales.

In freshwater fish production, the majority of farms are small-scale and hence have little negotiating power over the feed mills. There is strong competition among large and small feed millers. On the other hand, feed mills are at the mercy of farmers who stop farming when ex-farm prices are low and resume when fish prices rise. This was the case in 2013 with the tilapia and Asian seabass farmers in Malaysia and tilapia farmers in China. In Vietnam, yearly, pangasius farmers grapple with low ex-farm prices, sometimes at below the cost of fish production when supply is high.

In the last 5 years, the pangasius sector in Vietnam has seen both consolidation and integration. Processing companies have taken over smaller farms and feed mills with the objectives of building a complete supply chain from hatchery to production and processing and fulfilling sustainability and to facilitate the certification requirements of importers. Through feed subsidiaries, the processing company controls feed supply to their contract farmers. The effect is that independent feed producers are being squeezed out of the pangasius feed supply chain. The recent expansion into tilapia farming balances the lower demand for pangasius feeds for independent feed millers but these large integrators are also launching large-scale tilapia farming to stabilize their business (Merican, 2012, 2014a,b).

1.5 Extruded and functional feeds

A small number of feed companies such as Betagro (Thailand), Indian Bioler (India), Lucky and Uni President Enterprises (Taiwan) are producing extruded shrimp feeds. However, the field success of these feeds has been constrained by higher prices. Extruded feed costs 10–15% more in comparison to conventional pelleted feeds. However, extruded feeds can be functional feeds with the addition of probiotics and other additives which will further raise the selling price. According to A. Bhaskar (personal communication) farmers in India prefer extruded starter feeds with less fines. However, some issues with extruded feeds include the floating problem of 5–10% that has dampened demand for such feeds. This requires reworking of the formula and extruder settings to overcome this but, ultimately, the feed producer needs to regain the confidence of users with extruded feeds.

The use of autofeeder technology in shrimp farming, which originated in Thailand, is changing the types of feeds being used. The preferred physical characteristics of feeds for autofeeder have changed to avoid breakages in the feeder. Dust levels must be low and water stability is less important because shrimp pick up pellets quickly. In Thailand, where a significant number of farms are feeding shrimp using autofeeders, Wilson (2013) said that farmers use the large crumbles and small pellets only. In Malaysia, the preference is for extruded starter feeds instead of crumbles for the early culture period.

1.6   Rising production costs

Throughout the region, rising costs of raw materials to unprecedented levels posed a tremendous challenge to formulators. Since June 2012, prices rose by 20% for three major commodities: soybean meal, corn, and wheat. Fish meal prices have been on the rise too. Dong et al. (2013) reported that fish meal takes up 50% of the formulation costs for shrimp feeds in China. A shortage of animal protein sources was reported with the lifting of the EU ban on the use of animal by-products for aquafeed as suppliers diverted rendered raw materials to home markets with higher offer prices. Industry is of the opinion that it needs to depend on more plant proteins but soybean meal prices are tagged to fish meal prices and substitution does not result in considerable costs savings. In addition, there was currency weakness in most Asian countries in 2013. For the industry in Indonesia, this posed a significant challenge as the industry depends on imports for 80% of raw materials, according to the Aquafeed Division of the Feed Mill Association (Aqua Feed Division, GPMT). In addition, Indonesian producers also have the problem of increasing energy costs every 3 months when the government adjusts its fuel subsidy (Merican, 2014a,b).

In Vietnam, in replacing soybean meal with cheaper raw materials, formulators encounter issues with digestibility and antinutritional factors. In addition, replacements with local sources of raw materials, such as rice bran and local fish meal (50% CP), in pangasius feeds is difficult when there is no complete information on its nutritional requirements. In India, the use of fish meal has been impossible for fish feeds as feed millers work on small margins of 3% or lower and cannot absorb such wide fluctuations in ingredient prices for a long time (Merican, 2013).

1.7   Feed production and trends

Based on the available information, annual aquafeed production is around 30 million tonnes (Table 1.1). However, data on installed capacity gathered from various reports and industry sources, in several countries in Asia, showed that production is below

Table 1.1 Available information on aquafeed production in Asia in tonnes

Country	Shrimp feeds	Fish feeds		Details
		Freshwater fish	Marine fish	
China	1,500,000	17,950,000	1,125,000	*Source*: Dong personal communication (2014). Marine fish feeds calculated at FCR 1.5 for 750,000 tonnes production
Thailand	400,000	800,000	20,000	*Source*: Aramsiriwat (2013) and estimate from industry for marine fish feeds and shrimp feeds
Vietnam	478,000	2,200,000		*Source*: Industry estimates Figures in 2012 for fish feeds comprised 1.95 million tonnes of pangasius feeds (9% farm made) and 260,000 tonnes of tilapia and other freshwater fish, pelleted, and extruded.
Indonesia	320,000	1,100,000	100,000	Includes 50–100,000 tonnes of marine fish feeds. *Source*: Denny D. Indrajaja. GPMT
India	572,000	684,000		Vijay Anand and Umakanth (2014)
Taiwan	25,509	168,733	243,741	Published information in 2011 (Wu, personal communication). Freshwater feed was for tilapia and others and marine fish feeds for milkfish, grouper, and seabass
Malaysia	60,000	100,000	48,000	*Source*: Industry estimates
Bangladesh		1,070,000		A total in 2012 and 2–3% for shrimp and prawn (Mamun-Ur-Rashid et al., 2013)
Japan		415,941		All feeds including freshwater fish feeds, Japanese Fish Farming and Feed Association (Serge Corneillie, personal communication)
Korea		120,000		All feeds in 2013 from Alltech Global Feed Survey (2014)
Philippines		500,000		All feeds in 2013 from Alltech Global Feed Survey (2014)

capacity throughout Asia. In China, the shrimp feed production is only 50% of the production capacity at 2.5 million tonnes per year (tpy) in 2012. However, additional capacity is increasing at 100,000 tonnes annually (Dong et al., 2013). Similarly in Malaysia, shrimp feed production in 2013 is estimated to be less than 50% of production capacity and even at peak production of 110,000 tonnes in 2010, it was 84%

of capacity. In Thailand, Wilson (2013) indicated a usage rate of only 45% of the 2 million tpy installed capacity for shrimp feed in 2012.

Vijay Anand and Umakanth (2014) reported that, in India, the production capacity was 2.88 million tpy of fish and shrimp feed but only 1.23 million tonnes of feed was sold in 2013. This indicated that only 43% of installed capacity was utilized. In fish feed production, the most significant development was the increase in extrusion lines in India. Capacity increased 306% from 507,000 tpy to 1.5 million tpy in 2013 (Vijay Anand and Umakanth, 2014) but usage was only 44%. Feed producers would like to use excess capacity to produce floating pellets for the rohu fish, an Indian carp, but with little success.

As shrimp production is increasing in India and Indonesia, reports showed that most market leaders in shrimp feed production are utilizing almost 90–100% capacity to meet demand. In India, Avanti Feeds reported production at almost full capacity (90%) in 2013. In the open market for shrimp feeds in Indonesia the installed capacity is about 300,000 tpy and capacity usage is 80–100%. PT Matahari Sakti and PT CJ Jombang are utilizing all available capacity to meet demand. In comparison, PT Central Proteinaprima (CP Prima) reported using only 50–60% of its total installed capacity of 200,000 tpy at three feed mills (Kontan, 2013).

1.8   New capacity and new entrants

Despite the above, there are new entrants; fish feed mills diversifying into shrimp feed production such as Growel Feeds in India with five pellet mills. Growel has also expanded with two new extrusion lines for fish feed production. CP India has acquired Shree Vijay Aqua Feeds and will enter into the pelleted fish feed business. Similarly, in Indonesia, PT Central Proteinaprima (CP Prima) will be building up capacity in fish feed production with two new mills, to add to its three existing mills.

Both in fish and shrimp feed production, the strategy in India and Indonesia is to be closer to customers. In India, Avanti Feeds is expanding to meet the shrimp feed shortage in the east coast and south of India. In Indonesia, Java-based and market leaders, PT Matahari Sakti and PT CJ Jombang continue to build up capacity in shrimp feed production with new feed mills closer to shrimp farms in west Indonesia.

In Vietnam, the largest pangasius feed producer Viet Thang was acquired by Hung Vuong (HV) the second largest pangasius catfish integrator in Vietnam. The largest pangasius integrator, Vinh Hoan has diversified into marine fish feeds as well as increasing feed lines for pangasius feeds as it expands fish production areas. Similar moves are being taken by others such as CL-Fish.

1.9   Shrimp feed types

The several brands of shrimp feeds manufactured by Asian feed companies revolve around two species; *P monodon* or *Litopenaeus vannamei*. The specifications will differ by the protein levels; the range for monodon feeds is 40–41% crude protein (CP) for starter and crumble feeds and 38% CP for grower and finisher pellets. It is common for feed mills to use imported fish meal (origin Peru or Chile) for such premium feeds and this costs more. Farmers prefer such feeds when stocking density is high or after a successful harvest and they have better affordability level. *L. vannamei* feeds contain 34–36% for starter feeds to 32–34% for grower and finisher. However, some specifications for premium *L. vannamei* feed marketed in Vietnam have much higher crude protein from 42% to 43% in starter diets to 38–41% in finisher pellets.

The range of ingredients will depend on the type of feed (economy or premium). Aside from either local or imported fishmeal, the common raw materials include soybean meal, wheat flour, fish oil, lecithin, and cholesterol. With improvements in production technology, the use of wheat flour has been reduced and binder eliminated (Dong et al., 2013). In Thailand and most other countries, local fish meal (CP < 55%), when available, is commonly used in *L. vannamei* feed production. Imported fish meal (CP > 65%) is usually used for premium shrimp feeds and marine fish feed production. The challenge to formulate with less fish meal and also to replace fish meal and fish oil with more sustainable plant sources is ongoing. Among Asian aquafeed producers, R&D by the leading aquafeed producer, CPF, led to the announcement that it would produce shrimp feed with no fish meal by 2015.

1.10   Feed prices

There is a general trend of raising and bringing down prices in line with changes in raw material costs. Price increases range from two to three times/year to once/year. At the end of 2013, producers in Malaysia raised shrimp feed prices to as high as USD 1.2/kg and for tilapia feeds, USD 0.83/kg. In Vietnam, a one-time increase in shrimp feeds raised prices to USD 1.33–1.37/kg in 2013 but feed costs to the farmer can increase to as high as USD 1.66–1.90/kg with credit terms and high interest rates (Merican, 2014a,b). A feed producer in Thailand empathizes with the hardship of farmers facing EMS and did not increase shrimp feed prices. However, a one-time price increase, which requires approval from the Department of Fisheries (DOF), raised shrimp feed prices to USD 1.40–1.50/kg.

In Indonesia, Aqua Feed Division GPMT indicated that as far as possible increases are to compensate only for the change in exchange rate. The recent rise in feed costs brought up shrimp feed prices to USD 0.99/kg and USD 0.58–0.66/kg for fish feeds. Both in Vietnam and Indonesia, pangasius farmers grapple with low ex-farm prices, sometimes at below the cost of fish production when supply is high. Feed costs are VND10,000–11,000 (USD 0.47–0.52/kg) and feed conversion ratios (FCRs) range from 1.6 to 1.7. In India, pangasius farmers fluctuate between using pelleted and extruded feeds, the latter being more expensive, based on ex-farm fish prices. In 2013, ex-farm prices for the pangasius have been around INR60–65/kg (USD 0.97–1.0/kg) while feed cost were INR 29–30/kg (USD 0.48/kg).

1.11 Country developments

1.11.1 China

China is the world's leading aquaculture producer and thus consumer of aquafeeds. The sector showed CAGR of 8.6% over the 10-year period (2004–2013, Figure 1.1). The 2013 estimates on total aquafeed production were higher than the production of 18.6 million tonnes in 2012 (CAPPMA, 2014). This includes about 400,000 tonnes of moist feeds, mainly for eel production. Feed mills in China produce both extruded and pelleted fish feeds; 85% are pelleted and are mainly for carp, tilapia, and catfish. The 2012 production was a substantial increase (23%) from the official data for 2009 which showed 14.26 million tonnes of aquafeeds.

In the case of marine fish feeds, feed mills in China produced 900,000 tonnes of feeds for seabass, eel, pompano, and groupers in 2013, up from 500,000 tonnes in 2009. All seabass and pompano feeds are extruded while groupers are fed on extruded, pelleted feed as well as trash fish. According to Guangdong Hinter Biotechnology, part of the market leader, Guangdong Haid group, the yearly growth rate in fish feed production was 10% for the 2010–2013 and it is expected to increase at a rate of 10% in the next 3 years.

Dong et al. (2013) reported that shrimp feed production declined to 1.25 million tonnes in 2012, similar to that for 2009 at 1.22 million tonnes. The total production estimate rose to 1.5 million tonnes in 2013 and exports to Vietnam and Malaysia in 2013 were less than 10,000 tonnes of shrimp feed. Production in 2013 was contrary to an expected trend of lower feed volumes because of EMS. In its 2013 annual report, Tongwei indicated that sales of shrimp and specialty feed enjoyed a growth of 24.30% over 2012 (Tongwei Feeds, 2013), while Haid said that it supplied 300,000 tonnes in

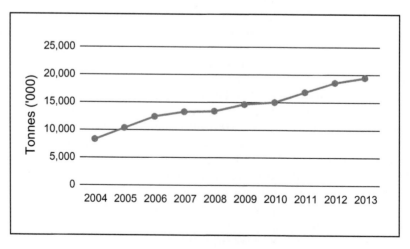

Figure 1.1 China's aquafeed production from 2004 to 2013.

2012. An expansion in shrimp feed production is expected as capacity is increasing at 100,000 tonnes annually (Dong et al., 2013).

China has about 1,000 aquafeed mills. Among the top four market leaders, Tongwei is fully integrated with farm, hatchery, and processing facilities with 40 feed mills while Haid is integrated with hatcheries. Haid also has 40 feed mills, located mainly in the coastal provinces (Guangdong, Fujian, Shandong, Zhejiang, Jiangsu, Liaoning, etc.). The smaller companies are Evergreen and CP; both are integrated with hatcheries, grow-out, and processing plants. Each has around 20 feed mills. Shrimp feed production is from 100 feed mills and 70% of production came from the four top feed producers (Dong et al., 2013) led by Evergreen and Haid.

The Ministry of Agriculture (MoA) has responsibility for the entire national administration and supervision of feed and feed additives. The feed industry is highly regulated as, vertically, the administrative lines run from the top MoA directly to the provincial feed working offices, city, and county feed working offices to township animal husbandry stations (Enting et al., 2010). According to industry sources, since 2013, there are more regulations adding to the five standards on feed products and feedstuffs in China. It is mandatory for feed mills to adhere to these regulations. There are also quality control standards which, according to industry, are on a par with global standards or even stricter.

As the industry is highly competitive, R&D is important for a producer to be ahead of competitors. Providing services is a marketing strategy introduced in 2006 and Guangdong Haid is an example with a 2,000-strong technical service team. This worked well as it increased its share of the shrimp market in 2013 (Dong et al., 2013).

1.11.2 Thailand

For several years until 2012, Thailand's aquafeed industry was dominated by shrimp feed production. EMS affected shrimp farming in late 2012 but Thongrod (2013) showed that shrimp feed sales had declined from the peak of 827,000 tonnes to 757,000 tonnes in 2011. A further decline occurred in 2012. Wilson (2013) also indicated a 10% decline from production in 2011 to only 740,000 tonnes. Furthermore, industry estimates a volume decline of 40% in 2013. The estimated capacity was more than 2.1 million tpy in 2013 with the completion of a new feed mill in 2013 (Wilson, 2013).

The fish feed sector is dominated by feeds for freshwater fish, namely the tilapia. Production of freshwater fish feeds has been rising at a rate of 7.4% per year from 2004 to 2013 (Aramsiriwat, 2013). Both pelleted and extruded feeds are produced for tilapia, Clarias and Pangasius catfish, snakehead, climbing perch, frog, and vegetarian fish feeds. Based on a request from fish farmers for lower-cost feeds, feed mills have developed a herbivorous feed with 15.5% crude protein and 10% crude fiber costing Thai Baht 8/kg (0.25USD/kg) for 5–6mm pellets (Bujel et al., 2001). An emerging sector is the marine fish feed but production is relatively small, estimated at 20,000 tonnes in 2013.

In Thailand, out of the total of 56 aquafeed mills, 13 were producing only shrimp feeds, 20 both shrimp and fish feeds, and 18 only shrimp feeds. According to Wilson

(2013), 85% of shrimp feed production in 2010 came from six companies: Charoen Pokphand Public Foods Limited (CPF), Thai Union Feed Mill (TFM), Grobest, Lee Feed Mills, Thailuxe Feeds, and Asian Feeds. CPF, TUF, and Asian Feeds are fully integrated in the shrimp sector. CPF and Gold Coin, through SyAqua, are involved in shrimp genetics and produce vannamei shrimp broodstock. All shrimp feeds are pelleted. Betagro, a major fish feed producer produced and marketed extruded shrimp feeds for a short period.

In the fish feed sector, the leading companies CPF, Thailuxe, and Betagro produce both pelleted and extruded fish feeds for various fish types. TFM is leading in the production and marketing of feeds for the seabass *Lates calcarifer* which it began to produce in 2007. This follows the change from farming marine shrimp to the seabass in brackish-water ponds because of diseases affecting the shrimp. The latest entrant to the industry is TRF feed mill which has a 96,000 tpy capacity and produces 65% shrimp feeds and 35% fish feeds. It will also consider production of feeds for the seabass, grouper, and cobia once the farming of these expands in Thailand (Merican, 2014b).

The aquafeed industry is tightly regulated by the DOF, which controls feed specifications for the various products, registration of facilities, and prices. DOF sets the basic quality parameters and monitors feed quality through inspections and feed audits. All feeds must be registered and are specific to the production facility and ingredients must be declared. Medicated feeds, including antibiotics, is prohibited. Feed mills are required to have GMP to operate, most have ISO9001, HACCP, and ISO14001 (Wilson, 2013). Some of the Thai feed producers have BAP certification.

Some recent issues facing the feed sector highlighted by Thongrod (2013) include the use of non-GMO soybean meal for production of seafood to EU markets and fish meal supply. Thailand produces 500,000 tpy of fish meal used for shrimp feeds and starter diets in the animal production industry. A small portion (15%) is exported. There is a 15% tax on imports of fish meal with 65% crude protein. By January 2015, there was to be a requirement that 50% fishmeal must be from sustainable fisheries.

1.11.3   Vietnam

The expansion in fish feed production in Vietnam was fuelled by the demand for the pangasius catfish in international markets. In 2000, pangasius fish production was 37,500 tonnes and this escalated to 1.35 million tonnes in 2011. In 2013, total fish feed production was only 1.9 million tonnes and included those for the tilapia, snakehead, grouper, and seabass (AAP, 2013). The demand for tilapia feeds is increasing and usage of tilapia feeds was estimated at 300,000 to 500,000 tonnes in 2013. The pangasius sector is very important and the Vietnam Association of Seafood Exporters and Producers has asked that the focus be on feeds with better FCRs; the target FCR is 1.3 from the current high range of 1.6 for it to be sustainable.

During the years from 2006 to 2010, CAGR for shrimp and fish feed production in Vietnam was 7.4% and 11.8%, respectively. Shrimp feed demand in 2011 declined by 9% when farms succumbed to EMS in 2010. Feed production increased in 2013 to 600,000 tonnes, which industry said was a result of balancing shrimp production with the conversion of extensive monodon shrimp farming to more intensive vannamei. However, although there was a recovery in shrimp farming in Vietnam, feed producers are of the opinion that there is still some uncertainty in feed demand.

In 2010, there were 106 feed mills for aquafeed production (Serene, personal communication). There is a clear division of the markets leaders in the fish and shrimp feed production. The top shrimp feed companies are Uni President (UPV), Charoen Pokphand (CP), Grobest, CJ Vina, and Sheng Long Bio-Tech International. Some have integrated upstream with hatcheries to provide post larvae to grow-out farms. Grobest is the feed supplier to Minh Phu, the largest integrated shrimp farm in Vietnam. All also produce extruded fish feeds for the pangasius, tilapia, and marine fish (mainly for the seabass and grouper). UPV exports both marine fish and shrimp feeds to Malaysia and India. UPV's imports of shrimp feed into India increased to 20,000 tonnes in 2012. Sheng Long which produces fish and shrimp feeds exports these feeds to Brunei, the Philippines, Indonesia, Malaysia, and Sri Lanka.

In 2012, in the fish feed market, the top 11 companies, led by Viet Thang and Proconco, produced 84% of pangasius feeds. However, in 2013, the second-largest pangasius integrator HV fully acquired Viet Thang and with its feed unit, Tay Nam Panga Feed Joint Stock became the largest fish feed producer with an estimated capacity of more than 500,000 tpy. Several of the feed mills are certified under Global Gap's Compound Feed Manufacturing standard.

The local industry predicted that the aquafeed industry in Vietnam will continue to grow. Recent additions to the industry included a new feed mill for aquafeed and animal feeds for Cargill, a fish feed extrusion plant for Sheng Long and De Heus added a new plant with a dedicated line for aquafeeds. Recent developments also include the acquisition of Tomboy Aquafeed JSC, shrimp, and marine fish feed producer by Nutreco and a joint venture between Anova Feeds, mainly in pangasius feed production with Ewos, Norway. As the marine fish farming develops and extruded feeds replace trash and farm-made feeds, industry expects more feed companies to produce feeds for Asian seabass, pompano, grouper, and cobia.

1.11.4 Indonesia

Indonesia's production of fish feeds, led by those for freshwater fish, has been on an upward trend in the past 3 years, from 905,000 tonnes in 2011, to 982,000 tonnes in 2012, and 1.1 million tonnes in 2013. Overall the fish feed sector grew at an annual rate of 7% during 2010–2013 and is expected to continue its expansion at an annual rate of 12% from 2014 to 2017, according to Aqua Feed Division GPMT.

The changeover to extruded floating pellets is relatively recent in the freshwater fish feed sector in Indonesia. According to Aquaculture Feed GPMT, the annual rate of growth for extruded feeds was 11% for 2011–2013 and it is expected to rise to 15% in 2014–2017. In contrast, pelleted fish feeds grew at an annual rate of 2% from 2011 to 2013 and are expected to continue rising at 8% in 2014 to 2017. The expansion in fish feed production is attributed to the industrialization of fish farming which also includes the national policy to farm large (800 g) pangasius catfish for export, farming the Clarias catfish in biofloc systems and a new strain of saline tilapia for the domestic market. The government has plans for the production of 1.2 million tonnes of the pangasius for export as well as for domestic consumption. Production was 651,000 tonnes in 2012 (Trobos Aqua, 2013).

Shrimp feed production had a CAGR of 13.4% for the period from 2006 to 2009 in Indonesia, based on estimates on feed production given by industry and Aquaculture Feed GPMT. Feed production declined in the years from 2010 to 2012, as producers faced crop losses from infectious myonecrosis virus or IMNV which affected farms from 2009. All shrimp feeds are pelleted. In 2008, production was 46% of capacity (Agrina, 2008). In 2012 and 2013, some feed mills reported full utilization of capacity while others, lower production, such as only 50–60%. The reason is that as logistic costs are extremely high in Indonesia, it is essential that feed supply is close to farms. In shrimp feed production, Aqua Feed Division GPMT expects an annual growth of 15% from 2014 to 2017.

Recovery in shrimp production was confirmed in 2013 when shrimp feed production rose 320,000 tonnes; up from a low of 250,000 tonnes in 2011. Farms are showing consistent increases in harvest volumes. High shrimp prices encourage farms to restart farming. Another driver for feed demand was the successful nationwide program to rehabilitate abandoned ponds. Feed millers also benefitted from the trend of increasing stocking density after each successful harvest by 10–20% and the use of higher-cost premium feeds.

The major feed company is the fully integrated PT Central Proteinaprima (CP Prima), producing both fish and shrimp feeds at six mills. It is involved in farming at the largest shrimp farm in Indonesia and in shrimp genetics and broodstock production, post larvae supply from its hatcheries and seafood processing. It is also part of the animal feed conglomerate, CP Indonesia. Shrimp feed production is mainly for the integrated operations which use more than 62,300 tonnes of feed annually. In 2014, CP Prima alone expected a 20% increase in shrimp feed sales.

The other large feed company with integrated operations is the aquafeed subsidiary of animal feed conglomerate PT Japfa Comfeed. PT Suri Tani Pemuka (STP) producing both fish and shrimp feeds. STP, a pioneer in marine fish feed production has marine fish hatcheries and has embarked on marine fish farming to push for a rapid expansion in marine fish farming. It has farms producing pompano, seabass, and hybrid groupers with an annual production of 2,200 tonnes (Trobos Aqua, 2013).

Other than these integrated companies, there are 12 companies supplying the open shrimp feed market. In 2013, production was led by PT CJ Jombang (CJ), PT Matahari Sakti (MS), PT Grobest Indomakmur, PT Mabar, PT Gold Coin, and PT Luxindo. With success in marketing their feeds in the last few years, CJ and MS have been increasing their shrimp feed sales and have recently expanded capacity with new feed mills. PT Sinta Prima and PT Cargill are leaders in the open market for fish feeds. In 2013, PT Sinta Prima began operations at a new feed plant (Kontan, 2013). PT MS is the first local company to export feeds. It exports 10% of its production; mainly tilapia and grouper feeds to Africa, Bangladesh, Brunei, and Malaysia.

The government has urged feed companies to reduce feed costs to match low selling prices for various freshwater fish and provide better profit margins for small-scale farmers. FCR for pelleted freshwater fish feeds ranges from 1.5 to 1.7 and for extruded feeds, 1.0 to 1.4. With the Clarias catfish, the association has managed to reduce FCR to 0.6–0.7 in high-density cultures with probiotics. New feed ingredients being tested are insect larvae *Hermetia illucens* and palm kernel meal. Major feed companies have announced expansion plans such as PT CP Prima increasing its pangasius production to 70 tonnes/month from 30 tonnes/month. PT Mabar Feed Mill is integrating into pangasius farming.

1.11.5 India

The rise in volumes of shrimp feed was most apparent in India, a result of the shift to farm the vannamei shrimp since 2009. Feed volumes rose 110% to 500,000–572,000 tonnes in 2013 from 272,000 tonnes in 2010. Meanwhile, the installed capacity of shrimp feed increased 241% from 390,000 tpy to 1.33 million tpy (Vijay Anand and Umakanth, 2014). During the period from 2006 to 2013, the CAGR was 19.5%.

The increase in demand is due to the opening of new areas for vannamei shrimp farming and conversion from farming monodon shrimp in traditional and extensive systems to the more intensive farming of the vannamei shrimp. Although the stocking density is regulated at $60 \, PL/m^2$, farmers attempt higher stocking density following successful harvests. Vannamei shrimp accounted for 85% of production in 2013 in contrast with 33% in 2010. This renewed interest in shrimp farming has led to fish feed producers investing in pelleting equipment for shrimp feed production.

In 2013, the installed capacity for fish feed production was 1.55 million tonnes, 322% of that estimated for 2010 at 408,000 tonnes (Vijay Anand et al., 2010). The growth is largely in the production of extruded fish feeds. The current production at 684,000 tonnes is only 44% of capacity but is a 59% increase of the production in 2010 (430,000 tonnes, Vijay Anand, personal communication). The rapid expansion was in anticipation of a large uptake of extruded feeds for various freshwater species. However, at present, the main market for floating feeds and sinking pelleted feeds is the pangasius while the Indian carps, rohu (*Labeo rohita*), catla (*Catla catla*), and mrigal (*Cirrhinus cirrhosus*) are fed pelleted feeds or farm-made feeds. The latter is a mash comprising de-oiled rice bran and groundnut cake, cotton seedcake, etc. placed in a perforated bag. With the availability of pelleted feeds, the freshwater aquaculture sector is slowly converting to sinking pellets.

Nearly all of the 26 feed mills operating in 2013 for fish feed production produce extruded fish feed which accounts for more than 95% of total fish feed production. Farmers also demand that pelleted feeds, now with 18% crude protein, also match that of extruded feeds with 27% crude protein. Feed producers would like to use excess capacity to produce floating pellets for the rohu, but with little success. Rohu has a higher ex-farm value as compared to the pangasius and higher profit margins as it is fed mainly with rice bran.

Shrimp feed production is led by Avanti Feeds, Charoen Pokphand India (CP), Waterbase and Godrej Aqua Feeds, Grobest, and Cargill. Avanti is the market leader for both vannamei and monodon shrimp feeds. Recently, both Avanti and CP have been expanding capacity and Avanti has three feed mills. In fish feed production, the leading companies are Growel and Indian Broiler (IB) with 10 other players with capacity ranging from 10 to 33 tph. Growel, which entered the industry in 2010, sold 110,000 tonnes in 2013 and has expanded its fish feed production to target the feed market in Andhra Pradesh, the main pangasius farming area. It started shrimp feed production in 2014. IB used some spare extruder capacity to develop extruded feeds for shrimp.

1.11.6 Bangladesh

The production of commercial fish feeds has increased at a rate of 32% during the years from 2008 to 2012 to 1.07 million tonnes. In addition, there was an estimated 0.4 million tonnes of farm-made feeds prepared by small village-based feed-making enterprises. Feeds are for the pangasius catfish, Nile tilapia climbing perch, and Indian major carps and for other freshwater fish. Production is mainly feeds for the pangasius catfish (60–65%) followed by that for the tilapia (35–45%). Sinking feeds dominate at 81% while extruded feeds only account for 19%, but the rate of extruded floating feeds has been increasing at 89% over the last 4 years. Using least-square demand projections, the authors projected that feed demand will rise to 1.8 million tonnes in 2015 (Mamun-Ur-Rashid et al., 2013).

There are about 100 feed mills producing aquafeeds but 60–70% of the market is dominated by 8–10 large producers. During the last 2 years, around 20 companies have expanded their capacities and have installed extruders for floating fish feed production. About 20 feed mills carry out toll milling for the large farms. Vertical integration is a recent trend and eight feed millers supply farms with feeds as well as mono sex tilapia fingerlings. Two large groups are fully integrated with a hatchery, grow-out farming, feed production, and processing. Raw materials for the processing also come from other farms (Mamun-Ur-Rashid et al., 2013).

1.11.7 Malaysia

In the last 3 years, due to EMS, the demand for shrimp feed declined to 60,000 tonnes in 2013, from a high of 126,000 tonnes in 2010. Demand was expected to decline further to 45,000 tonnes in 2014, according to an industry source. In fish feed production, feed producers face uncertainty in demand, linked to frequent fluctuations in ex-farm prices for tilapia and Clarias catfish. When ex-farm prices decline, farmers shift to using farm-made feeds, estimated at 60,000 tonnes in 2013. The concern is for the food safety and hygiene, as the common raw material for farm-made feeds is poultry waste. The government's target is to increase production of tilapia and catfish to substitute for the decline in fish supply from capture fisheries. The feed industry has to play its role and focus on production of lower-cost feeds. As such, it is expected that demand for pelleted feeds will increase at 20% during 2014–2017 but only 15% for extruded feeds.

The feed market is competitive for the four major feed producers in the country led by Star Feedmills, part of the Charoen Pokphand group, Cargill, Dindings Soya Multifeeds, and Gold Coin, as well as importers of feeds for marine fish and shrimp from Vietnam, Indonesia, and Taiwan. However, developments in marine fish farming in offshore cage farms are expected to increase the demand for extruded feeds. In the marine fish feed sector, feed demand is usually balanced by farmers changing species when demand and prices of a particular species falls.

2. Aquafeed in the Americas

2.1 Development of aquafeed production in the Americas

According to the FAO (FishStat Plus, 2014), total aquaculture production in the Americas within the 16 reported countries was approximately 2.7 million metric tones (MT) (Table 1.1). The top six producers in the Americas are Chile (818.7 MT) and Canada (132.3), mainly producing marine fishes such as salmon and trout, Brazil (685.7 MT) and the United States (167.9 MT) mainly producing freshwater fishes such as tilapia, tambaqui in Brazil and catfish in the United States, Ecuador (321.8 MT), and finally Mexico (137.9 MT) mainly produce white legged shrimp.

Aquaculture is a major economic activity in Chile. Among the diverse aquacultures practiced in Chile, Atlantic salmon (*Salmo salar*), followed by rainbow trout (*Oncorhynchus mykiss*) and coho salmon (*Oncorhynchus kisutch*) are the largest group representing over 80% of the harvest volume from the year 2000 onward (FAO, 2009). Until 2007, when the infectious salmon anemia virus hit, Chile experienced, over 15 years, a huge growth in its salmon aquaculture becoming the second-largest salmon and trout producer after Norway. In 2012 production levels returned to those achieved in 2006. Chile and Canada have been effectively farmed through good management practices including feeding equipment that allows a more efficient use of extruded sinking 46–50% protein feeds. Over the years the growth rates have been increasing steadily and the FCR has also been steadily decreasing (improving) annually through genetic improvement programs, management, and improved feeding systems, achieving a FCR of 0.9 to 1.3 (Tacon and Metian, 2008; Sarker et al., 2013; Bridson, 2014).

Nile tilapia (*Oreochromis niloticus*), Chitralada strain, and Tambaqui (*Colossoma macropomum*) farming in Brazil are typically produced, with 28–40% protein extruded floating feeds in reservoir cages and ponds achieving mean FCRs of 1.3 (Frota, 2004).

The aquaculture industry in the United States continues to be dominated by the production of channel catfish (*Ictalurus punctatus*), primarily under intensive conditions in earthen ponds. Feeding is done typically with extruded floating 28–35% protein feeds with FCRs ranging from 1.8 to 2.6.

Ecuador shrimp (*L. vannamei*) is mainly cultured in extensive systems, which means that stocking densities vary from eight to 15 shrimp/m^2, in earthen ponds that range from 6 to 20 hectares, achieving crops from 650 to 1,500 kg/ha. Mainly pelletized sinking 28–35% protein feeds are used with estimated FCRs ranging from 1.0 to 1.4, whereas in Mexico shrimp is cultured under higher densities varying from 25 to 40/m^2, achieving crops from 1,000 to 3,000 kg/ha. In August 2013 EMS caused large-scale die-offs of farmed-raised shrimp, reducing production to less than 50% in the states of Sonora, Sinaloa, and Nayarit. This situation is making tilapia farming the country's new aquaculture favorite. In 2015, Mexico's tilapia production is expected to reach 87,000 tons, and could reach 100,000 tons by 2016.

2.2 Country development

Although aquafeed production is not easy to quantify due to the lack of information, efforts should be made to estimate regional volumes through different mechanisms including aquaculture production combined with the typical FCRs of the individual cultured species and country. Under these calculations, the total compound aquafeed production in 2012 within the 16 reported countries has been estimated at 3.62 million MT (Table 1.2). This assessment refers specifically to complete diets made by the industrial-sized mills of commercial manufacturers and integrators. This value is 68% more than the tonnage quantified by Ziggers (2012), who reported for North America 0.28 million MT and for Latin America 1.88 million MT.

The five countries with the largest national outputs accounted for 82% of the tonnage in 2012, including Chile (981,800 MT), Brazil (837,800 MT), the United States (460,200 MT), Ecuador (434,800 MT), and Mexico (256,600 MT). The major cultivated species groups fed compound aquafeeds in 2012 were freshwater fish (40%), salmonids (34%), shrimp (26%), and other marine fish (0.2%) (see Table 1.3).

Table 1.2 Total 2012 aquaculture production by species in the Americas (values given in 1,000 MT)

Country	Shrimp and freshwater crustaceans	Salmonids (salmon, trout)	Freshwater fishes (tilapia, tambaqui, catfish)	Other marine fishes (cobia, snapper, flounder)	Total
Chile		817.7		0.4	818.1
Brazil	74.4	3.7	607.6		685.7
Ecuador	281.1		40.8		321.8
United States	44.9	37	167.9	1.9	251.7
Mexico	100.7	6.9	28.5	1.8	137.9
Canada		131.7	0.5		132.2
Colombia	8.9	6.1	74.5	0.2	89.7
Peru	17.8	24.8	4.8		47.4
Honduras	27		7.8		34.9
Costa Rica	3.0	0.8	23.4		27.2
Venezuela	19.6	0.3	6.3		26.1
Nicaragua	24.3				24.3
Cuba	3		21.6		24.6
Guatemala	12.3		5.6		17.9
Panama	6.7	0.2	0.3	0.2	7.5
El Salvador	1.2		4.1		5.3
TOTAL	624.9	1029.2	993.7	4.5	2652.3

Table 1.3 **Estimated 2012 market of compound aquafeed in the Americas (values given in 1,000 MT, as fed basis)[a]**

Country	Shrimp and freshwater crustaceans	Salmonids (salmon, trout)	Freshwater fishes (tilapia, tambaqui, catfish)	Other marine fishes (cobia, snapper, flounder)	TOTAL
Chile		981.2		0.6	981.8
Brazil	103.8	4.8	729.2		837.8
United States	89.5	48	319	3.7	460.2
Ecuador	337		97.8		434.8
Mexico	191.3	10.3	51.4	3.6	256.6
Canada		144.8	0.7		145.5
Colombia	13.4	8.6	119.2	0.3	141.5
Peru	28.5	31	8.6		68.1
Honduras	48.6		15.7		64.3
Costa Rica	4.9	1.1	43.2		49.2
Venezuela	35.2	0.4	11.3		46.9
Nicaragua	43.8				43.8
Cuba	5.2		34.5		39.7
Guatemala	20.9		9.8		30.7
Panama	10.8	0.3	0.5	0.4	12.0
El Salvador	1.9		7.4		9.3
TOTAL	934.8	1230.5	1448.3	8.6	3622.2

[a]Calculated amount of feed according to FAO (2014) aquaculture production report, Tacon and Metian (2008), and author-estimated FCRs.

The aquafeed industry in America is technologically well-developed, typically producing extruded feeds for fish and pelletized feeds for shrimp. Aquaculture feeds are generally composed of a mix of macro-ingredients such as wheat, soybean meal, fish meal, squid meal, krill meal, rapeseed cakes, wheat, fish oil, plant oils, rapeseed oil, corn, corn gluten, sorghum, poultry meal, hemoglobin meal, feather meal, sunflower cake, cottonseed meal, soybean lecithin, and additives such as amino acids, vitamins, and minerals, pre- and probiotics, enzymes, essential oils, and pigments, among others.

Compound salmonid aquafeeds in Chile and Canada are technologically developed. A major trend for practical diets has been the development of extruded high-energy diets, which have contributed to a reduction in Atlantic salmon farming, from 3.5 to 1.2 kg feed dry matter per kg of fish produced, during the last two decades (Austreng, 1994). Together with this trend, strong efforts have been made to steadily reduce the "fish in:fish out" ratio through the reduction of fish meal levels (17–20%) and fish oil to 12%. Alternative ingredients used today are typically rendered land–animal proteins and fats (24–30%), vegetable proteins (20–25%), and vegetable oils (5–12%). The performance of these diets has been maintained through the use of additives to increase the digestibility of the ingredients and the immunological response of the fish.

Our Brazilian compound aquafeed estimates resulted higher compared to the 653,000 MT reported by Zani (2014), who estimates a 2012 demand of 575,000 MT for compound feed for fish and 75,000 MT of shrimp compound feed. According to Zani (2014), a stimulus released by the "Official Harvest Plan" to encourage aquaculture production and a tendency to harmonize the requirements for environmental licensing, kept the dynamism established in their supply chain and will allow increased production in the following years.

Ecuador shrimp farming typically uses various levels of protein, beginning with 40% for the initial stages and 35% or 28% for grow-out stages depending on the stocking density and the season of the year. Several companies produce shrimp feeds in Ecuador, and they are making strong efforts to steadily reduce the inclusion levels of fishmeal and increase growth, reducing the cost of feed to optimize the FCR.

Acknowledgments

Zuridah Merican would like to thank Ir Denny Indradjaja, Chairman of Aqua Feed Division, Indonesian Feed Mills Association, Dong Qiufen, Guangdong Hinter Biotechnology, Vietnam, for their input in the preparation of this chapter as well as several industry members who requested anonymity.

References

AAP, 2008. Uncertainties ahead with raw materials. Aqua Cult. Asia Pac. 4 (1), 16–18.
AAP, 2009. Industry squeezed and high feed costs to remain. Aqua Cult. Asia Pac. 5 (1), 10–15.
AAP, 2010. Managing costs and improving feed performance. Aqua Cult. Asia Pac. 6 (1), 12–15.
Agrina, 2008. <http://agrina-online.com/show_article.php?rid=7&aid=1224>.
Alltech Global Feed Survey (2014). <http://www.alltech.com/sites/default/files/alltechglobal-feedsummary2014.pdf>. Alltech, Kentucky, USA.
Aramsiriwat, B. 2012. Thai Feed Mill Association, Agricultural Update on Thailand's Feed and Livestock Situation presented at 9th Southeast Asia US Agricultural Cooperators Conference (ACC2012), Phuket, Thailand, August 29, 2012.
Aramsiriwat, B. 2013. Thailand's Feed Situation, Thai Feed Mill Association, presented at VIV 2013, March 13, 2013, Bangkok, Thailand.
Austreng, E., 1994. Historical development of salmon feed. Annual report 1993, Institute of Aquaculture Research (AKVAFORSK), As-NLH, Norway, 32 pp.
Bhujel, R.C., Yakupitiyage, A., Warren, A., Turner, W.A., Little, D.C., 2001. Selection of a commercial feed for Nile tilapia Oreochromis niloticus/broodfish breeding in a hapa-in-pond system. Aquaculture 194 (2001), 303–314.
Bridson, P., 2014. Monterey Bay Aquarium Seafood Watch, Atlantic Salmon, Coho Salmon. Report from <www.seafoodwatch.org>.
CAPPMA – China Aquatic Products Processing and Marketing Alliance 2014. <http://www.cappma.org/hangyexinxi_article.php?big_class_id=7&small_class_id=17&article_id=16541>.

Dong, Q., Peng, Z., Zhang, S., Yang, Y., 2013. An overview of shrimp feed industry in China. International Aquafeed January–February, 22–24.

Enting, I., Wang, B., Zhang, X., van Duinkerken, G., 2010. The animal feed chain in China opportunities to enhance quality and safety arrangements. Minist. Agric., Nat. Food Qual. August.

FAO, 2009. Fish and feed inputs for aquaculture: practices, sustainability and implications. FAO Fisheries and Aquaculture Technical Paper No. 518.

FAO FishStat Plus, 2014. FAO Fisheries Department. Fishery Information. Data and Statistics Unit Fishstat Plus: Universal software for fishery statistical time series. Aquaculture production: quantities 2912. Vers. 230.

Flegel, T.W., Siripong Thitamadee, S., Sritunyalucksana, K., 2014. Update on Top Disease Threats including EMS/AHPND for Shrimp Cultured in Asia. Presented at The Aquaculture Roundtable Series (TARS 2014), Phuket, Thailand, August 20–21, 2014, pp. 50–59.

Frota, M.C., 2004. Presente do Egito. Globo Rural 226-AGO/04.

Kontan, 2013. Rising feed production at 12% in 2013. <http://industri.kontan.co.id/news/produksi-pakan-ternak-bisa-tumbuh-12-di-tahun-ini>.

Mamun-Ur-Rashid, M., Belton, B., Phillips, M., Rosentrater, K.A., 2013. Improving aquaculture feed in Bangladesh: from feed ingredients to farmer profit to safe consumption. WorldFish, Penang, Malaysia. Working Paper No. 2013-34.

Merican, Z., 2012. Squeezed from both sides in 2012. Aqua Cult. Asia Pac. 9 (1), 18–25.

Merican, Z., 2013. Driving barramundi in Thailand. Aqua Cult. Asia Pac. 9 (1), 15–16.

Merican, Z., 2014a. Selective growth in 2013. Aqua Cult. Asia Pac. 10 (1), 22–27.

Merican, Z., 2014b. A valued partner for growth. Aqua Cult. Asia Pac. 10 (3), 28–31.

Sarker, P.K., Bureau, D.P., Hua, K., Drew, M.D., Forster, I., Were, K., et al., 2013. Sustainability issues related to feeding salmonids: a Canadian perspective. Reviews in Aquaculture 5, 1–21.

Tacon, A.G., Metian, M., 2008. Global overview on the use of fish meal and fish oil in industrially compounded aquafeeds: trends and future prospects. Aquaculture 285, 146–158.

Thongrod, S., 2013. Current Status of Aquafeed Industry in Thailand. Thai Union Feed Mill Co., Ltd, Thailand.

Trobos Aqua, 2013. The market for catfish widens. <www.trobos.com>.

Tongwei Feeds, 2013. Homepage. <www.tongwei.com>.

Vijay Anand, P.E., Umakanth, R., 2014. Shaping India aquaculture further. Aqua Cult. Asia Pac. 10 (5), 28–32.

Vijay Anand, P.E., Cremer, M., Ramesh, G., 2010. Indian freshwater fish farming: moving to soy-based extruded fish feeds. Aqua Cult. Asia Pac. 5 (4), 25–29.

Wilson, T., 2013. An overview of the shrimp farming and shrimp feed manufacturing industries in Thailand. Presented at the World Aquaculture Society Meeting, Nashville, TN, February 21–25, 2013.

Zani, A., 2014. Brazilian Feed Industry Association. Sindiracoes, Setor de Alimentacao Animal. Bulletin May. 1–5.

Ziggers, D., 2012. Global survey: feed production reaches record of 873 million tonnes. <http://www.AllAboutFeed.net>.

Feed formulation software

2

A. Victor Suresh
United Research (Singapore) Pte. Ltd, Singapore

2.1 Introduction

The first record of using a computer to produce a feed formula was made in 1951 when F.V. Waugh (1951) published a paper titled "The Minimum-Cost Dairy Feed." Computers were just becoming commercially available during the early 1950s, so it is quite remarkable that solutions to feed formulation through computers were sought as early as 1951. Dent and Casey (1967) authored the first ever published book on computer-based formulation titled *Linear Programming and Animal Nutrition*. Using computers for practical feed formulation in the feed industry began in the 1970s when computers were more affordably priced for large enterprises. However, widespread use of computers for feed formulation began in the 1980s when personal computers became commercially available. In the present age, feed formulation is unthinkable without the use of computers.

Software for feed formulation has changed with changes in computer hardware and communication technology. A wide variety of software options is available. Capabilities and costs of the different options vary considerably. Costs have declined over time, while programs have become increasingly user-friendly.

This chapter aims to provide readers with a general understanding of the mechanics of modern formulation software in the feed sector, options available for those who are interested in buying formulation software, and some guidance on the practical use of the software in aquafeed formulation.

2.2 General overview of the formulation process in the feed industry

The feed industry relies on formulation to arrive at products that meet specific requirements for animal performance, physical characteristics, quality and functionality, and profitability of the feed manufacturing enterprise.

There are four types of formulated products used in the feed industry (AAFCO, 2000):

- Concentrate: A feed used with another to improve the nutritive balance of the total and intended to be further diluted and mixed to produce a supplement or a complete feed.
- Supplement: A feed used with another to improve the nutritive balance or performance of the total and intended to be: (i) fed undiluted as a supplement to other feeds; or (ii) offered

S. Nates (Ed): Aquafeed Formulation. DOI: http://dx.doi.org/10.1016/B978-0-12-800873-7.00002-6

as a free choice with other parts of the ration separately available; or (iii) further diluted and mixed to produce a complete feed

- Premix: A uniform mixture of one or more micro-ingredients with diluent and/or carrier. Premixes are used to facilitate uniform dispersion of micro-ingredients in a larger mix.
- Complete feed: A nutritionally adequate feed for animals. By specific formula it is compounded to be fed as the sole ration and is capable of maintaining life and/or promoting production without any additional substance being consumed except water.

A product may be formulated as a fixed recipe or as a dynamic formula depending on the feed manufacturer's requirements. A fixed recipe means that all specified ingredients for the product must be available every time the product is manufactured, and the proportions of the ingredients in the product do not vary. In contrast, dynamic formulation uses flexible options in ingredient usage depending on availability and cost. Fixed recipes are commonly used in the manufacture of premixes and specialty feeds like larval feeds while dynamic formulation is followed for making most of the products in the feed industry.

Formulation of products in the feed industry requires knowledge of the nutrient content of ingredients and target nutrient levels in the products. Analysis of ingredients for nutrient content prior to formulation and checking products for nutrient content after production are important for accurate formulation and details relevant to this aspect are given in Chapter 6.

While it is possible to arrive at formulas using calculations by hand, computers are invariably used to formulate feeds. A simple spreadsheet may serve the purpose for some, especially if the formula is a fixed recipe and simple with a few ingredients. A vast majority of formulation in the feed sector, however, is done using computer software that relies on a mathematical method called linear programming (LP).

2.3   LP-based feed formulation

LP was developed during the Second World War by the Russian mathematician Leonid Kantorovichto to provide solutions to military logistics and has been acknowledged as one of the most important contributions to economic management in the twentieth century. Essentially, LP is a method to achieve the best outcome (such as maximum profit or lowest cost) amidst a range of constraints that are defined mathematically. In the context of feed formulation, the constraints are product specifications that are defined as minimum and maximum levels of nutrients and ingredients that the product can have and the objective is to find the lowest cost at which various ingredients can be combined to make the product.

Least cost optimization or *least cost formulation* are the terms that are commonly used to refer to the LP-based formulation. Sometimes the terms are misinterpreted as if LP is used for arriving at a cheap feed of low quality. This is an incorrect perception because every quality variable that is measurable and additive in nature can be expressed within an LP system, and cost optimization is achieved only *after* reaching the set quality target. Knowing how to define and manage constraints correctly is a key trait distinguishing a successful feed formulator.

2.4 Essential components of LP-based feed formulation software

All LP-based feed formulation software has at least three databases: one each for nutrients, ingredients, and products.

1. Nutrients: Nutrients play the central role in the formulation system because they link both ingredients and products. The database may be a simple list of nutrients with corresponding units (Table 2.1) or have additional information such as the species to which the nutrients may be relevant. For example, astaxanthin is relevant for salmonids, but not for carp; and cholesterol is relevant for crustaceans but not for finfish. By the way, the term *nutrients* in a formulation system does not strictly refer to nutrients as in nutritional science but any measurable parameter that the feed processor wishes to optimize through formulation. Physical parameters such as color, smell, and density and functional parameters such as attractability and palatability may also be treated as nutrients in formulation systems.

2. Ingredients: A database on ingredients stores four vital pieces of information: ingredient names, availability of the ingredients for formulation, nutrient composition, and cost. A nutrient composition table of an ingredient in the database is commonly referred to as the *matrix*. The matrix is the actual nutrient composition of the ingredient in most cases (see Table 2.2 for an example). The matrix can also be used to reflect the effects an ingredient may have on the nutritional value of the feed. Enzymes, for example, improve the digestibility of feed ingredients, and this improvement can be captured in the form of a matrix (Table 2.3). The matrix values in this case do not reflect the nutrient analysis value of the enzyme but the effect that it will have on the feed at the given recommended dosage assuming that sufficient quantity of substrates is available for enzymatic action.

Table 2.1 List of nutrients commonly used in aquafeed formulation

Nutrient	Unit
Moisture	%
Crude protein	%
Crude fat	%
Crude fiber	%
Ash	%
Macro minerals (calcium, phosphorus, sodium, etc. Phosphorus is usually expressed on an available basis)	%
Trace minerals (zinc, manganese, iron, copper, selenium, etc.)	mg/kg
Amino acids (arginine, histidine, lysine, methionine, etc. The amino acids are usually expressed on a digestible basis)	%
Fatty acids and other lipid components (linoleic acid, linolenic acid, EPA, DHA, cholesterol, phospholipids, etc.)	%
Starch and nonstarch polysaccharides	%
Energy (usually expressed as digestible energy for aquafeeds)	kcal/kg or MJ/kg
Vitamins (vitamin A, thiamin, riboflavin, etc.)	IU/kg, mg/kg, or µg/kg

Table 2.2 Sample matrix value of soybean meal, 46% protein

Nutrient	Unit	Value
Dry matter	%	88.00
Crude protein	%	46.30
Crude fat	%	2.70
Crude fiber	%	4.50
Ash	%	7.20
Digestible energy	kcal/kg	3150.00
Calcium	%	0.35
Phosphorus	%	0.64
Available phosphorus	%	0.25
Sodium	%	0.03
Linoleic	%	1.25
Potassium	%	0.05
Linolenic	%	0.16
Arginine	%	2.70
Lysine	%	2.72
Methionine	%	0.61
Met + Cys	%	1.24
Threonine	%	1.82
Digestible arginine	%	2.34
Digestible lysine	%	2.30
Digestible methionine	%	0.52
Digestible Met + Cys	%	1.05
Digestible threonine	%	1.59
Choline	mg/kg	2706.00

Table 2.3 Sample matrix value of Pegazyme FP, a multienzyme product of Bentoli, Inc.

Nutrient	Unit	Value
Dry matter	%	98.00
Crude protein	%	2350.00
Digestible energy	kcal/kg	433000.00
Calcium	%	500.00
Available phosphorus	%	600.00
Digestible lysine	%	147.00
Digestible methionine	%	56.00
Digestible Met + Cys	%	102.00
Digestible isoleucine	%	123.00
Digestible threonine	%	73.00
Digestible tryptophan	%	21.00

Matrix values of ingredients may be constant, i.e. the formulator sets the nutrient values of the ingredients once and uses them repeatedly. Published values of ingredients in resources such as NRC books of nutrient requirements for various species may be used for this purpose. The values are referred to as *table values* or *book values*. The practice of relying on table values for matrix would likely result in inaccurate formulation because the nutrient composition of feed ingredients varies widely all the time. So, nutrient values should be updated based on the analysis of ingredients that are available for formulation. Since it is not practical to conduct an extensive analysis of every batch of every single ingredient received in a feed mill, feed manufacturers rely on analyzing a few components from which other components can be predicted. Typical proximate analysis for moisture, crude protein, crude fat, crude fiber, and ash serves this purpose. Crude protein value, for example, may be used for estimating amino acid composition by using an established relationship between crude protein and each amino acid. Near infrared spectroscopy (NIRS) enables fast and accurate proximate analysis of incoming ingredients. More details about the use of NIRS may be found in Chapter 6.

Manually updating all matrix values of ingredients every time the ingredients are analyzed can be a tedious and error-prone exercise. Automatic updating of matrix values is, therefore, desirable. By manually entering a few input values, for example, crude protein, it is possible for equations run by the software to arrive at calculated values of amino acids.

The cost of ingredients is in most instances the sum cost of goods, transport, and any applicable taxes. Some feed manufacturers include the cost of financing and shrinkage (reduction in ingredient quantity due to moisture loss, pests, process loss, etc.) in the cost as well. While the cost of an ingredient already bought and stored in the feed mill does not change, the economic value of the ingredient would fluctuate due to fluctuations in the cost of the same or alternative ingredients in the marketplace. Some feed manufacturers favor the use of *replacement cost*, the cost at which the ingredient would be brought in when it becomes exhausted, as the cost of the ingredient in formulation. Due to practical difficulties in arriving at the replacement cost, the challenges it will pose in the planning, procurement, and storage of ingredients, and problems it will create in the books of accounts, most use only the actual cost. Nevertheless, it would be a beneficial exercise to run formulas at replacement cost of ingredients to understand how ingredient use would change in the future and accordingly plan the current use and future procurement of ingredients.

3. Product specifications: Product specifications define the nutrient levels desired in the formula and the ingredient inclusion levels. Either a lower limit and/or an upper limit for each nutrient and ingredient are set (Table 2.4). The limits are commonly referred to as *constraints* and are the drivers of the solution. Each constraint has a potential cost associated with it, so constraints need to be set with careful consideration by the nutritionist.

Once all essential components of an LP-based feed formulation software are set up (Figure 2.1), formulation of products typically requires only the invoking of a key stroke or press of a button to produce the cost-optimized formula. If conflicting constraints or unrealistic targets are set within product specifications (for example, a formula with such a high protein target that none of the combinations of the ingredients can reach it) then the program would return an error called *infeasible solution*. When it occurs, sources of the errors must be identified and corrected.

The formulator should carefully examine the output of formulation software for any possible errors. Repeated exercise of optimization would be necessary in most cases to arrive at an acceptable formula that can be sent for production.

Table 2.4 **Sample product specifications for a tilapia grower feed**

	Minimum	Maximum
(A) Nutrient specifications		
Crude protein, %	28	
Crude fat, %	5.5	
Crude fiber, %		4
Ash, %		15
Digestible energy, kcal/kg	2600	
Digestible lysine, %	1.43	
Digestible methionine + cysteine, %	0.9	
Digestible threonine, %	0.74	
Linoleic acid, %	0.8	
Linolenic acid, %		
Available phosphorus, %	0.5	
Calcium, %		2.5
Sodium, %	0.4	
Potassium, %	1.0	
Magnesium, %	0.15	
Vitamin C, mg/kg	50	
Choline, mg/kg	500	
(B) Ingredient specifications, %		
Fishmeal, 62% protein		
Soybean meal, 46% protein		50
Rapeseed meal		10
Poultry by-product meal, 55% protein		25
Blood meal, spray-dried		5
Maize	20	30
Rice bran		15
Dicalcium phosphate		
Salt		1
Limestone		1
Fish oil	0.5	
Vegetable oil		
Vitamin premix	0.1	
Lysine HCl		
DL-methionine		
L-threonine		
Trace mineral premix	0.5	
Mold inhibitor	0.2	

Sophisticated feed formulation software provide several advanced tools for formulation efficiency and are discussed in detail elsewhere in this chapter, but one useful tool that comes with almost all feed formulation software is *shadow pricing*. For each ingredient that was rejected, the software would indicate how much the cost of the

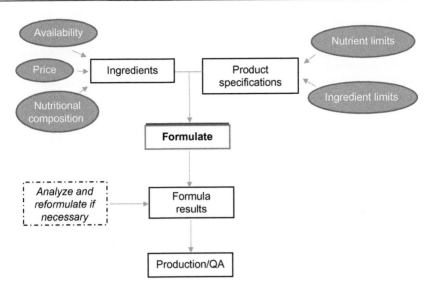

Figure 2.1 Overview of formulation process.

ingredient must fall before it can be included in the formula. The shadow cost of an ingredient or nutrient constraint is indicated by a value that shows what the impact of one unit change in the constraint will have on the formula cost. Ingredient buyers benefit from shadow cost data of ingredients to understand the real value of ingredients and negotiate prices, while formulators and nutritionists use the marginal price change information to gauge what constraints cost the formula the most and critically review them for possible cost reductions.

2.5 Software options

A wide variety of feed formulation software is presently available. Capabilities and costs of the different options vary considerably. Setting up of a software package for use involves considerable time and money. Users should, therefore, have a good understanding of their needs before selecting an appropriate software. The following are key criteria for evaluating available options and coming up with a decision:

1. User type and needs: Is the user a researcher, a consultant, a feed manufacturer with one feed plant or a feed manufacturer with several feed plants or a supplier to the feed manufacturer? How big and complex will the nutrient, ingredient, and product specification databases be? Will the user need to integrate the formulation software with other applications, for example, lab data systems, bag label generation, ingredient inventory systems, etc.? Special needs such as traceability must be considered as well.
2. Data inputs required to run a formulation system: Formulation software generates formulas based on the data provided by the feed company. The primary data input in any formulation

system are the ingredient composition data (such as the proximate composition values typically obtained from the lab) which drive various computational models to derive other nutrient values such as amino acid composition. The feed company should have access to reliable and accurate input data. It should also ideally have a system in place to update the data periodically and dynamically. If the company does not have a reliable data source, it will have to find out from the software providers whether such data can be procured.

3. Compatibility of the formulation software with the existing hardware and software setups: Computer hardware and software keep evolving and compatibilities change over time. A formulation company that offers products across multiple platforms is preferable to one that is limited to one platform. More importantly, the company should constantly be upgrading its products to take advantage of new capabilities offered by developments in information technology. Most formulation software are designed to run on a standard, store-bought desktop or laptop. Only when a high level of data security is needed, or highly complex problems such as large multi-blends with multiple price sets, multi-plant optimizations and associated logistics are to be solved, will an advanced computer system be required.

4. Post-sales support services offered by the formulation company: This includes personnel training, troubleshooting, customization, help desk, etc. It would be diligent to find out whether the formulation company has on its staff sufficient resources to support its customer base. This will be especially important in countries and companies that have limited IT resources.

5. Integration capacity with other company software, especially manufacturing and accounting: Formulation is only one step in the feed business that converts raw materials into finished feed. Capacity and ease of integration of the software are essential if formula changes need to be implemented rapidly throughout the organization.

6. Hardware requirements: As indicated earlier, most formulation software are designed to run on a standard, store-bought desktop or laptop. Most would run within a Microsoft Windows operating systems (OS). Options to run within Apple iOS are limited, so users of Apple computers will be required to partition the hard drives of their computers to run the software. Enterprises with multiple users accessing a common database will require to set up a client–server system with Internet connectivity and appropriate firewall-based security for remote access. The arrival of cloud computing has resulted in formulation being offered as a service. Users buy the service and connect via the internet to access databases stored in the cloud and perform formulation.

7. Cost: Software to formulate feed costs from absolutely free to a few hundred dollars to several thousand dollars depending on the features available and the number of user licenses required. Established companies charge for services like set up and training and annual fees for software updates. Cloud-based formulation as a service is typically charged as an annual recurring fee rather than as a one-time licensing fee for the software.

Commercial LP-based feed formulation software programs:

1. Microsoft Excel with the Add-In tool called Solver is a widely available option that can be used for feed formulation by teachers, students, and small-scale feed-mixing operations. The tool allows one to find an optimal value for a formula in one cell which is called the objective cell subject to constraints, or limits, on the values of other formula cells on a worksheet. Decision variables are defined in a group of cells and participate in computing the formulas in the objective and constraint cells. Solver adjusts the values in the decision variable cells to satisfy the limits on constraint cells and produces the result in the objective cell. A customized spreadsheet for feed formulation was developed by Thomson and Nolan (2001). The spreadsheet was available for download from the host institution of the authors, the University

of New England, Australia for a number of years. Unfortunately it is no longer available, but another version that is similar in function and named Windows User-Friendly Feed Formulation for Poultry and Swine is available for download from the University of Georgia's extension services website (extension.uga.edu). A detailed description of a predecessor of this application is presented in Pesti and Miller (1993). Using a spreadsheet tool with Solver is sufficient for teaching and learning and running occasional formulas, but frequent formulation of multiple formulas in a real feed mill environment will be difficult to manage.

2. There are inexpensive software packages like WinFeed (www.winfeed.com) that provide easy-to-operate interfaces to manage feed formulation for the purposes of teaching and small-scale feed mixing. Whether the companies involved in the development of the software have continued interest in updating the software and providing the necessary support need to be ascertained before buying the software.

3. Software programs such as Feedsoft (www.feedsoft.com) provide user-friendly interfaces and options to manage multiple clients and plants as well as cloud services to provide efficient tools for consultants and small feed companies that value ease of use.

4. Enterprise-level feed formulation solutions for large feed operations with multiple locations and thousands of formulas are offered by a handful of companies, namely Adifo (Bestmix software; www.adifo.be), A Systems (Allix^2 software; www.a-systems.fr), Feed Management Systems (Brill software; www.feedsys.com), and Format International (Format software; www.formatinternational.com). While these companies also offer single-user versions of formulation software for consulting nutritionists and small feed companies, their strength is in providing advanced features and options for various formulation and related functions (such as inventory management, quality control, and finance), data safety and security, and integration with information systems throughout an enterprise. The software packages typically provide the following features and options:

   a. Optimizing a single formula or multiple formulas at a time: Called Single Blend (or Single Mix) or Multi Blend (or Multi Mix), software companies provide options to optimize formulas one at a time or several formulas all at once. Single blend allows formulas to be optimized one at a time. When using single blend, it is assumed that all the ingredients available for use in the formula exist without any limitation in quantity. This is not a reasonable assumption as all ingredients are only available in finite quantities for feed production in a given location at any point in time. The alternative is to optimize all the formulas simultaneously and let the ingredients be used by the formulas in which those ingredients are valued the highest.

   b. Effective formulation of concentrates, supplements, and premixes: Designing and formulating concentrates, supplements, and premixes is not as straightforward as designing and formulating complete feeds. When a nutritionist wants to design a concentrate, she needs to consider the nutrient contribution of supplements that will be offered to the animal and vice versa. Advanced formulation software allow for the optimization of both components so that the resulting product will accurately serve the market needs. Sometimes, a product may have more than one stage of production and it is desirable that the feed formulation software provides options to optimize, save, and use formula from one stage as an ingredient to be used in the second stage.

   c. Integration with laboratory data and automated ingredient value recalculating models: Advanced formulation software provide in-built tools to facilitate integration with laboratory systems to stream in data on ingredient quality and automatically recalculate nutrient values of ingredients based on laboratory data. The equations that drive the recalculations can be encrypted to protect the intellectual property of the organization that developed the equations.

d. Sensitivity analysis: Feed formulators require tools to tell them why certain ingredients available in the feed mill were not used in a formula by the LP program and what sort of changes in formula cost would occur if nutrient specifications are altered or cost of the ingredients change. Formulation software packages include *basic* sensitivity analysis that provide what are known as *shadow prices* of ingredients that are rejected and nutrient levels that reach constraining levels (minimum/maximum) specified in the formula. In the case of nutrients, shadow price is the value by which the cost of the feed would be increased or decreased if the constraint were to be increased or reduced by one unit. For example, if the shadow price of crude protein is $0.05 in a formula with crude protein specification of 24% minimum, reduction of crude protein to 23% would result in a formula cost reduction of $0.05. In the case of ingredients, shadow prices indicate how much the price of an ingredient should be lower for inclusion in the formula. For example, if the price and shadow price of an ingredient are $0.3 and $0.03, respectively, then the price of the ingredient should become $0.27 to be included in the formula. Advanced software packages offer options for more detailed sensitivity analysis through what is known as *parametric analysis* to not only show the shadow prices but also how inclusion levels would change if an ingredient's cost would change in a wide range of values (Figure 2.2). Using this tool would help make decisions on the use of ingredients already in the feed mill's warehouse as well as about future purchases.

e. Stochastic programming: Variation in the nutrient composition of ingredients is common and unavoidable, and it imposes the risk of not achieving targeted nutrient levels in feed formulation. A common approach to managing this risk is to provide a margin of safety by either undervaluing the ingredients (for example, a fishmeal with an average crude protein value of 62% may be treated as a fishmeal with a crude protein value of 61.8%) or over-formulating the feed (for example, a feed with a target crude protein value of 32% may be formulated with a target crude protein value of 32.5%). However, this approach imposes a cost. *Stochastic programming* (also referred to as chance-constrained

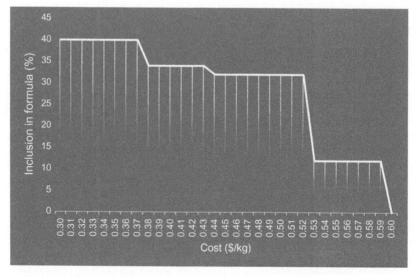

Figure 2.2 Parametric price map of an ingredient.

programming) provides a nonlinear approach to feed formulation by taking into account the variability of nutrients and the probability of meeting the nutrient requirement (Roush, 2004). Some advanced feed formulation software provide this option.

2.6 Conclusion

The current feed industry is provided with wide-ranging options in software for feed formulation. Selection of a suitable program depends on careful analysis of the needs and cost–benefit. LP-based feed optimization has become extremely affordable. However, use of even the simplest and easiest to use program requires substantial time and effort in setting up the program to work effectively. Users require periodic updating of knowledge with respect to new tools provided by software companies to enhance usability and functionality of feed formulation software to serve demands for increasing efficiency in not only formulation but also throughout the organization in functions such as purchasing, regulatory compliance, production, quality control, and sales.

Acknowledgments

I am grateful to George W. Chamberlain and Jan Van Eys for teaching me the basics of feed formulation. Benoit Chesnais and Arpad Zsok reviewed the first draft of the chapter and provided valuable feedback.

References

AAFCO, 2000. Official Publication 2000. Association of American Feed Control Officials Incorporated. Atlanta, Georgia.

Dent, J.B., Casey, H., 1967. Linear Programming and Animal Nutrition. Crosby Lockwood, London, p. 128.

Pesti, G.M., Miller, B.R., 1993. Animal Feed Formulation: Economics and Computer Applications. Chapman & Hall, New York, NY, p. 166.

Roush, W.B., 2004. Stochastic programming: addressing risk due to variability. Aqua Feeds Formul. Beyond 1 (3), 6–8.

Thomson, E., Nolan, J., 2001. UNEForm: a powerful feed formulation spreadsheet suitable for teaching or on-farm formulation. Anim. Feed Sci. Technol. 91, 233–240.

Waugh, F.V., 1951. The minimum-cost dairy feed. J. Farm Econ. 33, 299–310.

Understanding the nutritional and biological constraints of ingredients to optimize their application in aquaculture feeds

3

Brett Glencross
Ridley Agriproducts, Narangba, Queensland, Australia

3.1 Introduction

The identification, evaluation, and development of ingredients are key steps in the development of effective formulation strategies for all aquaculture feeds. Traditionally there has been much reliance on the use of fish meal and fish oil in aquaculture diets and because of this intensive aquaculture has been perceived as a net fish user rather than producer (Naylor et al., 2009). However, in addition to alleviating concerns about the reliability of aquaculture as a food provider, and also the long-term sustainability of aquaculture as an industry, the use of alternative ingredients also empowers the formulator with additional options. These options include improving the technical qualities of pellets and also introducing certain nutrients and nutraceuticals to increase the value of the resultant compound diet in which the ingredients are included.

Both protein meals and oils can generally be divided into those of plant or animal origin and many have considerable potential to supply the required dietary nutrients for fish. Clearly the optimization of the use of these resources in aquaculture diets depends on a detailed understanding of the chemical composition of these products and the consequences of feeding these products and their influence on each specific species being fed.

In assessing the value and potential of a range of plant protein resources there has been considerable research on the use of these resources in the diets of aquaculture species (Gatlin et al., 2007; Hardy, 2011). Soybean meal is the most widely produced and used plant protein source in aquaculture diet formulations, and has been applied with considerable success in diets for a wide range of species (Kaushik et al., 1995; Refstie et al., 1998; Glencross et al., 2004b). However, there are a range of other plant protein sources like field peas, lupins, and rapeseed meals and each of these grains has been shown to provide some value as a potential aquaculture feed ingredient (Burel et al., 2000, 2001; Booth et al., 2001; Glencross et al., 2003a,b).

The use of plant protein resources in fish diets can also introduce a suite of problems. Not only does the use of high levels of plant proteins increase the potential for inducing an essential amino acid limitation, but most plant protein resources also

S. Nates (Ed): Aquafeed Formulation. DOI: http://dx.doi.org/10.1016/B978-0-12-800873-7.00003-8

contain a variety of antinutritional (biologically active) factors (ANF). The influence of these ANF on fish can be considerable and varied (Francis et al., 2001).

Rendered animal meals are another protein resource stream that has been widely used in aquaculture diet formulations, with considerable success (Williams et al., 2003a; Glencross et al., 2011b). However, in some regions (e.g., Europe) there is limited use of these protein resources due to legislation and policy based on human health risk concerns (EU, 2001; Woodgate and Van Der Veen, 2004). However these concerns have never been substantiated in aquaculture species and there is still widespread use of these resources throughout most of the world. Additionally, there is substantial production of animal-derived oils from some sectors and these too have substantial potential as a feed resource (Turchini et al., 2009).

In some respects, with the modern trend toward using high-nutrient-dense diets in aquaculture, there is arguably a greater need in identifying and adopting alternatives to the use of fish oils than there is for fish meals. Although there has been much attention on the use of alternatives to fish oil use in aquaculture diets over recent decades, the consolidation of information on this issue remains a high priority for the aquaculture industry (Turchini et al., 2009).

This chapter examines the present state of knowledge of the use of a range of plant- and to a lesser extent animal-derived protein and oil resources in diets for aquaculture species. Consistent with an established framework for ingredient evaluation practices, first a characterization of the origins and the composition of key protein meal and oil commodities is presented. This is intended to detail the chemistry of each ingredient and the variations that occur between the different varieties and processing forms (Glencross et al., 2007a,b). Secondly, the digestibility of those ingredients and what role the characterization traits play in that digestible value, forms a second critical part of the assessment. Thirdly, the influence that different ingredients play on palatability (as assessed by feed intake) follows, and then fourthly, the ability of an animal to use those digested nutrients for growth (utilization) (Glencross et al., 2007a,b). Another critical step in the feed formulation and manufacture process is the influence that those ingredients play in the physical properties of a pellet in which they are included, often referred to as the ingredient's functional qualities (Thomas and van der Poel, 2001). Therefore this chapter frames the assessment of different ingredients according to this framework and how it applies to a range of aquaculture species.

3.2 Characterizing ingredients

3.2.1 Origin and processing

The origin (species and region) and processing used to produce different ingredients has a critical impact on their nutritional value. As an example, in lupin meals the efficiency of the dehulling processes used to produce a kernel meal has a clear impact on the protein content of the resultant meal (Drew et al., 2007). There can also be distinct effects of where certain protein meals are obtained from (region) and the conditions in which they were produced, as shown by Glencross et al. (2008) with genotype and growing season effects also impacting on the composition. Similar processing effects can be found with most grain protein sources and additionally with rendered animal meals, substantial differences can be observed in composition and nutritional values subject to species from which they are produced and also how the meals are dried (e.g., ring-dried, drum-dried, or spray-dried). Therefore, in addition to collecting chemical composition data, obtaining as much background data on origin and processing can substantially improve our understanding of different control points on the nutritional value of different ingredients.

3.2.2 Chemical composition of protein meals

3.2.2.1 Protein and amino acids

The crude protein content of the different protein resources considered in this chapter varies markedly between each resource (Table 3.1). Notably, the level of protein among different plant protein resources, like soybean meals, also varies according

Table 3.1 **Proximate specifications of key plant protein meal resources and fish meal (all values g/kg as received basis unless otherwise specified)**

Ingredient	Dry matter (DM)	Protein	Lipid	Ash	CHO (carbohydrate)	Gross energy (MJ/kg DM)
Anchovetta (fish) meal	926	668	75	186	4	21.8
Krill meal	917	687	129	117	0	21.5
Narrow-leaf lupin kernel meal[a]	910	402	70	33	497	20.7
Yellow lupin kernel meal[b]	937	496	67	41	410	21.0
White lupin kernel meal[c]	922	455	137	36	405	23.1
Lupin protein concentrate	937	716	81	25	115	23.4
Soy protein concentrate	920	871		4	14	66.0
Field pea meal	903	257	12	28	703	18.6
Expeller canola meal	938	380	136	59	425	22.0
SE canola meal	889	432	30	74	464	19.7
SE soybean meal	890	503	12	88	397	19.2
Full-fat soybean meal	909	416	196	53	336	23.4
Poultry offal meal	919	559	109	147	104	21.3
Blood meal	887	844	1	16	0	20.4
Meat and bone meal	950	560	88	324	27	19.3
Wheat	905	117	20	9	760	16.7
Wheat gluten	924	656	40	7	133	21.2

[a]*L. angustifolius.*
[b]*L. luteus.*
[c]*L. albus.*
SE, solvent extracted.
Source: Data derived from Glencross (2014), unpublished.

to whether the grain is hulled or dehulled, full-fat, or extracted and even according to the method of oil extraction used (Hardy, 2011). A similar level of variability is also observed of canola meals between oil extraction method used (Glencross et al., 2004a) and in lupin meals according to the method of protein concentration used (Drew et al., 2007). Solvent-extracted soybean meal has among the highest crude protein content of the plant protein meals though this is similar to that of yellow lupin kernel meals. The protein content of the field pea meal is generally among the lower of the potential plant protein meals considered for use in aquaculture diets. Processing of different grains to produce concentrates and isolates (including glutens in this category) tends to result in very high protein levels in the different products, often higher than that of fish meal. Protein content of most rendered animal meals is typically marginally lower than that seen in fish meal (500–600 g/kg).

The amino acid composition of the different ingredients also varies considerably (Table 3.2). In comparison to fish meal, which is usually considered as the ideal amino acid source for fish, most plant protein meals are relatively limited in their lysine and methionine content. Of the plant protein resources considered in this chapter the lysine content was highest in soybean protein concentrate and lowest in wheat gluten. However, the relative (% of protein) level of lysine was highest in canola but still lowest in wheat gluten. By contrast, most rendered animal meals have an amino acid profile closer to that observed of fish meal. Notably, plant protein meals tend to be largely devoid of taurine, which has been identified as an important amino acid in the diet of some species (Jirsa et al., 2014).

3.2.2.2 Lipids

The amount of lipids in plant protein ingredients varies considerably among the different resources (Table 3.1). Full-fat soybean meal has considerably greater levels of lipid than most other plant protein meals. Most lupin kernel meals are low in lipid (<100 g/kg) while rapeseed/canola meal, when processed by expeller extraction, can have reasonably high lipid contents at about 10–14%, but when solvent-extracted is generally of a similar lipid content to both pea meals and the solvent-extracted soybean meals. Those protein-concentrated grains, whether lupin, soy, or cereal in origin tend to have among the lower levels of residual fat in the meal. Rendered animal meals also have variable lipid levels, but typically less so than that of plant protein meals, and generally range from 100 to 150 g/kg.

The fatty acid composition of the lipid content of each of the meals also varies considerably among each of the resources (Table 3.3). Notably, the fatty acids are often examined on a proportional basis of the total fatty acids. Therefore consideration needs to be given to not only the levels of the specific fatty acids, but also the total amount of lipid in each meal. Notably, each of the plant proteins has relatively low levels of saturated fatty acids (SFA) in contrast to fish oil and rendered animal meals. The levels of monounsaturated (MUFA) and polyunsaturated fatty acids (PUFA) in plant protein lipids can vary considerably. The highest levels of monounsaturates are found in rapeseed and olive lipids, with the highest levels of PUFA found in linseed and soybean lipids. PUFA in each of the plant protein meals are predominated by

Table 3.2 Amino acid composition (g/kg as received basis) of range of ingredient resources

	Fish meal	Poultry	Blood	Krill	Wheat	Soybean	NL kernel[a]	YL kernel[b]	Canola	Wheat gluten	SPC
Alanine	45	43	80	45	5	22	14	16	26	19	28
Arginine	41	45	34	44	7	36	51	53	26	26	50
Aspartic acid	69	52	94	73	8	60	42	49	39	25	77
Cysteine	10	11	7	8	4	10	6	13	7	20	10
Glutamic acid	94	82	70	86	44	92	75	115	67	299	123
Glycine	39	61	38	34	6	20	17	19	24	25	27
Histidine	13	12	56	14	1	9	10	12	12	13	17
Isoleucine	29	25	4	30	5	22	16	18	19	28	28
Leucine	56	46	120	51	10	39	28	38	34	53	52
Lysine	55	36	81	44	5	32	17	21	28	11	37
Methionine	21	14	20	22	2	7	4	5	15	15	9
Phenylalanine	30	25	80	32	6	24	17	19	19	43	35
Proline	41	49	36	34	27	42	19	24	26	115	31
Serine	30	35	49	32	7	27	21	26	20	40	38
Taurine	5	3	0	5	0	0	0	0	3	0	0
Threonine	32	27	46	32	5	20	15	17	20	21	28
Tyrosine	23	18	24	27	4	18	16	15	15	27	24
Valine	33	27	71	31	6	23	15	15	21	28	30
Tryptophan	6	6	12			6	3				
sum AA	674	617	924	644	151	509	388	476	421	807	644

[a] *L. angustifolius.*
[b] *L. luteus.*
SPC, soy protein concentrate.
Source: Data from Glencross (2014), unpublished.

Table 3.3 Fatty acid composition (% of total fatty acids) of the key plant and animal oils

Fatty acid	Marine oils				Terrestrial animal oils			Terrestrial plant oils					
	Anchovy	Jack Mackerel	Capelin	Menhaden	Bovine	Ovine	Avian	Soybean	Palm	Canola	Linseed	Olive	Corn
SFA													
14:0	8	6	8	8	4	2	1	1	1				
16:0	18	17	11	19	25	18	22	8	44	6	5	11	11
18:0	6	4	1	4	19	16	6	3	5	3	4	4	2
20:0	4							1		1			
22:0	1												
MUFA													
16:1n-7	11	8	11	9	4	2	6	2	1	1		1	
18:1n-9	15	25	17	13	36	38	38	21	39	54	16	76	24
20:1n-9	3	4	19	2	1	1	1	2		2			
22:1n-11	2		15	1						1			
PUFA													
18:2n-6	1	4	2	2	3	13	20	57	11	20	19	7	58
18:3n-3	1	1	1	1	1	4	1	7		13	56	1	1
18:4n-3	3	2	2	3			1						
LC-PUFA													
20:4n-3	1	1											
20:4n-6	1	1		1									
20:5n-3	12	13	5	11									
22:5n-3	2	2	1	2									
22:6n-3	12	8	3	9									
SFA	36	26	20	31	48	35	29	13	50	9	9	15	13
MUFA	30	37	62	25	41	41	44	24	40	57	16	77	24
PUFA	5	6	4	5	4	17	22	64	11	33	75	8	59
LC-PUFA	26	25	8	23	0	0	0	0	0	0	0	0	0
n-3	30	26	11	26	1	4	2	7	0	13	56	1	1
n-6	2	5	2	2	3	13	20	57	11	20	19	7	58

Source: Data derived from Glencross (2009).

linoleic acid (18:2n-6), though appreciable levels of linolenic acid (18:3n-3) are also found in some products, such as linseed lipids. All plant lipids (and also lipids of rendered animal meals) are essentially devoid of long-chain polyunsaturated fatty acids (LC-PUFA), which are an important feature of fish meals.

3.2.2.3 Carbohydrates

Plant protein meals are typified by possessing higher levels of carbohydrates than most other feed ingredients, and with the exception of chitin in some crustacean meals and nominal amounts of glycogen, carbohydrates are largely absent in animal meals. The carbohydrate content can also be quite different among the many plant protein resources used in feeds. Typically carbohydrates are classified as either starch or nonstarch polysaccharides (NSPs). The NSPs primarily form the structural polysaccharides of the seed, though some are considered nonstructural, whereas starch is present largely as an energy reserve in the seed for the germinating plant embryo. While starch has two forms (amylose and amylopectin), there is a much greater degree of complexity in the range of NSPs present.

The precise composition of the NSPs differs between the plant protein resources. The hemicellulose content of the NSPs is generally proportionally greater in lupin kernel meals than in other resources such as field peas, rapeseed, and soybeans in which the cellulose content comprises a greater proportion of fiber (Petterson et al., 1999). A further group of polysaccharides, the pectins, are comprised primarily of β-(1,4)-galactan, which itself is comprised of subunits of L-rhamnose, L-arabinose, D-galactose, and galacturonic acid. The polysaccharide group of the lignins is prevalent in soybean meals, with considerably lower levels in the lupin kernel and field pea meals (Glencross et al., 2008). Both pectins and lignins have been shown to have direct effects on the digestibility of protein within plant protein ingredients (Glencross et al., 2012).

3.2.2.4 Antinutritional factors

ANFs, also referred to as biologically active substances, are essentially evolutionary developments by plants to enable some level of defense against being eaten (Francis et al., 2001). In this sense these ANFs are essentially a variety of chemical defense mechanism being employed by plants. However, the variety of ANFs found in the different plant species, let alone their seeds, varies quite widely, both in diversity of antinutrient type and relative concentration (Table 3.4).

3.2.2.4.1 Alkaloids

Alkaloids are generally bicyclic, tricyclic, or tetracyclic derivatives of the molecule quinolizidine (Petterson et al., 1999). Data on the influence of alkaloids on fish, show that alkaloids are generally considered a feeding deterrent because of their bitter taste (Glencross et al., 2006; Serrano et al., 2011, 2012). While the alkaloids are found primarily in the legumacae family (peas and beans), high levels are found in some varieties of lupins (Glencross et al., 2006). Present levels of alkaloids in commercial lupin varieties, such as Lupinus angustifolius are usually less than 200 mg/kg.

Table 3.4 Antinutrient levels (all values mg/kg) in a range of plant protein resources

	Alkaloids	Phytate	Tannins	Trypsin inhibitor	Oligosaccharides	Polyphenolics
L. angustifolius (cv Mandelup) kernel meal	33	5,222	0	9,222	85,556	2,889
L. angustifolius (cv Myallie) kernel meal	43	4,839	0	5,161	62,366	2,796
L. angustifolius (cv Merrit) kernel meal	11	5,761	0	1,957	64,130	2,391
L. angustifolius (cv Belara) kernel meal	11	5,889	0	6,444	88,889	2,778
L. luteus (cv Wodjil) kernel meal	143	7,253	0	2,857	102,198	2,418
L. albus (cv Kiev mutant) kernel meal	231	4,545	568	6,250	78,409	15,341
Whole soybean	292	8,202	0	22,247	40,449	4,719
Solvent-extracted soybean meal	44	8,889	0	10,000	62,222	4,556
Solvent-extracted canola meal	22	12,043	1,505	10,860	11,828	15,269
Whole canola (cv Surpass 501TT)	22	10,440	549	9,451	12,088	14,835
Whole field pea (cv Laura-Dunwa)	11	5,435	7,826	8,152	33,696	9,565
Dehulled field pea (cv Laura-Dunwa)	11	6,154	989	10,989	36,264	2,527

Source: Data derived from Glencross (2014), unpublished.

3.2.2.4.2 Glucosinolates

Glucosinolates in their own right have little biological activity. The actual antinutritional components are the breakdown products of glucosinolates, such as isothyocyanates, nitriles, thiocyanate anions, and vinyloxazolidinethiones and all have some goitrogenic activity. Effectively these compounds induce hypothyroidism in most vertebrate animals and lead to reduced levels of the thyroid hormones triiodothyronine (T_3) and thyroxine (T_4). The breakdown of glucosinolate to these products is achieved via the enzyme myrosinase which cleaves the glucosinolate to produce glucose and a variety of subsequent bioactive compounds. The modes of action of these bioactive compounds is closely involved with the synthesis of T_3 and T_4, notably thiocyanate anions compete with the active transport of iodine to the thyroid while vinyloxazolidinethiones block the coupling of subunits of the precursors to T_4. The consequence of the hypothyroidic response in fish is usually manifested by a reduced metabolic rate leading to lethargy, low appetite, and subsequently poor growth (Burel et al., 2000).

3.2.2.4.3 Lectins

Lectins, also known as hemagglutinins, are proteins that possess a specific affinity for carbohydrates. These proteins cause the agglutination of erythrocytes, hence their alternative name. The primary mode of the antinutritional action of lectins is their ability to reduce the absorption of nutrients from the gastrointestinal tract (Francis et al., 2001). Notably, lectins being proteins are heat labile and can be inactivated by precooking of meals.

3.2.2.4.4 Oligosaccharides

The oligosaccharides are generally α-galactosyl homologs of sucrose. Oligosaccharides also contain significant amounts of the raffinose, stachyose, verbascose, and sucrose families. Of these raffinose has a single galactose moiety linked to a sucrose molecule, while stachyose has two and verbascose three (Petterson, 2000). High levels of raffinose oligosaccharides have been reported to present some negative nutritional effects, some of which may be applicable to fish (Glencross et al., 2003a,b). These include: (i) interference with the digestion of other nutrients, (ii) osmotic effects of oligosaccharides in the intestine, and (iii) anaerobic fermentation of the sugars resulting in increased gas production. Other studies examining ethanol-soluble carbohydrates (most likely to be oligosaccharides) from soybean meals on Atlantic salmon, have also shown some antagonistic effects (Refstie et al., 1998).

3.2.2.4.5 Phytate

The molecule inositol hexaphosphate and salt ions of this molecule are commonly referred to as phytate. Phytate is strongly negatively charged at all pH values usually encountered in feeds. Consequently, phytate has been known to complex with proteins at acidic pH values and also polyvalent ions, such as zinc, at intestinal pH values. This has been reputedly attributed to a reduced availability of these nutrients to animals when fed diets with a significant phytate content. It has also been suggested that high dietary calcium levels can exacerbate the complexation of zinc with phytate. Other significant effects that have been attributed to phytate include depressed growth, depressed feed intake, reduced protein utilization, and depressed thyroid function (Petterson, 2000). The commercial use of exogenous enzyme supplements has made considerable improvements to the utilization of phytase by both pigs and poultry. The key to this is the use of the enzyme phytase (EC 3.1.3.8), which cleaves the phosphate units from the inositol base. Studies have indicated that there may be potential for phytase use with fish diets (Carter and Hauler, 2000; Mwachireya et al., 1999). Interestingly, improved feed intakes have also been observed of Atlantic salmon when fed diets containing phytase (Carter and Hauler, 2000).

3.2.2.4.6 Protease inhibitors

Protease inhibitors are specific substances that have the ability to inhibit the proteolytic activity of certain digestive enzymes. A range of protease inhibitors have been identified from a variety of plant meals. Notably, soybean is a prominent plant meal with a substantial protease inhibitor content and it has been shown to impact on fish (Krogdhal et al., 1994). Five different types of protease inhibitors have been identified in the seeds of this plant, which has trypsin inhibitor levels of about 60,000 mg/kg DM in unprocessed seed and about 3,400 mg/kg DM in processed meal (White et al., 2000). The mode of action of protease inhibitors is primarily through either the competitive or allostearic binding of the molecule to the digestive enzyme to render it inactive. Similar to lectins, being proteins, protease inhibitors are heat-labile and can be inactivated by precooking of meals.

3.2.2.4.7 Saponins

Saponins are plant glycosides with a steroid or triterpenoid structure as part of the molecule. Similar to alkaloids, saponins are also a bitter-tasting molecule. This means that their primary antinutritional basis is as a feeding deterrent. An additional effect attributable to saponins is an increase in the permeability of the small intestine mucosal cells (Fenwick et al., 1991). Saponins have also been implicated in a series of nutritional pathologies in salmonids associated with the use of soybean meals in particular (Bureau et al., 1998; Urán et al., 2008, 2009).

3.2.2.4.8 Tannins

Tannins are a group of polyphenolic compounds that bind to other proteins to either inhibit their activity in the case of digestive enzymes or to prevent their digestion, in the case of most other proteins. There are two tannin subgroups, those being either the hydrolyzable or condensed (nonhydrolyzable) forms. The condensed tannins have been reported to be able to precipitate proteins, particularly the digestive enzymes. Tannins can also form cross-linkages between proteins and other macromolecules and render them unavailable for digestion (Griffiths, 1991). These inhibitory facets, in conjunction with an astringent taste, constitute the antinutritional characteristics of tannins (Petterson, 2000).

3.2.2.5 Contaminants

Like all biological products, plant and animal protein meals (and oils) can suffer from contamination with undesirable chemicals. Residues of heavy metals, chemicals, molds, and mycotoxins can contaminate meals causing a significant reduction in their nutritional value to animals (Van Barneveld, 1999). Among the different chemical residues to be considered are preemergent herbicides, selective herbicides, fungicides, insecticides used during crop growth phases and at preharvest, and postharvest insecticides, antibiotics, and anthelmintics, organochlorides, and polychlorinated biphenyls. Management and even detection of each of these different contaminants can at times be difficult and costly and the risk of occurrence is generally low.

3.3 Chemical composition of oils

In addition to the protein meals that are produced from some plant varieties and also rendered animal products, another key commodity produced, in some cases for which some plants are specifically grown for, is their oil- and lipid-bearing qualities. The attributes of the lipid products produced by these different products is quite varied. This is most easily examined on a basis of the functional chemical classes found in most oils.

3.3.1 Lipid classes

Plant lipids typically constitute a range of lipid classes including neutral lipids, polar lipids, and sterols. For the most part, particularly in oilseeds such as canola and soybeans, the predominant lipid class is that of neutral lipids constituting primarily the triacylglycerols. In association with the triacylglycerols are lower amounts of diacylglycerols and monoglycerides. The amount of these later neutral lipids varies with the level of biosynthetic activity of the plant and also the quality of the oil extracted from the seed and how much damage it has undergone. In rendered oils, such as tallow and poultry oil, they also are predominantly triacylglycerides with only minor fractions of other neutral lipid classes.

Polar lipids are also present in some plant oils, usually as phosphatidylcholine or phosphatidylinositol. Typically, high levels of phosphatidylcholine are found in products such as lecithin and some of the gums associated with oil extraction. Generally these are not included in most oil products produced from plant resources, but residual levels of phosphatidylcholine or phosphatidylinositol may be found.

The sterol classes found in plants differ somewhat to those found in most animal systems. Typically these products are of the phytosterol group and generally have limited value as a nutrient in fish. Indeed there is some speculation over the potential antinutritional potential of some of these sterols because some possess estrogenic activities.

3.3.2 Fatty acids

Fatty acids comprise a length (chain) of carboxyl groups with the introduction of double (alkene) bonds introducing levels of unsaturation along the chain. A nomenclature designating the number of and position of the first of such bonds from the carboxyl terminus of the chain is established with the notation of n – or omega. On this basis a fatty acid chain with 18 carboxyl units, two unsaturated bonds with the first of those bonds, six bonds from the carboxyl terminus of the chain would be denoted as 18:2n-6.

Fatty acids can also generally be grouped according to whether they are SFA, MUFA, PUFA, or LC-PUFA. In most plant oils the fatty acid chains are 12 carboxyl units or more in length and can possess up to four (4) unsaturated bonds. In contrast some animal, fungal, and algal oils the fatty acids can possess up to six (6) unsaturated bonds. The position of these bonds tends to conform to the pattern of n-3, n-6, n-9, n-11. There are exceptions such as some of the fatty acids that possess n-5 and n-7 bonds.

In plant oils, such as those from rapeseed, soybean, and palm the common fatty acids tend to be 16:0, 16:1n-7, 18:0, 18:1n-9, 18:2n-6, and 18:3n-3. Those making up the greatest concentration are usually 16:0, 18:0, 18:1n-9, 18:2n-6, and 18:3n-3. However, the relative proportions of the fatty acids is generally unique to that of each particular plant species, with variations also seen within cultivars.

The major difference in fatty acid composition between the plant and terrestrial animal oils and those of marine origin (fish and krill oils) is the lack of appreciable levels of the long-chain polyunsaturates (LC-PUFA) such as eicosapentaenoic (EPA) or docosahexaenoic (DHA) acids. This difference is important because of the essential need by most aquaculture species for a certain level of these LC-PUFA in their diet.

3.4 Digestibility, palatability, and utilization value of plant protein meals

A wide range of plant protein commodities have been reported in the literature as being used as alternatives to fish meal in aquaculture diets for virtually all aquaculture species (Gatlin et al., 2007; Hardy, 2011). Key among the ingredients used has been soybean meals. More recently, this has further progressed to the development and evaluation of protein concentrates and isolates from soybean meals and some other ingredients. There are a range of other plant protein meals that are of interest as potential alternatives. Among those resources that are considered here are soybean, field peas, lupins, and rapeseed meals.

The approach taken in this review to evaluating each of these feed grain commodities has been to identify the digestible value of the primary plant protein resources and also the variety of resources of that grain available. In addition, maximum inclusion limits and factors influencing their inclusion have also been identified where possible along with some details on antinutritionals and recent findings on each respective plant protein resource following the convention posed by Glencross (2009).

3.4.1 Soybean

Soybean meals constitute one of the largest volumes of both plant protein meals and feed ingredient resources available in the world. Not only is the volume (some 177 million tonnes per annum in 2011) the largest, but a considerable diversity of products is also available. A range of processed soybean products (including protein concentrates and protein isolates) have been used by the aquaculture feed sector and have been evaluated in a range of species. Examples of digestibility values of some of the products in a range of species are presented in Table 3.6. There has been a huge amount of work evaluating the use of soybean products in a range of species. The following is a small spectrum of what is available and highlights some of the key tangibilities of this ingredient.

Kaushik et al. (1995) evaluated the digestible value of a wide range of soybean meals of various forms in rainbow trout (*Oncorhynchus mykiss*) (Table 3.5). From this study it was shown that all of the soybean meals had very high protein digestibilities. The extrusion of the meals had little influence on their digestible energy value. It could be argued that a single extrusion has some benefit, but that there is little additional benefit with further extrusion. The comparison of soybean meal and soy flour showed a minor improvement in the digestibility of both protein and energy. This was concomitant with the decrease in particle size of the ingredient being examined. This was attributed to an increased availability of both the protein and carbohydrate content from the meals. The concentration of the protein content was clearly seen to improve the overall value of soybean meals with an increase in digestible energy content without any loss in protein digestibility.

The work of Kaushik et al. (1995) further demonstrated that up to 62% of the diet could be comprised of a soy protein concentrate without loss of feed intake or performance of the fish. In contrast, the inclusion of soy flour at 42% of the diet resulted in a deterioration of growth, though feed intake was relatively uninfluenced. It was suggested that, at the higher inclusion level of soy flour, the NSP content of this ingredient was causing problems with nutrient absorption by the fish. In addition, the influences of oligosaccharides and saponins were also implicated. Work by Glencross et al. (2011a) observed a similar response of rainbow trout to solvent-extracted soybean meals in extruded diets (Figure 3.1). This was despite the diets being formulated to an equivalent digestible protein and energy basis. In this more recent study a minor decline in feed intake was seen at the 40% inclusion level of the soybean meal and the feed conversion was unaffected suggesting that it was this decline in feed intake that was the primary problem.

The relative nutritional value of a range of soy protein products when included in diets for Atlantic salmon (*Salmo salar*) on an equivalent protein basis was examined by Refstie et al. (1999). In this study solvent-extracted soybean meal was compared against an oligosaccharide reduced soybean meal, a soy protein concentrate, and a soy protein isolate. Significant improvements to organic matter and protein digestibilities were observed with increased level of processing of the soybean meals. Digestibility estimates of soybean meals presented by Refstie et al. (2006) were similar to those of Kaushik et al. (1995) (Table 3.5). Interestingly, recent data on the comparability of the different fecal collection methods used by different laboratories has shown some marked differences, in some cases with greater effect than that seen by fish species (Glencross et al., 2004b; Glencross, 2011). Some improvements to lipid digestibility were observed with removal of oligosaccharides and some of the other NSPs, but this was not consistent with the greatest levels of NSP removal as exhibited from the protein isolates treatment (Refstie et al., 1998, 2006).

Work by Krogdhal et al. (1994) demonstrated that the protease inhibitors, notably trypsin inhibitor from soybeans, had a clear negative impact on the intestinal trypsin activities and protein and amino acid digestibilities of rainbow trout. In a second

Table 3.5 Digestibility of a range of soy protein products when fed to different fish species

Ingredient	Ingredient composition					Method	Apparent digestibility		Species	Data source
	Dry matter	Protein	Lipid	Ash	CHO		Protein	Energy		
Full-fat soybean meal (double extrusion)						Settlement	97.2	86.7	O. mykiss	Kaushik et al. (1995)
Full-fat soybean meal (single extrusion)						Settlement	97.7	85.1	O. mykiss	Kaushik et al. (1995)
Soy flour						Settlement	95.1	80.7	O. mykiss	Kaushik et al. (1995)
Soy protein concentrate	939	590	54	79	277	Settlement	106.9	85.6	O. mykiss	Glencross et al. (2004)
Soy protein concentrate	939	590	54	79	277	Settlement	90.1	101.2	S. salar	Glencross et al. (2004)
Soy protein concentrate						Settlement	96.1	83.3	O. mykiss	Kaushik et al. (1995)
Soy protein isolate	938	893	13	47	47	Settlement	97.8	93.1	O. mykiss	Glencross et al. (2004)

Ingredient						Method			Species	Reference
Soy protein isolate	938	893	13	47	47	Settlement	97.4	117.4	*S. salar*	Glencross et al. (2004)
Soybean meal	893			47		Settlement	92.8	76.8	*O. mykiss*	Kaushik et al. (1995)
Soybean meal (solvent extracted)	890	503	12	88	397	Stripping	86.8	75.1	*O. mykiss*	Glencross and Hawkins (2004)
Soybean meal (solvent extracted)	890	503	12	88	397	Settlement	96.2	81.0	*P. auratus*	Glencross and Hawkins (2004)
Soybean meal (solvent extracted)	896	500	17	86	397	Stripping	103.4	65.5	*L. calcarifer*	Glencross (2011)
Soybean meal (solvent extracted)	909	518	47	69	365	Settlement	99.0	83.3	*O. mykiss*	Glencross et al. (2004)
Soybean meal (solvent extracted)	896	500	17	86	397	Stripping	97.0	70.5	*O. mykiss*	Glencross (2011)
Soybean meal (solvent extracted)	890	503	12	88	397	Settlement	79.4	59.2	*P. auratus*	Glencross et al. (2004)
Soybean meal (solvent extracted)	921	457	58	76	358	Settlement	94.4	88.0	*P. hypophtalamus*	Hien et al. (2010)
Soybean meal (solvent extracted)	909	518	47	69	365	Settlement	94.4	89.0	*S. salar*	Glencross et al. (2004)
Soybean meal (solvent extracted)	907	521	19	69	1000	Stripping	91.1	72.2	*S. salar*	Refstie et al. (2006)

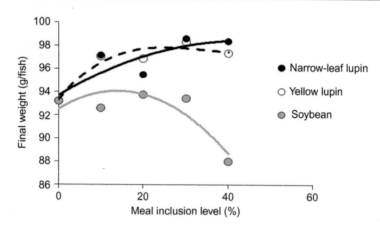

Figure 3.1 Growth response of *O. mykiss* to increasing inclusion of narrow-leaf lupin kernel meal, yellow lupin kernel meal, and solvent-extracted soybean meal, when each is included in diets formulated to equivalent digestible protein and energy specifications.
Source: Data derived from Glencross et al. (2011a).

study by this group Olli and Krogdahl (1995) examined the influence of some of the alcohol-soluble components of soybeans on the digestibility of lipids by Atlantic salmon. These workers noted that there was a significant reduction in the total digestibility of lipid and also some specific fatty acids, notably the saturates. It was not clear from this study, but it was implied, that these effects were due to the influence of soybean oligosaccharides and possibly also saponins, both of which were found in the alcohol-soluble components of soybeans. Additional complications on digestive function have been observed in Atlantic salmon and some other species through the use of soybean meals. A condition referred to as distal enteritis, where degradation of the villi of the distal intestine is accompanied by an inflammatory response, occurs (Baeverfjord and Krogdahl, 1996; Urán et al., 2008, 2009). This has tended to limit soybean meal inclusion in the diet of these species to less than 10%.

3.4.2 Lupin

Lupin production is substantially lower than that of many other plant protein ingredients, with about 2 million tonnes produced globally each year. There are numerous works that have studied the nutritional value of lupin meals and products when fed to a variety of aquaculture species. There are close to 200 different species of lupins, but of these there are really only four species which have been evaluated in aquaculture species. These are *Lupinus albus*, *L. angustifolius*, *Lupinus luteus*, and *Lupinus mutabilis* (Burel et al., 2000; Glencross and Hawkins, 2004; Glencross et al., 2010a).

Assessment of the digestibility characteristics of different species and processing forms of lupins are shown in Table 3.6. Protein digestibilities of lupin products are typically high (>90%), whereas energy digestibility ranges from 50% to 100%+ subject to the removal of the NSP fraction of the grain. Substantial variation in digestibility of both protein and energy has been shown to be affected by the efficiency

Table 3.6 Digestibility of a range of lupin protein products when fed to different fish species

| Ingredient | Ingredient composition | | | | | | Method | Apparent digestibility | | Species | Data source |
	Dry matter	Protein	Lipid	Ash	CHO			Protein	Energy		
Lupin protein concentrate	942	690	93	31	186		Settlement	101.0	86.6	*O. mykiss*	Glencross et al. (2004)
Lupin protein concentrate	942	690	93	31	186		Settlement	108.7	105.9	*S. salar*	Glencross et al. (2004)
Lupin protein isolate	926	810	125	30	35		Settlement	98.6	93.8	*O. mykiss*	Glencross et al. (2004)
Lupin protein isolate	926	810	125	30	35		Settlement	96.9	104.5	*S. salar*	Glencross et al. (2004)
Narrow-leaf lupin kernel meal	916	412	64	35	489		Stripping	96.1	73.4	*L. calcarifer*	Glencross (2011)
Narrow-leaf lupin kernel meal	885	415	53	33	499		Settlement	97.2	70.5	*O. mykiss*	Glencross et al. (2004)
Narrow-leaf lupin kernel meal	916	412	64	35	489		Stripping	92.2	69.6	*O. mykiss*	Glencross (2011)
Narrow-leaf lupin kernel meal	910	411	60	32	497		Stripping	93.1	62.4	*O. mykiss*	Glencross and Hawkins (2004)
Narrow-leaf lupin kernel meal	910	411	60	32	497		Settlement	99.1	62.4	*P. auratus*	Glencross and Hawkins (2004)
Narrow-leaf lupin kernel meal	885	415	53	33	499		Settlement	130.4	69.6	*S. salar*	Glencross et al. (2004)
Narrow-leaf lupin kernel meal	905	425	75	34	466		Stripping	70.5	67.9	*S. salar*	Refstie et al. (2006)
Narrow-leaf lupin protein concentrate	937	754	153	23	70		Stripping	86.0	100.4	*L. calcarifer*	Glencross (2011)

(Continued)

Table 3.6 (Continued)

Ingredient	Ingredient composition					Method	Apparent digestibility		Species	Data source
	Dry matter	Protein	Lipid	Ash	CHO		Protein	Energy		
Narrow-leaf lupin protein concentrate	937	754	153	23	70	Stripping	97.5	88.8	*O. mykiss*	Glencross (2011)
Narrow-leaf lupin protein concentrate	926	783	110	29	78	Stripping	96.8	94.6	*S. salar*	Refstie et al. (2006)
White lupin kernel meal	922	455	137	36	405	Stripping	88.3	64.0	*O. mykiss*	Glencross and Hawkins (2004)
White lupin kernel meal	922	455	137	36	405	Settlement	100.1	60.9	*P. auratus*	Glencross and Hawkins (2004)
Yellow lupin kernel meal	913	567	67	39	327	Stripping	81.2	82.7	*L. calcarifer*	Glencross (2011)
Yellow lupin kernel meal	913	567	67	39	327	Stripping	92.5	78.5	*O. mykiss*	Glencross (2011)
Yellow lupin kernel meal	937	496	55	38	410	Stripping	95.3	64.9	*O. mykiss*	Glencross and Hawkins (2004)
Yellow lupin kernel meal	937	496	55	38	410	Settlement	97.7	69.5	*P. auratus*	Glencross and Hawkins (2004)
Yellow lupin kernel meal	909	537	77	44	342	Stripping	79.4	69.7	*S. salar*	Refstie et al. (2006)
Yellow lupin protein concentrate	931	819	112	29	40	Stripping	98.6	113.0	*L. calcarifer*	Glencross (2011)
Yellow lupin protein concentrate	931	819	112	29	40	Stripping	96.5	90.2	*O. mykiss*	Glencross (2011)
Yellow lupin protein concentrate	932	811	55	32	102	Stripping	107.7	92.3	*S. salar*	Refstie et al. (2006)

of seed coat removal (dehulling), but above 50% dehulling efficiency it has been shown that there is not much improvement in protein digestibility, but the energy digestibility keeps improving and the protein concentration in the meal also increases (Figure 3.2) (Glencross et al., 2007a,b). Additionally, there is substantial variation in protein digestibility among kernel meals of different genotypes of the *L. angustifolius*

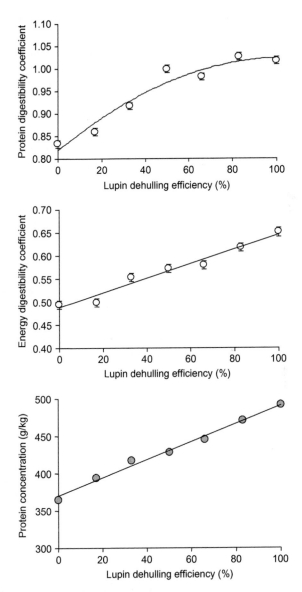

Figure 3.2 Effect of increasing efficiency of lupin dehulling (removal of seed coat) on the digestibility of protein and energy of the ingredient when fed to *O. mykiss*.
Source: Data from Glencross et al. (2007a,b).

species, with protein digestibility coefficients being documented to range from 65% to 100%+ (Glencross et al., 2008). Although variation due to genotype, production season characteristics, and processing state have been documented, studies by Glencross et al. (2003a,b, 2008) demonstrated that protein digestibility was intrinsically linked to the protein and lignin content within the meal.

Among kernel meals of all four species of lupin (*L. albus*, *L. angustifolius*, *L. luteus*, and *L. mutabilis*) digestibility when fed to rainbow trout (*O. mykiss*) has been shown to be relatively consistent (Glencross and Hawkins, 2004; Glencross et al., 2010a). The digestibility of protein from each of these lupin kernel meals was generally equal to or better than for the soybean meal. The digestibility of protein concentrates and isolates from *L. angustifolius*, *L. luteus*, and *L. mutabilis* was also compared to soybean meal (Glencross et al., 2010a). In each case the protein digestibility from each species was about the same (90–93%), though energy digestibility varied from 58% to 84% subject to the amount of NSP removed. A commensurate improvement in energy digestibility was also seen with the increase in removal of NSP with the production of protein isolates from different lupin species, though limited gains in protein digestibility occurred as they were already arguably maximal anyway (Glencross et al., 2004b, 2007b).

Another salient difference between lupin meals and soybean meals is the level of phosphorus digestibility observed in rainbow trout. High phosphorus digestibilities have been obtained from kernel meals of *L. angustifolius* (100%), *L. albus* (100%), and *L. luteus* (100%), while the soybean meal had a significantly poorer phosphorus digestibility (45%) (Glencross and Hawkins, 2004). This finding has important implications for the development of phosphorus-limiting diets, which are sometimes required in freshwater aquaculture farming systems.

Inclusion of lupin kernel meals in diets on growth and feed utilization parameters has also demonstrated good results. Glencross et al. (2011a) showed that up to 40% of kernel meal from either *L. angustifolius* or *L. luteus* could be included in extruded diets for rainbow trout and not have any negative impact on growth (Figure 3.1). These marginal improvements in growth with the use of lupin kernel meals were underpinned by increases in feed intake relative to the fish-meal-based control used in this study, thereby providing good evidence for the beneficial effects on palatability by these ingredients. Similar improvements in feed intake and growth with the use of lupin kernel meals have been observed in other species like Asian sea bass (*Lates calcarifer*) and Atlantic salmon (*S. salar*) (Glencross et al., 2011b; Salini and Adams, 2014). In contrast to soybean meal, studies looking for distal enteritis in lupin-meal-fed fish have found that this feed grain does not induce this pathology in Atlantic salmon (Refstie et al., 2006).

A nonnutritional feature of lupins is their effect on the feed extrusion process. The NSP content of lupins has highly hygroscopic properties meaning a higher amount of water is required to be added to the mash during extrusion. This, in combination with the protein fraction of the lupins leads to a stronger, denser pellet (Figure 3.3). The higher amount of water added also has the benefits of reducing wear and depreciation of feed extrusion equipment (Glencross et al., 2010b).

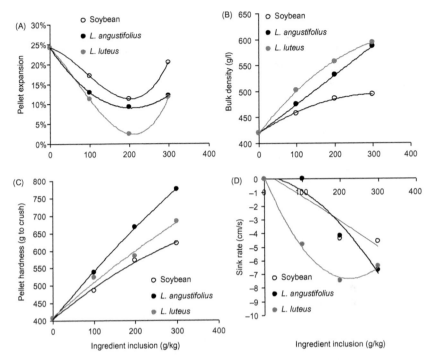

Figure 3.3 Influence of solvent-extracted soybean meal, *L. angustifolius* kernel meal or *L. luteus* kernel meal on the technical properties of pellets produced using extrusion processing. Represented are (A) pellet sink rate; (B) pellet hardness; (C) pellet bulk density; (D) pellet expansion.
Source: Data from Glencross et al. (2010).

3.4.3 Rapeseed

Rapeseed, also known as canola, are names used to describe the plants *Brassica napus* and *Brassica campestris*. The name "canola," is actually a combination of two words – "Canadian" and "oil," based upon particular varieties developed by the Canadians but now used throughout the world. These varieties of the plant are notable in that the seeds yield oil with less than 2% erucic acid, and the air-dried, oil-free meal contains less than 30 micromoles of glucosinolates per gram (or 30 mmol/kg) (Anderson-Hafermann et al., 1993). Although the primary product from canola is its oil content, canola meal is also a valuable protein resource for use in the animal feed industries. Like soybean and lupins, the carbohydrate content of rapeseed is largely devoid of starch.

The digestibilities of rapeseed meals have been evaluated in several studies across a range of species. Burel et al. (2000) examined the nutritional value of both solvent-extracted and heat-treated rapeseed meals when fed to rainbow trout (Table 3.7). These authors observed that there was a level of consistency in the protein digestibility of both rapeseed meals (90.9% and 88.5%). However, the heat-treated rapeseed

Table 3.7 Digestibility of a range of rapeseed protein products when fed to different fish species

Ingredient	Ingredient composition						Method	Apparent digestibility (%)		Species	Data source
	Dry matter	Protein	Lipid	Ash	CHO			Protein	Energy		
Canola meal (expeller extracted)	916	388	133	53	426		Stripping	99.6	68.8	*O. mykiss*	Glencross (2011)
Canola meal (expeller extracted)	916	388	133	53	426		Stripping	63.3	59.7	*L. calcarifer*	Glencross (2011)
Canola protein Concentrate	483	483	33	59	425		Settlement	74.4	64.9	*P. auratus*	Glencross et al. (2004)
Canola protein isolate	971	935	34	57	−26		Settlement	97.6	84.7	*O. mykiss*	Mwachireya et al. (1999)
Commercial canola meal	892	407	76	123	394		Settlement	88.1	55.6	*O. mykiss*	Mwachireya et al. (1999)
Expeller Canola meal	898	381	136	66	418		Settlement	89.2	62.1	*P. auratus*	Glencross et al. (2004)
Expeller canola meal – 120°C	898	381	136	66	418		Settlement	61.4	34.3	*P. auratus*	Glencross et al. (2004)

Expeller canola meal – 150°C	898	381	136	66	418	Settlement	37.4	27.4	*P. auratus*	Glencross et al. (2004)
Galactosidase-treated canola meal	981	446	55	100	399	Settlement	84.2	56.1	*O. mykiss*	Mwachireya et al. (1999)
Heat-treated rapeseed meal	915	433	9	82	391	Settlement	88.5	70	*O. mykiss*	Burel et al. (1998)
Methanol-ammoniated canola meal	882	443	43	144	370	Settlement	83.8	41.4	*O. mykiss*	Mwachireya et al. (1999)
Phytase-treated expeller canola meal	898	381	136	66	418	Settlement	97.3	61.1	*P. auratus*	Glencross et al. (2004)
Phytase-treated canola meal	990	487	47	116	350	Settlement	84.4	51.5	*O. mykiss*	Mwachireya et al. (1999)
Sieved canola meal	895	415	75	125	385	Settlement	85.8	52.3	*O. mykiss*	Mwachireya et al. (1999)
Solvent extract rapeseed meal	937	431	48	79	379	Settlement	90.9	76.4	*O. mykiss*	Burel et al. (1998)
Solvent-extracted canola meal	962	431	22	86	461	Settlement	82.1	46.1	*P. auratus*	Glencross et al. (2004)

meal (70%) had poorer energy digestibility than the solvent-extracted rapeseed meal (76%). This is probably a direct reflection of the markedly lower fat content (43 g/kg DM cf. 9 g/kg DM) of the heat-treated meal. In most cases, the relative digestibility of the energy of each of the ingredients was a direct response to the protein content of the ingredient and the relative protein digestibility of that ingredient.

Mwachireya et al. (1999) also evaluated the nutritional value of a range of rapeseed products derived from the physical, enzymatic, and chemical processing of a commercially produced meal (Table 3.6). The products ranged in protein concentration from 407 g/kg DM in the unprocessed commercial meal to a protein content of 935 g/kg DM in the canola protein isolate (CPI). the energy content of each of the meals was relatively constant, though substantially enriched by the production of a protein isolate. The antinutritional content of the meals was also reduced with processing. Digestibility of the protein of each of the canola meals varied only minimally. The greatest improvement in protein digestibility was observed with the development of the CPI.

Glencross et al. (2004a) examined the influence of the oil extraction method and heat damage on the digestibility of rapeseed meals when fed to red sea bream (*Pagrus auratus*). There was a significant difference in protein digestibility between the solvent-extracted (83%) and expeller-extracted (94%) meals and the energy digestibility was even more divergent (44% vs. 62%). The application of heat (120°C or 150°C for 30 min) to the expeller-extracted meal resulted in a direct reduction in both protein and energy digestibilities. Protein digestibilities dropped from 94% to 51% to 23% with increasing heat, demonstrating a direct impact of heat damage on the ability of fish to digest protein from this feed grain. A deterioration in digestible and/or utilization value of rapeseed meals has been linked to the occurrence of Mailliard reactions reducing the reactive lysine in the meal through glycosylation of the amino residue on the lysine side chain (Oste, 1984).

Some of the earliest work on the effects of rapeseed meals on growth of fish was that by Higgs et al. (1982). In this study these workers evaluated the potential of two varieties of rapeseed meals and also that of a rapeseed protein concentrate (RPC) when fed to Chinook salmon (*Oncorhynchus tshawytscha*) (Table 3.8). This work was notable in that it presented some of the earliest work on the use of protein concentrates being used in fish diets. The performance of fish fed diets with the RPC was good. In most cases the growth rate of fish fed these diets was equivalent or superior to that of fish fed the fish-meal-based reference diet. There were also differences observed between the two varieties of canola meals. The protein content of the "Candle" rapeseed meal was marginally higher than that of the "Tower" rapeseed meal (410 g/kg cf. 381 g/kg). The levels of glucosinolates introduced into the test diets by each variety were also different, with the higher protein variety of "Candle" containing about 663 mg/kg and the "Tower" variety containing 557 mg/kg. At lower inclusion levels of either rapeseed meal, no deterioration in growth performance of the fish was observed. However, with the highest inclusion levels of the "Tower" meal a decline in growth was noted, but such a decline with a similar inclusion of "Candle" meal was not observed. A second study by Higgs et al. (1983) focused on the protein quality of another rapeseed variety "Altex," also when fed to Chinook salmon. In this study Higgs et al. (1983) fed isoenergetic diets, varying in protein content of 290,

Table 3.8 Evaluation of different varieties of canola meal and a CPC when fed to juvenile Chinook salmon, *O. tshawytscha*

	Reference	T1	T2	T3	C1	C2	C3	B1	B2	B3
Ingredients (g/kg)										
Cottonseed meal	94	–	–	–	–	–	–	–	–	–
Canola meal (Tower)	–	120	159	297	–	–	–	–	–	–
Canola meal (Candle)	–	–	–	–	129	171	320	–	–	–
RPC (Bronowski)	–	–	–	–	–	–	–	80.6	106	199
Fish meal	489	489	489	407	489	489	407	489	489	407
Dried whey	24	24	–	–	24	–	–	24	–	–
Wheat germ meal	43	43	–	–	43	–	–	43	–	–
Wheat middlings	96	61	84	10	70	96	30	110	149	130
DL–Methioine	4.3	3.2	2.5	2.2	2.6	3.4	2.0	4.5	4.6	4.9
Other ingredients	249.7	259.8	265.5	283.8	242.4	240.6	241	248.9	251.4	259.1
Diet proximate composition										
Crude protein (g/kg DM)	548	539	538	536	553	556	534	558	558	549
Crude fat (g/kg DM)	181	170	172	170	164	167	168	162	160	175
Isothiocynates (mg/kg DM)	0	46	61	113	68	90	168	0	0	0
Oxazolidnethione (mg/kg DM)	0	21	28	52	18	23	44	0	0	0
Total glucosinolates (mg/kg DM)	0	67	89	165	86	113	212	0	0	0
Growth performance										
Specific growth rate (%/day)	1.8	1.8	1.9	1.9	1.8	1.8	1.8	1.9	2.0	2.1
Nitrogen retention (%)	30.0	31.0	32.0	25.0	31.0	33.0	30.0	20.0	27.0	35.0
Food conversion ratio (g/g)	5.6	5.3	4.8	6.3	5.6	4.8	5.3	5.6	5.3	4.2

NFE, nitrogen-free extractive.
Source: Data derived from Higgs et al. (1982).

390, and 490 g/kg. The canola meal was included to provide 0%, 11.5%, and 23% of the total dietary protein, essentially replacing the fish meal content of the diet. These workers suggested that the nutritive value of canola and fish meal protein were similar when evaluated in the 290 and 390 g/kg protein diets. However, when the canola meal was included in the 490 g/kg protein diet a decrease in growth performance of the fish was observed. Primarily this was a response to deterioration of feed intake by the fish. From this study Higgs et al. (1983) concluded that "Altex" rapeseed meal could comprise up to 25% of the dietary protein content without adversely influencing the protein quality of the diet. However, this limit was contingent on the total glucosinolate level of the diet being kept below 2.65 mmol/kg. These two studies by Higgs et al. (1982, 1983) provided good support that canola meals could be useful ingredients when included in diets for fish. They identified that glucosinolates were a potential problem with high inclusion levels. There was also some indication that there was inherent variability between different meal varieties, but that the development of protein concentrates offered some of the best potential.

A study by Teskeredžić et al. (1995) examined the inclusion of normal and dephytinized RPC when fed to rainbow trout. Three different sources of RPC were evaluated as partial or total replacements of fish meal in practical, isonitrogenous (430 g/kg) and isoenergetic (21.6 MJ/kg GE) diets. The three sources of RPC included undephytinized RPC, undephytinized solvent-treated control RPC, and dephytinized RPC. Each was included in test diets at 19%, 39%, and 59% of the total diet, which effectively allowed replacement of 33%, 66%, and 100% of the dietary protein content. Growth, survival, feed intake, and feed efficiency of the fish were unaffected with the inclusion of either undephytinized RPC or dephytinized RPC at up to 66% replacement of the fish meal. Although 100% replacement of the fish meal did not influence feed intake of the fish, it did significantly reduce growth and accordingly feed efficiency.

A study by Forster et al. (2000) examined the potential of the exogenous enzyme phytase to improve the nutritional value of a canola protein concentrate (CPC). In this study improvements in the digestibility of protein in the diets were observed with the inclusion of CPC in the diet. However, no improvements in the digestibility of dietary energy content were observed with inclusion of CPC. The resultant effects of these changes in nutritional value on the growth performance of the fish were minimal, suggesting, if anything, only a slight deterioration in growth. The addition of phytase to the diet resulted in significant improvements in the digestibility of phytate, and at the highest inclusion level of phytase, improvements in phosphorus digestibility were also noted.

Most notable about the use of rapeseed/canola meals was the absence of any recent data on its utilization in feeds. Given the progress in feed formulation technology and diet specifications in the past 10 years, it would be prudent to revisit the use of this ingredient.

3.4.4 Field peas

Field peas, although legumes from the same family as lupins and soybeans, differ considerably in their composition. Notably pea meals have a lower protein content (~250 g/kg), a very low fat content (~10 g·kg), and a high carbohydrate content

(700 g/kg), which is predominantly starch. There has also been some production and evaluation of pea protein concentrates (PPC). These products have a protein level more consistent with soybean meals and some lupin kernel meals, with the added advantage that their carbohydrate content contains significant levels of starch in contrast to the NSP of the other grains.

Booth et al. (2001) also examined the nutritional value of a range of plant legumes including field peas, faba beans, chick peas, and vetch when fed to silver perch (*Bidyanus bidyanus*) (Table 3.9). In this study each of the grains was also evaluated in whole-seed meal and kernel meal forms. Notably, the dehulling process realized only a minor increase in protein content of the meal from about 25.5% to 27.7%. This level of increase in protein content was consistent among the legume meals evaluated. The fat content of each of the legume meals, field peas included, was low, with most containing less than 2% fat and no increase in fat content was observed with dehulling. The protein digestibility of the field pea meals was exceeded only by that of Faba bean meals. Improvements in protein digestibility of field peas were observed with dehulling. This effect was consistent across all of the legume meals examined, excepting chick pea meals. Energy digestibilities of each of the plant legumes were improved with dehulling of the grains. However, digestibility of the Faba bean meals was an exception to this, with a moderate decrease in energy digestibility with dehulling. Notably, the carbohydrate content of each of the meals examined was predominated by starch. Therefore it is possible that the fish are obtaining significant energetic value from this nutrient.

Carter and Hauler (2000) evaluated the use of a field PPC, *L. angustifolius* kernel meal, and a soybean meal when fed to Atlantic salmon. Each ingredient was included in the diet to constitute 40% of the total dietary protein. The highest protein digestibilities were those observed from the diets with PPC (93.8%), followed by the soybean meal diets (92.0%) and the *L. angustifolius* kernel meal (91.3%). The apparent energy digestibilities of the PPC diet (83.1%) were slightly lower than the diet that included soybean meal (86.4%), with the diet containing *L. angustifolius* kernel meal having the lowest energy digestibilities (80.5%). Of note were the phosphorus digestibilities, the highest of which was the diet containing the *L. angustifolius* kernel meal (46.7%), followed by the PPC (41.6%). The diet containing soybean meal (27.5%) had considerably poorer phosphorus digestibility than even the reference diet (39.8%).

The use of PPC has also been examined in Atlantic salmon (Overland et al., 2009). Using two different types of concentrates (35% protein and 50% protein), these researchers found that either PPC made no impact on the growth or feed conversion relative to the fish meal reference diet, but the soybean meal diet used in the same experiment had a poorer growth and higher feed conversion. Digestibility of the diets used in the growth study showed reduced lipid digestibility in the soybean-fed fish, but no impact of either of the PPCs. Protein digestibility was unaffected by any treatment.

3.4.5 Cereals and glutens

Most cereal grains are notable in that they possess very low levels of protein and are generally used for their starch properties, not their protein. However, some cereals are

Table 3.9 Digestibility of a range of legume and cereal protein products when fed to different fish species

Ingredient	Ingredient composition					Method	Apparent digestibility		Species	Data source
	Dry matter	Protein	Lipid	Ash	CHO		Protein	Energy		
Chickpea kernel meal	–	242	50	–	–	Settlement	81.2	60.2	*B. bidyanus*	Booth et al. (2001)
Chickpea whole seed	–	208	47	–	–	Settlement	84.8	53.6	*B. bidyanus*	Booth et al. (2001)
Faba bean kernel meal	–	313	13	–	–	Settlement	96.6	58.8	*B. bidyanus*	Booth et al. (2001)
Faba bean whole seed	–	277	13	–	–	Settlement	91.6	62.2	*B. bidyanus*	Booth et al. (2001)
Field pea – extruded	909	260	4.5	33	612	Settlement	87.9	68.9	*O. mykiss*	Burel et al. (1998)
Field pea kernel meal	–	277	10	–	–	Settlement	88.1	67	*B. bidyanus*	Booth et al. (2001)
Field pea whole seed	–	255	11	–	–	Settlement	83.3	54.5	*B. bidyanus*	Booth et al. (2001)
Vetch kernel meal	–	323	9	–	–	Settlement	87.7	81.8	*B. bidyanus*	Booth et al. (2001)
Vetch whole seed	–	309	9	–	–	Settlement	74.9	55.5	*B. bidyanus*	Booth et al. (2001)
Barley	1000	151	44	21	784	Stripping	152.7	54.7	*L. calcarifer*	Glencross et al. (2010)
Corn	1000	52	26	18	905	Stripping	149.9	43.2	*L. calcarifer*	Glencross et al. (2010)
Faba beans	1000	380	63	36	521	Stripping	104.3	61.6	*L. calcarifer*	Glencross et al. (2010)
Oats	1000	135	91	25	749	Stripping	98.0	52.4	*L. calcarifer*	Glencross et al. (2010)
Sorghum	1000	138	39	15	808	Stripping	109.9	53.7	*L. calcarifer*	Glencross et al. (2010)
Tapioca	1000	7	3	4	986	Stripping	n/a	58.0	*L. calcarifer*	Glencross et al. (2010)
Triticale	1000	205	26	20	749	Stripping	110.7	57.3	*L. calcarifer*	Glencross et al. (2010)
Wheat	1000	196	31	15	758	Stripping	100.2	65.2	*L. calcarifer*	Glencross et al. (2010)
Wheat gluten	910	838	9	8	146	Settlement	103.6	95.8	*P. auratus*	Glencross and Hawkins (2004)
Wheat gluten	910	838	9	8	146	Stripping	101.0	90.7	*O. mykiss*	Glencross and Hawkins (2004)
Wheat gluten	1000	710	46	8	236	Stripping	121.9	107.3	*L. calcarifer*	Blyth et al. (2015)

processed to isolate their protein content, which has, in addition to good nutritional properties, some useful functional properties as well.

A study by Glencross et al. (2012) examined the digestibilities of a range of cereal grains when fed to barramundi (*L. calcarifer*) in extruded feeds. Among the cereal grains were wheat, oats, sorghum, triticale, and barley. Significant variability in the digestible energy and starch value were seen among the cereal grains (Table 3.9). Ingredient starch digestibility ranged from 18% to 96%. Protein digestibilities were all very high, often exceeding 100%. Energy digestibilities ranged from 43% for corn to 65% for wheat. However, among the different starch ingredients a strong negative correlation was observed between the ingredient amylopectin content and starch digestibility of that ingredient. It was noted from this study that overall, barramundi have limited ability to digest starch and that there was an effect of inclusion level on the ability of the fish to digest starch.

In addition to the study by Glencross et al. (2012) on cereal grain digestibilities, these authors also examined the digestibility of wheat gluten when fed to barramundi (Table 3.7) (Blyth et al., 2015). Further studies on rainbow trout and red sea bream (Glencross and Hawkins, 2004) also confirm that protein digestibility of wheat gluten is generally 100%+ in all observances. Energy digestibility of this ingredient ranges from 91% to 100%+.

What was not found was a clearly defined study examining the serial inclusion of wheat (or corn gluten) on performance by a fish species. This work would be useful to help establish intake and utilization boundaries for this ingredient (Glencross et al., 2007a,b).

3.4.6 Rendered animal meals

By-products from animal slaughtering have long been used as feed ingredients in aquaculture. Traditionally unused parts of the animal, such as offal, were minced and fed and over time this evolved to being combined with dry materials to make wet pastes incorporating such animal by-products. The modern strategy though is to process and dry the by-products to improve their shelf life and utility. However, like all processing effects and input (species and organ) origin effects, there are also substantial differences in the qualities of rendered animal by-products.

Bureau et al. (2000) thoroughly examined the nutritional value of a broad range of rendered animal protein sources when fed to rainbow trout (Table 3.10). These authors found that there were some clear effects of drying method on the protein digestibility of blood meals and that energy digestibility is strongly linked to protein digestibility in all rendered products. Spray drying produced the best-quality blood meals, while those drying methods using high temperatures resulted in poorer protein digestibility. Similarly, drying also had a clear impact on feather meals, with steam drying producing a better-quality product. The effects of drying on the digestibility of the meat and bone meals were less clear. Drying temperatures ranging from 125°C to 138°C and drying times of 25–60 min largely produced protein and energy digestibilities of equal qualities. A similar result was also seen with the poultry by-product meals, though higher processing temperatures did reduce the protein digestibility in that case.

Table 3.10 Digestibility of a range of rendered animal protein products when fed to different fish species

Ingredient	Ingredient composition				Method	Apparent digestibility		Species	Data source
	Dry matter	Protein	Lipid	Ash		Protein	Energy		
Anchovetta meal	931	749	87	161	Stripping	83.6	79.2	*S. salar*	Refstie et al. (2006)
Fish meal A	1000	744	75	162	Stripping	89.1	80.7	*L. calcarifer*	Blyth et al. (2015)
Fish meal B	1000	721	85	158	Stripping	96.1	82.0	*L. calcarifer*	Blyth et al. (2015)
Blood meal	1000	953	1	20	Stripping	84.1	70.8	*L. calcarifer*	Blyth et al. (2015)
Blood meal – blood plasma, spray dried	965	938	–	17	Settlement	99	99	*O. mykiss*	Bureau et al. (1999)
Blood meal – blood cells, spray dried	922	775	–	110	Settlement	96	93	*O. mykiss*	Bureau et al. (1999)
Blood meal – coagulated, steam dried	931	884	–	69	Settlement	84	79	*O. mykiss*	Bureau et al. (1999)
Blood meal – ring dried	895	945	25	16	Settlement	88	88	*O. mykiss*	Bureau et al. (1999)
Blood meal – rotaplate dried	944	914	35	19	Settlement	82	82	*O. mykiss*	Bureau et al. (1999)
Blood meal – whole blood, spray dried	913	914	–	57	Settlement	96	92	*O. mykiss*	Bureau et al. (1999)
Blood meal – whole blood, spray dried	915	940	–	23	Settlement	97	94	*O. mykiss*	Bureau et al. (1999)

Feather meal – disk dried	932	876	67	26	Settlement	81	80	*O. mykiss*	Bureau et al. (1999)
Feather meal – long hydrolysis	903	830	109	26	Settlement	81	86	*O. mykiss*	Bureau et al. (1999)
Feather meal – ring dried	872	870	81	24	Settlement	81	78	*O. mykiss*	Bureau et al. (1999)
Feather meal – steam dried	914	824	108	30	Settlement	87	76	*O. mykiss*	Bureau et al. (1999)
Meat and Bone – beef+pork, 128°C, 30 min	950	503	112	317	Settlement	88	82	*O. mykiss*	Bureau et al. (1999)
Meat and bone – mixed source 125°C, 30 min	963	590	116	228	Settlement	83	68	*O. mykiss*	Bureau et al. (1999)
Meat and bone – mixed source, 133°C, 40 min	959	532	169	229	Settlement	92	88	*O. mykiss*	Bureau et al. (1999)
Meat and bone – mixed source, 138°C, 60 min	931	532	115	270	Settlement	87	76	*O. mykiss*	Bureau et al. (1999)
Meat and bone – reduced ash	953	577	188	147	Settlement	87	73	*O. mykiss*	Bureau et al. (1999)
Meat and bone – mixed source, 132°C, 25 min	968	553	150	250	Settlement	88	82	*O. mykiss*	Bureau et al. (1999)
Poultry by-product meal – 132°C, 40 min	972	647	136	154	Settlement	91	87	*O. mykiss*	Bureau et al. (1999)
Poultry by-product meal – 138°C, 30 min	940	692	114	157	Settlement	89	83	*O. mykiss*	Bureau et al. (1999)

Bureau et al. (2000) also evaluated the potential of a range of both feather meals and meat and bone meals as protein sources in diets for rainbow trout. These authors demonstrated that it was possible to include up to 15% feather meal in the diet without affecting growth and feed conversion. The incorporation in a diet of up to 24% meat and bone meals was also possible without affecting growth but resulted in a significant reduction in feed efficiency relative to reference diet. No significant differences were observed among the different feather meals and meat and bone meals used.

Williams et al. (2003a) examined the potential of two different meat and bone meals to replace fish meal in diets for barramundi. In a series of farm and laboratory experiments these authors were able to demonstrate that it was possible to replace all of the fish meal without impact on growth and only minor impairments on feed conversion. The meat and bone meals used were a high-ash variety (26% ash, 52% protein) and low-ash variety (14% ash, 60% protein). These authors also followed-up this growth performance study by assessing the organoleptic characteristics and fatty acid composition of the fish (Williams et al., 2003b). Fish fed the meat-meal-based diets were reported as being sweeter and firmer than those fish fed a fish-meal-based reference diet. Interestingly, scores for "fishy" flavor were also higher for those fish fed the meat-meal-based diets. Overall it was reported that complete replacement of all marine protein sources from the diet did not detract from the sensory value of the fish.

3.5   Nutritional value of plant and animal oils to aquaculture species

The inclusion of alternative oils to fish oil, such as plant oils like rapeseed or soybean oil, or animal oils like poultry oil or tallow, to aquaculture diets is not a new concept (Turchini et al., 2009). However, rational consolidation of information on this issue remains a high priority for the aquaculture industry (Tacon and Metian, 2008; Naylor et al., 2009). However, the replacement of fish oil in diets of fish, particularly marine fish, is likely to present further complications. Most fish have shown an essential requirement for some of the PUFA as they lack the enzymes required to desaturate 18 carbon chains any further than one unsaturated bond. Although some fish have a capacity to elongate and further desaturate the PUFA 18:2n-6 and 18:3n-3 to the LC-PUFA (18:4n-3, 20:4n-6, 20:5n-3: EPA, 22:4n-6, 22:5n-3, 22:6n-3: DHA), most marine fish species have shown a limited capacity to synthesize these LC-PUFA and hence need some or all of these fatty acids in their diet in particular amounts (Kanazawa et al., 1979).

3.5.1   Plant oils

Plant oils are one potential option for the replacement of fish oils in diets for fish (Turchini et al., 2009). Compared to fish oils, plant oils are notable in their deficiency in any of the LC-PUFA. The fatty acid composition varies substantially among the different plant oils (see Table 3.3). Soybean oils contain high levels of the PUFA linoleic (18:2n-6, LOA) and linolenic (18:3n-3, LNA) fatty acids. Similar to soybean oils, rapeseed oils are also deficient in any of the LC-PUFA. While rapeseed oils have appreciable quantities of both LOA and LNA, the proportions that they represent within the oil are substantially different from most other plant oils, with rapeseed oils containing a greater content of LNA than most other oils, other than linseed oil. This difference is a potential advantage, in that this brings a greater concentration of the n-3 fatty acids that are important for fish growth and flesh quality attributes, albeit as a short-chain n-3 PUFA. There have been a lot of studies in this area over the past decade and a few reviews that cover this topic of replacement of fish oils (Glencross, 2009; Turchini et al., 2009). Studies on salmonids predominate this area, reflective of the high reliance of this sector on high-lipid diets. Additionally, studies on marine fish are also prevalent as these species appear to have a sensitivity to manipulation of the LC-PUFA content of their diet, reflective of their inability to synthesize these fatty acids (Kanazawa et al., 1979; Glencross, 2009).

Different oils have different digestibilities, and this is largely affected by the fatty acid composition of each oil. Austreng et al. (1979) determined the digestibilities of soybean oil and two other fish oil types and their specific fatty acid components. Typically soybean oils were also digested as fish oils. The authors observed that there was a trend observed toward increasing digestibility with decreasing fatty acid chain length and also an increase in digestibility with increasing level of fatty acid chain unsaturation. Typically the LC-PUFA were completely digested. Interestingly, the LOA digestibilities from fish oils were considerably lower than those from the soybean oil. It was suggested that this effect was attributable to interactions among the different fatty acids within the oils.

Hardy et al. (1987) examined the potential of a range of alternative oils in diets fed to Atlantic salmon where herring oil was replaced with either menhaden, tallow, or soybean oil. The findings of this study showed that the dietary lipid source did not affect growth or the proximate composition of the fillets. However, some changes in the composition of the fatty acids in the fillet lipids were observed. Furthermore, evaluation of the organoleptic characteristic fillets by a trained sensory panel indicated that no taste differences were observed among any of the treatments. These findings somewhat contrasted those of Dupree et al. (1979) who noted changes in the acceptability of channel catfish that had been fed corn oil in comparison to diets using solely fish oil (menhaden oil). Hardy et al. (1987) summated that it would be feasible to use any of these oil resources without compromising the growth of Atlantic salmon.

One of the earlier studies that evaluated the nutritional value or influence of canola oils on fish was that of Thomassen and Rosjo (1989). In this study, these researchers incrementally replaced the fish oil with soybean oil, low-erucic acid rapeseed oil, or high-erucic acid rapeseed oil in diets that were postcoated with oil to a total diet oil content of about 19%. This study was unusual in that despite claiming that growth was unaffected by the treatments, no growth or feed utilization data were provided. However, clear differences in fatty acid composition of each of the diets were noted and these differences were reflected in the fatty acids of the tissues.

Kalogeropoulos et al. (1992) examined the substitution of fish oil for soybean oil in diets fed to a marine species, the gilthead sea bream (*Sparus auratus*). In this study soybean oil was increased in 2% of the diet increments to replace the added fish oil, a total of 12% of the diet. The soybean oil inclusion in the diets also had a notable influence of diet fatty acid composition of the fish. Growth of fish fed the diets showed no influence of the soybean oil content until greater than 50% of the fish oil had been replaced by the soybean oil. Deterioration in the feed efficiency was also noted at the same point. Changes in the tissue composition of total fat and fatty acids were also noted in some treatments. Those treatments that had the greatest amounts of soybean oil had lower tissue fat levels. A similar study by Glencross et al. (2003b) examined the use of canola oils (crude and refined) and soybean oil in diets for red sea bream (*P. auratus*). Similarly, these authors incrementally replaced fish oil, but in contrast no effect on growth or feed utilization was observed with complete replacement of the fish oil by either refined canola oil or soybean oil. This may have been reflective of the fact that a high level of fish meal was used in the diets, supplying a significant level of LC-PUFA from the residual lipids in the meal. However, the crude canola oil used, which had substantial tannin levels and other gums and residuals, did result in a significant decline in growth and feed conversion at high levels of replacement. As with other studies the composition of the fatty acids in the fish reflected that of their diets and as such there was a decline in the LC-PUFA levels and a rise in the PUFA.

Thompson et al. (1996) evaluated the potential of sunflower oil when fed to Atlantic salmon and although they monitored the growth of the fish, the key parameter they were examining was the immune response capabilities of the fish. Their basic premise was that the change in the n-3 and n-6 fatty acid ratio in the diet would deteriorate the immune response of the fish. In contrast to the study reported by Hardy et al. (1987), a deterioration in growth was observed with the full substitution of sunflower oil for fish oil, even in a diet containing 70% fish meal. Consistent with most other studies the fatty acid composition of the fish also changed to reflect that of their diet. Although there were no significant differences in any of the nonspecific immune parameters measured, there were significant immunological advantages conferred on the fish fed the fish oil diets, such as a higher number of B-cells when placed in an immune-challenge test. These workers concluded that Atlantic salmon fed diets with a high n-6 content, such as sunflower or soybean oil, may be less resistant to infection than those fed a high n-3 diet, such as fish oil.

Karalazos et al. (2007) reported a study on the use of rapeseed oil and the interactive effects of dietary fat and protein contents on growth, body, and fatty acid composition of Atlantic salmon (*S. salar*). Using a series of six diets, each formulated to provide either 39% protein and 32% lipid or 34% protein and 36% lipid and within each dietary protein/fat level, rapeseed oil was blended with fish oil to provide 0%, 30%, or 60% of the added oil. At the end of the experiment growth was significantly affected by the inclusion of rapeseed oil, with growth of fish improving with increasing replacement. The results of this study supported that moderate inclusion of rapeseed oil was effective in Atlantic salmon culture with no negative effects on growth and feed conversion.

3.5.2 Terrestrial animal oils

Terrestrial animal oils (like beef tallow, pork lard, and poultry oil) are another oil resource with potential for use as replacers of fish oil in diets for fish. Similar to rendered protein meals, these products are produced as a by-product from the production of other meat sources. As these oils are derived from other fed animals, the composition of these oil sources also tends to reflect the composition of what those animals are fed. In contrast to most plant oils though, terrestrial animal oils are typified by higher levels of saturates.

Greene and Selivonchek (1990) evaluated a range of plant and animal oils and fats when included in diets fed to rainbow trout. Included in their study were salmon oil, soybean oil, linseed oil, poultry oil, pork lard, and beef tallow. In this study growth of the fish was greatest from the fish fed the pork lard and lowest by those fed the beef tallow, however there were no significant differences in growth among any of the treatments. Similarly, feed utilization efficiency was also unaffected by oil/fat type used. None of the different lipids resulted in any significant differences in the proximate composition of the whole fish, though lipid content of fillets from the soybean–oil-fed treatment was significantly higher than that of any of the other treatments. As expected, the fatty acid composition of the fish was a reflection of that of their respective diets and in most cases there was a concomitant decrease in the levels of the LC-PUFA, such as EPA and DHA, with increasing use of different oils and fats.

More recently, Emery et al. (2014) replaced poultry by-product oil with beef tallow at increments up to 50% of the added oil in diets fed to Atlantic salmon. At the end of the study the authors observed no impact on growth performance or feed conversion. However, tallow inclusion did reduce lipid and fatty acid apparent digestibilities and had a significant effect on the fatty acid composition of the fish, with a decrease in n-6 fatty acid content in tissues and an increase in the n-3:n-6 fatty acid ratio in fillet.

3.6 Processing effects of ingredients

One of the important features to consider with the use of ingredients, in addition to their nutritional values, is their properties in terms of feed processing. Irrespective of how nutritionally good an ingredient might be, if it cannot be incorporated effectively into a bound pellet, then its value as a useful ingredient is diminished. While steam pellet press processing is widely used in pig and poultry sectors for feed manufacture, it is only still used to make shrimp feed among the aquaculture sector. More commonly used in the twenty-first century for fish feed production is extrusion technology. Extrusion technology differs from pellet press technology in that it relies on plasticizing the ingredients through a combination of heat, shear, and compressional forces to change the chemistry of the combined product. Usually this results in increased melt/gelatinization of starches, causing their expansion and the cross-linking of proteins within the matrix. This produces a more strongly bound, but porous pellet. In some cases the porous nature of the pellet is used to flood it with lipids/oils using top coating or vacuum infusion to increase the overall lipid content of the diet. By contrast, pellet press technology uses compressional forces to bind things together, although in modern pellet presses a preconditioner is often used to hydrate glutens and induce a range of potential protein cross-linking to occur as well.

The influence of a range of plant protein ingredients on the extrusion process and the resultant pellets was studied by Glencross et al. (2010b). Using a base reference diet of fish meal and wheat, increasing levels (10%, 20%, and 30%) of *L. angustifolius* and *L. luteus* kernel meals and solvent-extracted soybean meal were added to each of the diets. The addition of each of the ingredients had some unusual properties on the expansion of the pellets (Figure 3.3A). Notably this reduction in expansion increased the bulk density of the pellets produced (Figure 3.3B). This bulk density had direct effects on the hardness (Figure 3.3C) and the sink rates of the pellets (Figure 3.3D). These authors also showed that at low inclusion levels (~10%) *L. angustifolius* kernel meal had significantly higher water retention capacity than either soy or *L. luteus* kernel meal.

Overland et al. (2009) examined the impact of using PPC on extrusion parameters for feed for Atlantic salmon. These authors observed that inclusion of PPC increased the torque and resultant specific mechanical energy during extrusion relative to that seen with a similar inclusion of soybean or the fish meal reference diet. Substantially less water was also required than during extrusion of the soybean. The inclusion of PPCs resulted in pellets that expanded less, were harder, and more durable than those made using either fish meal or with 20% inclusion of soybean.

In a study examining the use of a range of ingredients in extruded diets for barramundi, the authors also collected data on the influence of those ingredients used on the physical properties of the pellets produced (Glencross et al., 2011b). The study examine the inclusion of a range of plant and rendered animal products (and combinations) in extruded diets all produced to the same digestible protein and energy specifications, but with variable starch levels as wheat was used as the filler in this case. The study showed that the inclusion of either *L. angustifolius* kernel meal or *L. luteus* kernel meal at 30% significantly constrained pellet expansion and increased pellet bulk density and pellet hardness. These ingredients also induced much higher mash water retention than the fish-meal-based reference. Rapeseed (canola meal) inclusion at 20% resulted in expansion characteristics similar to those achieved with the lupin products, despite the lower inclusion level. And despite a similar bulk density (571 g/l), the hardness of the pellets with 20% canola was much lower than that of the lupin pellets. The other plant protein used in this study was wheat gluten, which was also included at 20%. While pellet expansion with the gluten was not that different relative to the fish-meal-based reference (62% vs. 64%), the bulk density was dramatically higher (729 g/l vs. 528 g/l). This density was also coupled with a hardness twice that of the lupin pellets and almost three times that of the fish meal reference. This hardness, despite the expansion, also resulted in a pellet that leaked oil and had a very fast sink rate. Poultry offal meal was also used at 30% inclusion. This ingredient had no effect on expansion, marginally increased bulk density (587 g/l vs. 528 g/l), reduced the pellet hardness and did improve the water retention in the mash. Overall, it could be argued that poultry offal meal behaved relatively similar to the fish meal in the extrusion process, but that each of the plant protein meals introduced a range of complexities that provide some potential advantages and some potential disadvantages depending on what the objective outcome of the pelleting process is.

References

Anderson-Hafermann, J.C., Zhang, Y., Parsons, C.M., 1993. Effects of processing on the nutritional quality of canola meal. Poult. Sci. 72, 326–333.

Austreng, E., Skrede, A., Eldegard, A., 1979. Effect of dietary fat source on the digestibility of fat and fatty acids in rainbow trout and mink. Acta Agric. Scand. 29, 119–126.

Baeverfjord, G., Krogdahl, A., 1996. Development and regression of soybean meal induced enteritis in Atlantic salmon, *Salmo salar* L., distal intestine: a comparison with the intestines of fasted fish. J. Fish Dis. 19, 375–387.

Blyth, D., Tabrett, S.J., Glencross, B.D., 2015. A study of the effects of faecal collection method and acclimation time on the digestibility of diets and ingredients when fed to juvenile barramundi (*Lates calcarifer*). Aquaculture Nutr. 21, 248–255. http://dx.doi.org/10.1111/anu.12159.

Booth, M., Allan, G.L., Frances, J., Parkinson, S., 2001. Replacement of fishmeal in diets of silver perch: VI. Effects of dehulling and protein concentration on the digestibility of four Australian grain legumes in diets for silver perch (*Bidyanus bidyanus*). Aquaculture 196, 67–85.

Bureau, D.P., Harris, A.M., Cho, C.Y., 1998. The effects of purified alcohol extracts from soy products on the feed intake and growth of Chinook salmon (*Oncorhynchus tshawystcha*) and rainbow trout (*Oncorhynchus mykiss*). Aquaculture 161, 27–43.

Bureau, D.P., Harris, A.M., Cho, C.Y., 1999. Apparent digestibility of rendered animal protein ingredients for rainbow trout (*Oncorhynchus mykiss*). Aquaculture 180, 345–358.

Bureau, D.P., Harris, A.M., Bevan, D.J., Simmons, L.A., Azevedo, P.A., Cho, C.Y., 2000. Feather meals and meat and bone meals from different origins as protein sources in rainbow trout (*Oncorhynchus mykiss*) diets. Aquaculture 181, 281–291.

Burel, C., Boujard, T., Corraze, G., Kaushik, S.J., Boeuf, G., Mol, K.A., et al., 1998. Incorporation of high levels of extruded lupin in diets for rainbow trout (*Oncorhynchus mykiss*): nutritional value and effect on thyroid status. Aquaculture 163, 325–345.

Burel, C., Boujard, T., Tulli, F., Kaushik, S., 2000. Digestibility of extruded peas, extruded lupin, and rapeseed meal in rainbow trout (*Oncorhynchus mykiss*) and turbot (*Psetta maxima*). Aquaculture 188, 285–298.

Burel, C., Boujard, T., Kaushik, S.J., Boeuf, G., Mol, K.A., Van der Geyten, S., et al., 2001. Effects of rapeseed meal glucosinolates on thyroid metabolism and feed utilization in rainbow trout. Gen. Comp. Endocrinol. 124, 343–358.

Carter, C.G., Hauler, R.C., 2000. Fish meal replacement by plant meals in extruded feeds for Atlantic salmon, *Salmo salar* L. Aquaculture 185, 299–311.

Dupree, H.K., Gauglitz, E.J., Hall, A.S., Houle, C.R., 1979. Effects of dietary lipids on the growth and acceptability (flavor) of channel catfish (*Ictalurus punctatus*). In: Halver, J.E., Tiews, K. (Eds.), Proceedings of the World Symposium on Finfish Nutrition and Fishfeed Technology, Hamburg, 20–23 June 1978, vol. II, Berlin, pp. 88–103.

Drew, M.D., Borgeson, T.L., Thiessen, D.L., 2007. A review of processing of feed ingredients to enhance diet digestibility in finfish. Anim. Feed Sci. Technol. 138, 118–136.

Emery, J., Smullen, R., Turchini, G., 2014. Tallow in Atlantic salmon feed. Aquaculture 422, 98–108.

EU Regulation, 2001. 999/2001 of the European Parliament and of the Council Laying Down Rules for the Prevention, Control and Eradication of Certain Transmissible Spongiform Encephalopathies, 31. Official Journal of the European Communities., L147.

Fenwick, G.R., Price, K.R., Tsukamato, C., Okubo, K., 1991. Saponins. In: Dmello, J.P.F., Duffus, C.M., Duffus, J.H. (Eds.), Toxic Substances in Crop Plants. The Royal Society of Chemistry, Cambridge, pp. 285–328.

Forster, I., Higgs, D.A., Donsanjh, B.S., Rowshandeli, M., Parr, J., 2000. Potential for dietary phytase to improve the nutritive value of canola protein concentrate and decrease phosphorus output in rainbow trout (*Oncorhynchus mykiss*) held in 11C fresh water. Aquaculture 179, 109–125.

Francis, G., Makkar, H.P.S., Becker, K., 2001. Anti-nutritional factors present in plant-derived alternate fish feed ingredients and their effect in fish. Aquaculture 199, 197–227.

Gatlin, D.M., Barrows, F.T., Brown, P., Dabrowski, K., Gaylord, T.G., Hardy, R.W., et al., 2007. Expanding the utilization of sustainable plant products in aquafeeds: a review. Aquacult. Res. 38, 551–579.

Glencross, B.D., 2009. Exploring the nutritional demand for essential fatty acids by aquaculture species. Rev. Aquacult. 1, 71–124.

Glencross, B.D., 2011. A comparison of the diet and raw material digestibilities between rainbow trout (*Oncorhynchus mykiss*) and barramundi (*Lates calcarifer*) – implications for inferences of digestibility among species. Aquacult. Nutr. 17, e207–e215.

Glencross, B.D., Hawkins, W.E., 2004. A comparison of the digestibility of several lupin (*Lupinus* spp.) kernel meal varieties when fed to either rainbow trout (*Oncorhynchus mykiss*) or red seabream (*Pagrus auratus*). Aquacult. Nutr. 10, 65–73.

Glencross, B.D., Curnow, J.G., Hawkins, W.E., 2003a. Evaluation of the variability in chemical composition and digestibility of different lupin (*Lupinus angustifolius*) kernel meals when fed to rainbow trout (*Oncorhynchus mykiss*). Anim. Feed Sci. Technol. 107, 117–128.

Glencross, B.D., Hawkins, W.E., Curnow, J.G., 2003b. Evaluation of canola oils as alternative lipid resources in diets for juvenile red seabream, *Pagrus auratus*. Aquacult. Nutr. 9, 305–315.

Glencross, B.D., Hawkins, W.E., Curnow, J.G., 2004a. Nutritional assessment of Australian canola meals. I. Evaluation of canola oil extraction method, enzyme supplementation and meal processing on the digestible value of canola meals fed to the red seabream (*Pagrus auratus*, Paulin). Aquacult. Res. 35, 15–24.

Glencross, B.D., Carter, C.G., Duijster, N., Evans, D.E., Dods, K., McCafferty, P., et al., 2004b. A comparison of the digestive capacity of Atlantic salmon (*Salmo salar*) and rainbow trout (*Oncorhynchus mykiss*) when fed a range of plant protein products. Aquaculture 237, 333–346.

Glencross, B.D., Hawkins, W.E., Evans, D., McCafferty, P., Dods, K., Jones, J.B., et al., 2006. Evaluation of the influence of the lupin alkaloid, gramine when fed to rainbow trout (*Oncorhynchus mykiss*). Aquaculture 253, 512–522.

Glencross, B.D., Booth, M., Allan, G.L., 2007a. A feed is only as good as its ingredients – a review of ingredient evaluation for aquaculture feeds. Aquacult. Nutr. 13, 17–34.

Glencross, B.D., Hawkins, W.E., Vietch, C., Dods, K., McCafferty, P., Hauler, R.C., 2007b. Assessing the effect of dehulling efficiency of lupin (*Lupinus angustifolius*) meals on their digestible nutrient and energy value when fed to rainbow trout (*Oncorhynchus mykiss*). Aquacult. Nutr. 13, 462–470.

Glencross, B.D., Hawkins, W.E., Evans, D., Rutherford, N., McCafferty, P., Dods, K., et al., 2008. Variability in the composition of lupin (*Lupinus angustifolius*) meals influences their digestible nutrient and energy value when fed to rainbow trout (*Oncorhynchus mykiss*). Aquaculture 277, 220–230.

Glencross, B.D., Sweetingham, M., Hawkins, W.E., 2010a. A digestibility assessment of Pearl lupin (*Lupinus mutabilis*) meals and protein concentrates when fed to rainbow trout (*Oncorhynchus mykiss*). Aquaculture 303, 59–64.

Glencross, B.D., Hawkins, W.E., Maas, R., Karopoulos, M., 2010b. Influence of the incorporation of various inclusion levels, lupin species and varieties of *L. angustifolius* kernel meal in extruded diets on pellet physical characteristics. Aquacult. Nutr. 16, 13–24.

Glencross, B.D., Rutherford, N.R., Hawkins, W.E., 2011a. A comparison of the growth performance of rainbow trout (*Oncorhynchus mykiss*) when fed soybean, narrow-leaf or yellow lupin kernel meals in extruded diets. Aquacult. Nutr. 17, e317–e325.

Glencross, B.D., Rutherford, N.R., Jones, J.B., 2011b. Fishmeal replacement options for juvenile barramundi (*Lates calcarifer*). Aquacult. Nutr. 17, e722–e732.

Glencross, B.D., Blyth, D., Tabrett, S.J., Bourne, N., Irvin, S., Fox-Smith, T., et al., 2012. An examination of digestibility and technical qualities of a range of cereal grains when fed to juvenile barramundi (*Lates calcarifer*) in extruded diets. Aquacult. Nutr. 18, 388–399.

Greene, D.H.S., Selivonchek, D.P., 1990. Effects of dietary vegetable, animal and marine lipids on muscle lipid and hematology of rainbow trout (*Oncorhynchus mykiss*). Aquaculture 89, 165–182.

Griffiths, D.W., 1991. Condensed tannins. In: Dmello, J.P.F., Duffus, C.M., Duffus, J.H. (Eds.), Toxic Substances in Crop Plants. The Royal Society of Chemistry, Cambridge, pp. 180–201.

Hardy, R.W., 2011. Utilisation of plant proteins in fish diets: effects of global demand and supplies of fishmeal. Aquacult. Res. 41, 770–776.

Hardy, R.W., Scott, T.M., Harrell, L.W., 1987. Replacement of herring oil with menhaden oil, soybean oil, or tallow in the diets of Atlantic salmon raised in marine net-pens. Aquaculture 65, 267–277.

Hien, T.T.T., Phuong, N.T., Cam Tu, T.T., Glencross, B.D., 2010. Assessment of methods for the determination of digestibilities of feed ingredients for Tra catfish, *Pangasianodon hypopthalamus*. Aquacult. Nutr. 16, 351–358.

Higgs, D.A., McBride, J.R., Markert, J.R., Dosanjh, B.S., Plotnikoff, M.D., Clarke, W.C., 1982. Evaluation of Tower and Candle rapeseed protein concentrate as protein supplements in practical dry diets for juvenile Chinook salmon (*Oncorhynchus tshawytscha*). Aquaculture 29, 1–31.

Higgs, D.A., Fagerlund, U.H.M., McBride, J.R., Plotnikoff, M.D., Dosanjh, B.S., Markert, J.R., et al., 1983. Protein quality of Altex canola meal for juvenile Chinook salmon (*Oncorhynchus tshawytscha*) considering dietary protein and 3,5,3′-triiodo-L-thyronine content. Aquaculture 34, 213–238.

Jirsa, D., Davis, D.A., Salze, G.A., Rhodes, M., Drawbridge, M., 2014. Taurine requirement for juvenile white seabass (*Atractoscion nobilis*) fed soy-based diets. Aquaculture 422, 36–41.

Kalogeropoulos, N., Alexis, M.N., Henderson, R.J., 1992. Effects of dietary soybean and cod-liver oil levels on growth and body composition of gilthead bream (Sparus aurata). Aquaculture 104, 293–308.

Kanazawa, A., Teshima, S.I., Ono, K., 1979. Relationship between essential fatty acid requirements of aquatic animals and the capacity for bioconversion of linolenic acid to highly unsaturated fatty acids. Comp. Biochem. Physiol. 63B, 295–298.

Karalazos, V., Bendiksen, E.A., Dick, J.R., Bell, J.G., 2007. Effects of dietary protein, and fat level and rapeseed oil on growth and tissue fatty acid composition and metabolism in Atlantic salmon (*Salmo salar* L.) reared at low water temperatures. Aquacult. Nutr. 13, 256–265.

Kaushik, S.J., Cravedi, J.P., Lalles, J.P., Sumpter, J., Fauconneau, B., Laroche, M., 1995. Partial or total replacement of fishmeal by soybean protein on growth, protein utilization, potential estrogenic or antigenic effects, cholesterolemia and flesh quality in rainbow trout, *Oncorhynchus mykiss*. Aquaculture 133, 257–274.

Krogdhal, A., Lea, T.B., Olli, J.J., 1994. Soybean proteinase inhibitors affect intestinal trypsin activities and amino acid digestibilities in rainbow trout (Oncorhynchus mykiss). Comp. Biochem. Physiol. 107A, 215–219.

Mwachireya, S.A., Beames, R.M., Higgs, D.A., Dosanjh, B.S., 1999. Digestibility of canola protein products derived from the physical, enzymatic and chemical processing of commercial canola meal in rainbow trout, *Oncorhynchus mykiss* (Walbaum) held in fresh water. Aquacult. Nutr. 5, 73–82.

Naylor, R.L., Hardy, R.W., Bureau, D.P., Chiu, A., Elliott, M., Farrell, A.P., et al., 2009. Feeding aquaculture in an era of finite resources. Proc. Natl. Acad. Sci. 106, 15103–15110.

Olli, J.J., Krogdahl, Å., 1995. Alcohol soluble components of soybeans seem to reduce fat digestibility in fish-meal-based diets for Atlantic salmon, *Salmo salar* L. Aquaculture Res. 26, 831–835.

Oste, R.E., 1984. Effect of Maillard reaction products on protein digestion. *In vivo* studies in rats. J. Nutr. 114, 2228–2234.

Overland, M., Sorensen, M., Storebakken, T., Penn, M., Krogdahl, A., Skrede, A., 2009. Pea protein concentrate substituting fish meal or soybean meal in diets for Atlantic salmon (*Salmo salar*) – effect on growth performance, nutrient digestibility, carcass composition, gut health, and physical feed quality. Aquaculture 288, 305–311.

Petterson, D.S., Harris, D.J., Rayner, C.J., Blakeney, A.B., Choct, M., 1999. Methods for the analysis of premium livestock grains. Aust. J. Agric. Res. 50, 775–787.

Petterson, D.S., 2000. The use of lupins in feeding systems. Asian Australas. J. Anim. Sci. 13, 861–882.

Refstie, S., Storebakken, T., Roem, A.J., 1998. Feed consumption and conversion in Atlantic salmon (*Salmo salar*) fed diets with fish meal, extracted soybean meal or soybean meal with reduced content of oligosaccharides, trypsin inhibitors, lectins and soya antigens. Aquaculture 162, 301–312.

Refstie, S., Svihus, B., Shearer, K.D., Storebakken, T., 1999. Nutrient digestibility in Atlantic salmon and broiler chickens related to viscosity and non-starch polysaccharide content in different soyabean products. Anim. Feed Sci. Technol. 79, 331–345.

Refstie, S., Glencross, B., Landsverk, T., Sørensen, M., Lilleeng, E., Hawkins, W., et al., 2006. Digestive function and intestinal integrity in Atlantic salmon (*Salmo salar*) fed kernel meals and protein concentrates made from yellow or narrow-leafed lupins. Aquaculture 261, 1382–1395.

Salini, M.J., Adams, L.R., 2014. Growth performance, nutrient utilization and digestibility by Atlantic salmon (*Salmo salar* L.) fed Tasmanian grown white (*Lupinus albus*) and narrow-leafed (*L. angustifolius*) lupins. Aquaculture 426, 296–303.

Serrano, E., Storebakken, T., Penn, M., Øverland, M., Hansen, J.Ø., Mydland, L.T., 2011. Responses in rainbow trout (*Oncorhynchus mykiss*) to increasing dietary doses of lupinine, the main quinolizidine alkaloid found in yellow lupins (*Lupinus luteus*). Aquaculture 318, 122–127.

Serrano, E., Storebakken, T., Borquez, A., Penn, M., Shearer, K.D., Dantagnan, P., et al., 2012. Histology and growth performance in rainbow trout (*Oncorhynchus mykiss*) in response to increasing dietary concentration of sparteine, a common alkaloid in lupins. Aquacult. Nutr. 18, 313–320.

Tacon, A.G.J., Metian, M., 2008. Global overview on the use of fish meal and fish oil in industrially compounded aquafeeds: trends and future prospects. Aquaculture 285, 146–158.

Teskeredžić, Z., Higgs, D.A., Dosanjh, B.S., McBride, J.R., Hardy, R.W., Beames, R.M., et al., 1995. Assessment of undephytinized and dephytinized rapeseed protein concentrate as sources of dietary protein for juvenile rainbow trout (*Oncorhynchus mykiss*). Aquaculture 131, 261–277.

Thomas, E., van der Poel, A.F.B., 2001. Functional properties of diet ingredients: manufacturing and nutritional implications. In: van der Poel, A.F.B., Vahl, J.L., Kwakkel, R.P.

(Eds.), Advances in Nutritional Technology 2001. Proceedings of the First World Feed Conference, Wageningen Pers, Utrecht, The Netherlands, 7–8 November, pp. 109–122.

Thomassen, M.S., Rosjo, C., 1989. Different fats in feed for salmon: influence on sensory parameters, growth rate and fatty acids in muscle and heart. Aquaculture 79, 129–135.

Thompson, K.D., Tatner, M.F., Henderson, R.J., 1996. Effects of dietary (n-3) and (n-6) polyunsaturated fatty acid ratio on the immune response of Atlantic salmon, *Salmo salar* L. Aquacult. Nutr. 2, 21–31.

Turchini, G.M., Torstensen, B.E., Ng, W.K., 2009. Fish oil replacement in finfish nutrition. Rev. Aquacult. 1, 10–57.

Urán, P.A., Gonçalves, A.A., Taverne-Thiele, J.J., Schrama, J.W., Verreth, J.A.J., Rombout, J.H.W.M., 2008. Soybean meal induces intestinal inflammation in common carp (*Cyprinus carpio* L.). Fish Shellfish Immunol. 25, 751–760.

Urán, P.A., Schrama, J.W., Rombout, J.H.W.M., Taverne-Thiele, J.J., Obach, A., Koppe, W., et al., 2009. Time-related changes of the intestinal morphology of Atlantic salmon, *Salmo salar* L., at two different soybean meal inclusion levels. J. Fish Dis. 32, 733–744.

Van Barneveld, R.G., 1999. Physical and chemical contaminants in grains used in livestock feeds. Aust. J. Agric. Res. 1999 (50), 807–823.

White, C.E., Campbell, D.R., McDowell, L.R., 2000. Effects of dry matter content on trypsin inhibitors and urease activity in heat treated soya beans fed to weaned piglets. Anim. Feed Sci. Technol. 87, 105–115.

Williams, K.C.K., Barlow, C.G., Rodgers, L.J., Ruscoe, I., 2003a. Potential of meta meal to replace fishmeal in extruded dry diets for barramundi, *Lates calcarifer* (Bloch). I. Growth performance. Aquacult. Res. 34, 23–32.

Williams, K.C., Patterson, B.D., Barlow, C.G., Rodgers, L., Ford, A., Roberts, R., 2003b. Potential of meat meal to replace fish meal in extruded dry diets for barramundi *Lates calcarifer* (Bloch). II. Organoleptic characteristics and fatty acid composition. Aquacult. Res. 34, 33–42.

Woodgate, S., Van Der Veen, J., 2004. The role of fat processing and rendering in the European Union animal production industry. Biotechnol. Agron. Soc. Environ. 8, 283–294.

Nutrient requirements

4

César Molina-Poveda
GISIS, Guayaquil, Ecuador

4.1 Introduction

One of the most important aspects in aquaculture is the nutrition, as for optimum development of organisms, it is essential to have all the necessary nutrients, in terms both of quantity and quality. Nutrients are important for species in culture to stay alive, be healthy and to grow. Even though nutritional principles are similar for all animals, the required level of nutrients varies between species. Animals in production systems need a diet that is properly balanced for the specific requirements of species; however it must also be considered that nutritional requirements are affected by the growth rate, growing conditions, and environmental factors. Knowledge of these allows the formulation of well-balanced and cost-effective feeds. This chapter summarizes information concerning the fate of energy and nutritional requirements of fish and shrimp.

4.2 Proteins and amino acids

Proteins are large, complex organic nitrogen compounds found in cells of all animals and plants that are indispensable for growth and maintenance of life. They provide amino acids for a variety of roles in living organisms and are often classified by these biological roles, which are summarized in Table 4.1.

Proteins are comprised of amino acids linked with peptide bonds and cross-linked between chains with sulfhydryl and hydrogen bonds. Although there are about 300 amino acids in natural sources, only 20 make up most proteins, each with different physical and chemical properties. The success of the synthesis of a particular protein will be dependent on the availability of the required amino acids.

4.2.1 Protein quality and evaluating protein

Protein quality refers to how closely the composition of the absorbed essential amino acids (EAA) of the protein (feedstuff) matches the dietary amino acids required by the animal and how digestible the protein source is.

Apart from chemically measuring amino acids and their availability within feed proteins, there are biological methods of evaluating protein quality:

a. *Specific growth rate (SGR)*
   The SGR of an animal is defined as the increase in weight per unit time and this is a fairly sensitive index of protein quality; being proportional to the supply of EAA. A higher

S. Nates (Ed): Aquafeed Formulation. DOI: http://dx.doi.org/10.1016/B978-0-12-800873-7.00004-X

Table 4.1 Classification of proteins by biological function

Biological function	Example
Structural: provide strength and structure	a. Glycoprotein – cell membranes and cell wall in plants b. Keratins – epithelial cells in the outermost layers of the skin, feathers c. Collagen – fibrous connective tissue (bone, cartilage) d. Elastin – elastic connective tissue (ligaments)
Contractile: muscle contraction; cell division	Myosin and actin are proteins needed for muscles contraction
Enzymes: accelerate biological reactions	Glutathione peroxidase which serves to protect cellular tissues against oxidative damage
Transport	Hemoglobin and hemocianin transport oxygen throughout the body. Fatty acid and lipids are mobilized by serum albumin and lipoproteins. Transferrin and ceruloplasmin transport iron and copper, respectively
Regulatory: regulate the functioning of other proteins	Insulin and glucagon regulate the activity of specific enzymes in the body
Storage: provide storage of essential nutrients	Egg albumin, casein (milk protein), zein (corn protein)
Protection: protect cells or the organism from foreign substances	Immunoglobulins recognize and breakdown foreign molecules. Antibodies and blood coagulation – fibrinogen, thrombin

value indicates that the protein is more digestible and that the absorbed EAA profile is more favorable for protein synthesis.

The following equation is used to calculate daily SGR:

$$\text{SGR} = (\ln W_f - \ln W_i \times 100)/t$$

where:

$\ln W_f$ = the natural logarithm of the final weight
$\ln W_i$ = the natural logarithm of the initial weight
t = time (days) between $\ln W_f$ and $\ln W_i$

b. *Protein efficiency ratio (PER)*

Due to its simplicity, PER has been used extensively to compare feed proteins. The better-quality proteins will result in a higher PER.

PER = Body Wt Gain in grams/Protein consumed in gram

c. *Protein conversion efficiency (PCE)*

PCE is defined as the ratio between the protein content of a test animal at the start and end of the experiment compared to the total protein fed.

PCE = [(PCf – PCi)/Protein fed] × 100

d. *Net protein utilization*

Apparent net protein utilization (ANPU) is the ratio of feed digestible protein deposited to proteins in carcass. This figure is affected by the AAE available within the tested protein, but is majorly affected by the level of limiting amino acids within the protein. This value can be determined by the following formula:

$$ANPU = (PCf - PCi)/(\text{Protein fed} \times \text{Digestibility coefficient})$$

True net protein utilization (TNPU) compares the body protein of animals receiving a test protein to animals receiving a protein-free diet for the same period of time to determine the change in carcass protein. The formula for TNPU is:

$$TNPU = (PCf - PCi) - (PCFf - PCFi)/(\text{Protein fed} \times \text{Digestibility coefficient})$$

where:

PCf: final carcass protein
PCi: initial carcass protein
PCFf: final carcass protein of protein-free diet
PCFi: initial carcass protein of protein-free diet

e. *Biological value (BV)*

The BV of a protein is the absorbed protein percentage from the feed that is retained in the body and is therefore available for incorporation into the proteins within the organism to support the body's needs. While more balanced the amino acids raise the BV of a protein source.

The following equation concerns the BV of a specific feed.

$$BV = [(\text{Feed N} - \text{Fecal N} - \text{Urinary N} - \text{Gills N})/(\text{Fed N})] \times 100$$

The parameters therefore refer to nitrogen (N) from that food protein that has been absorbed by or retained in the body. BV assumes protein is the only source of nitrogen and measures the proportion of this N absorbed by the body which is then excreted.

Nonetheless, some of the N excreted in the feces and urine that is produced in the body is known as "endogenous nitrogen". BV is lower in older animals relative to growing animals. A BV of 70 or greater can support growth as long as adequate energy is available.

True biological value (TBV) compares body N of animals receiving a test protein to animals receiving a N-free diet but the N retained in the body is compared to the N intake rather than absorbed N. Low values indicate lower utilization.

$$TBV\% = [\text{Food N} - (\text{Fecal N} - \text{Metabolic Fecal N}) - (\text{Urine N} - \text{Metabolic Urine N}) - (\text{Gills N} - \text{Metabolic Gills N})/(\text{Fed N})] \times 100$$

4.2.2 Protein requirements of aquatic species

Protein levels in aquaculture feed are generally from 25% to 60% crude protein (CP).

Most variation among aquatic species can be associated with whether the animals are: (i) coldwater or warmwater; (ii) freshwater or marine; (iii) carnivorous or omnivores; and (iv) finfish or crustaceans. Warm periods and tropical climates require less protein and carbon and vice versa. Carnivorous fish need 40–50% protein while omnivorous fish need 25–35% (Wilson, 2002). The protein requirements of fish (from 25–60% CP) decrease with age and the low energy requirements of fish result in higher percentage protein requirements. However, the grams of protein required per gram of tissue protein synthesis are fairly similar to terrestrial animals (NRC, 2011).

Protein requirements should be carefully examined as species in culture do not possess a real protein requirement, but require a balanced mixture of amino acids, hence nutritional requirements, expressed as a percentage of the diet, are influenced by the protein source, in addition to the age and physiological state of the animal, also the protein source used in the diet can lead to an overestimation of protein requirement (Mambrini and Guillaume, 2001).

Proteins are the most important components of the body of animals, representing 65–85% of the weight of fish and shrimp and their quality is determined by the profile of EAA (Jauncey, 1982). Since fish and shrimps have a limited capacity to synthesize protein at the rate which is required to promote the maximum growth from carbon skeleton, protein must be procured through the feed which is consumed. Although in basic qualitative nutrient needs there are similarities between fish and domesticated farm animals, however most fish species require high-protein-content diets. In general, protein levels in successful fish feed range from 23% to 55% to accommodate for differences in the physiological needs of different fish (Table 4.2). Most warmwater fish have protein needs similar to channel catfish, thus protein levels of 30–36% will probably be adequate for most warmwater fish diets. Maximum growth and optimum utilization were achieved when carp were fed on diets containing 38–35% CP (Ogino and Saito, 1970; Köprücü, 2012). Larumbe-Moran et al. (2010) studied the effect of four protein levels 20%, 30%, 40%, and 50% on the performance of *Oreochromis niloticus* fry cultured under different salinities, 0, 15, 20 and 25 ppt. Authors concluded that the tilapia cultured at salinities up to 15 ppt had the best growth at 30%, whereas those cultured at 20 or 25 ppt salinity had the best growth at 40% dietary protein.

The optimal dietary protein level required for maximal growth in farmed fish is reported to be slightly higher than that of terrestrial farm animals in terms of grams of protein intake per kilogram body weight gain. However, Wilson and Halver (1986) declared that warm-blooded animals such as chicks retain less protein per energy unit as compare with young fish. This is because fish have a lower maintenance energy requirement.

In crustaceans the optimum protein level is determined based on growth data with different levels of dietary protein provided, which is defined as the minimum amount of this nutrient necessary for the maximum growth. Some laboratory researches have shown wide variations in the requirement reported for shrimp by the different researchers within the same species and stage of development. Thus, Pedrazzoli et al. (1998) established a requirement for *Litopenaeus vannamei* of 40%, Cousin et al. (1991) 30%, while Aranyakananda and Lawrence (1993) found a requirement of 15% (Table 4.3). The optimal level of protein for juvenile *Penaeus monodon* (0.5–1.8 g) is about 40%, while for broodstock it is 50–55% (Lee, 1971).

Protein requirements can be affected by many factors, including the species, feeding habits, the phases of the life cycle, reproduction, dietary aspects such as quality protein and protein/energy ratio, and abiotic factors such as salinity and temperature. Bomfim et al. (2008) evaluated the effect of decreasing dietary crude protein (32%,

Table 4.2 Protein requirement (% feed) of some selected fish species

Species	Size range (g)	Requirement (%)	References
African catfish (*Clarias gariepinus*)	10.9–67.0	43	Ali and Jauncey (2005)
Atlantic salmon (*S. salar*)	57–207	45	Lall and Bishop (1977)
	80–211	55	Grisdale-Helland and Helland (1997)
Artic charr (*Salvelinus alpinus*)	2.6–31.9	39	Gurure et al. (1995)
Basa catfish (*Pangasius bocouti*)	100	25	Somboon and Semachai (2010)
Blue tilapia	0.016–2.5	36	Winfree and Stickney (1981)
(*Oerochromis aurea*)	0.4–9.7	55	Davis and Stickney (1978)
	7.5	34	Winfree and Stickney (1981)
Cachara catfish (*Pseudoplatystoma reticulatum*)	8.2–101.9	50	Henrique Gomes Cornélio et al. (2014)
Channel catfish	0.02–0.2	55	Winfree and Stickney (1984)
(*I. punctatus*)	7.0–29.0	32–36	Garling and Wilson (1976)
	14–100	35	Page and Andrews (1973)
	20–517	24	Li and Lovell (1992)
	27–354	24	Robinson and Li (1997)
	114–500	25	Page and Andrews (1973)
	594–1859	24	Li and Lovell (1992)
Chinook salmon (*Oncorhynchus tshawytscha*)	1.5–7.6	40–55	DeLong et al. (1958)
Coho salmon (*Oncorhynchus kisutch*)	14.5–21.6	40	Zeitoun et al. (1974)
Common carp (*C. carpio*)	1.0–2.0	38	Ogino and Saito (1970)
	4.3–9.0	31	Takeuchi et al. (1979a,b,c)
Gilthead sea bream (*S. aurata*)	2.6–18	40	Sabaut and Luquet (1973)
Golden shiner (*Notemigonus crysoleucas*)	0.2–1.3	29	Lochmann and Phillips (1994)
Gold fish (*Carassius auratus*)			Lochmann and Phillips (1994)
Grass carp (*C. idella*)	0.2–0.6	41–43	Dabrowski (1977)
	7.1–49.3	35	Köprücü (2012)
	2.4–8.0	23–28	Lin (1991)
Grouper (*Epinephelus malabaricus*)	3.8–9.2	48	Chen and Tsai (1994)
	9.2–53	44	Shiau and Lan (1996)

(Continued)

Table 4.2 (Continued)

Species	Size range (g)	Requirement (%)	References
Hybrid red tilapia	2.9–8.4	24	Shiau and Huang (1990)
(*O. niloticus* × *O. aurea*)	21	28	Twibell and Brown (1998)
Malabar grouper	17	55	Tuan and Williams (2007)
(*E. malabaricus*)			
Milkfish (*Chanos chanos*)	0.04–0.18	40	Lim et al. (1979)
Mozambique tilapia	1.8–10.3	40	Jauncey (1982)
(*O. mossambica*)			
Nile tilapia (*O. niloticus*)	0.012	45	El-Sayed and Teshima (1992)
	0.45–10.3	45	Abdel-Tawwab et al. (2010)
	0.84	40	Siddiqui et al. (1988)
	0.8	28	Bomfim et al. (2008)
	4.0–10.0	30	Wang et al. (1985)
	19.5–45.2	35	Abdel-Tawwab et al. (2010)
	23.5–55.9	27.5–35	Wee and Tuan (1988)
	40–64.7	35	Abdel-Tawwab et al. (2010)
	40	30	Siddiqui et al. (1988)
Rainbow trout	2.0–10.0	40	Satia (1974)
(*Oncorhynchus mykiss*)	10–31.3	24	Kim et al. (1991)
	61–491.8	42–51	Austreng and Refstie (1979)
Rainbow trout	6.5–20.5	40	Zeitoun et al. (1974)
(*Salmo gairdneri*)			
Red drum (*S. ocellatus*)	4.1–28	35–44	Daniels and Robinson (1986)
	2.0–27	40	Serrano et al. (1992)
Red Snapper (*Lutjanus*	8.0–76.3	40–43	Abbas et al. (2005)
argentimaculatus)			
Redbelly tilapia	1.6–3.4	35	Mazid et al. (1979)
(*Tilapia zillii*)			
Sunshine bass (*Morone*	125–550	41	Webster et al. (1995)
chrysops × *M. saxatilis*)			
White sturgeon (*Acipenser*	145–300	40	Moore et al. (1988)
transmontanus)			

31%, 30%, 29%, 28%, and 27%) with supplementation of amino acids, based on the ideal protein concept on Nile tilapia fingerlings' performance. The diets were formulated to be iso-caloric and digestible iso-lysine, with a constant amino acids to lysine minimum ratio. The reduction in the CP content of the diet did not affect the SGR, feed intake, digestible lysine intake, body protein content, fat and protein deposition rates, and nitrogen retention efficiency of fish. The authors concluded that the CP level of the diet for Nile tilapia fingerlings (0.80 g) can be reduced from 32% to 28%.

Table 4.3 **Protein requirement (percentage of feed) of cultivable shrimp species**

Species	Stage	Requirement (%)	References
L. vannamei	Protozoea	30	Durruty et al. (2000)
	Mysis	50	Durruty et al. (2000)
	Postlarvae	20–25	Velasco et al. (2000)
	Postlarvae	30–35	Colvin and Brand (1977)
	Juvenile	32	Kureshy and Davis (2002)
	Juvenile	40	Pedrazzoli et al. (1998)
	Juvenile	30	Cousin et al. (1991)
	Juvenile	>36	Smith et al. (1985)
	Juvenile	15	Aranyakananda and Lawrence (1993)
L. setiferus	Protozoea	30	Durruty et al. (2000)
	Mysis	60	Durruty et al. (2000)
	Postlarvae	50	García et al. (1998)
	Juvenile	28–32	Andrews et al. (1972)
	Juvenile	30	Lee and Lawrence (1985)
	Juvenile	30	Taboada et al. (1998)
Litopenaeus schmitti	Postlarvae	60	García and Galindo (1990)
	Juvenile	28–33	Galindo et al. (2002)
Litopenaeus stylirostris	Postlarvae	44	Colvin and Brand (1977)
	Juvenile	30	Colvin and Brand (1977)
Farfantenaeus brasiliensis	Juvenile	18–28	Hidalgo et al. (2000)
Farfantepenaeus aztecus	Juvenile	40	Venkataramiah et al. (1975)
	Juvenile	51	Zein-Eldin and Corliss (1976)
Farfantepenaeus paulensis	Juvenile	25–35	Ramos and Andreatta (2011)
Farfantepenaeus notialis	Juvenile	45	Galindo et al. (2003)
Farfantepenaeus duorarum	Postlarvae	50	García et al. (1998)
Farfantepenaeus californiensis	Postlarvae	44	Colvin and Brand (1977)
	Juvenile	35	Colvin and Brand (1977)
M. rosenbergii	Postlarvae	40	Millikin et al. (1980)
	Postlarvae	15	Boonyaratpalin and New (1982)
	Postlarvae	35	Balazs and Ross (1976)
	Juvenile	27	Stanley and Moore (1983)
Penaeus indicus	Juvenile	30–40	Bhaskar and Ali (1984)
	Juvenile	43	Colvin (1976)
Penaeus merguiensis	Juvenile	50–55	Aquacop (1978)
	Juvenile	34–42	Sedgwick (1979)
P. monodon	Postlarvae	55	Bages and Sloane (1981)
	Juvenile	40	Aquacop (1977)
	Juvenile	40	Alava and Lim (1983)
M. japonicus	Postlarvae	45–55	Teshima and Kanazawa (1984)
	Juvenile	52–57	Deshimaru and Yone (1978b)
	Juvenile	54	Deshimaru and Kuroki (1974b)
	Juvenile	50	Teshima et al. (2001)

Handwritten annotations: "Amino Group NH₂", "α Carbon (Cα)", "sounds like Kiral", "↳ α Carbon AKA chiral carbon", "↳ optical activity", "Carboxylic Acid"

4.2.3 Amino acids are the basic unit of protein

Amino acids (AA) are the building blocks of proteins. Most natural polypeptide chains contain between 50 and 2,000 amino acid residues and are commonly referred to as *proteins*. An amino group is attached to a carbon of a carboxylic acid with basic structure R–CHNH$_2$–COOH where the R group is hydrogen (glycine) or a carbon chain of variable length and branching points.

In animal tissues, the amino groups are always attached to the α carbon. Hence they are called α-amino acids. Animals cannot use any other type of amino acid to make tissue protein with some exceptions. The general structure of an amino acid has both a basic amine group and an acidic carboxylic acid group. There is an internal transfer of a hydrogen ion from the –COOH group to the –NH$_2$ group, leaving an ion with both a negative and a positive charge. This is called a zwitterion, a compound with no overall electrical charge, but which contains separate parts that are positively and negatively charged. The properties of AA result in D- and L-forms, that is, they rotate the plane of polarized light as in D- and L-forms of carbohydrates (mirror images). If the R group is hydrogen (glycine), no isomers can exist. Most AA in nature are L-form and particularly those in animal tissues. Plants and microorganisms may synthesize D-amino acids and fish have the ability to utilize D-amino acids to build tissues (Kim et al., 1992a; Deng et al., 2011). Many synthetic AA may be a mixture of D- and L-forms and thus may not be completely available for use as an AA supplement in shrimp feeds.

The AA content and the sequence of a specific protein are determined by the sequence of the bases in the gene that encode that protein. The chemical properties of the AA determine the biological activity of the protein. Each AA has unique characteristics arising from the size, shape, charge, hydrophobicity, and reactivity properties of its R group. As a result, the R group of AA exerts a profound effect on the structure and biological activity of proteins. Classification of the 20 protein AA according to the functional group on the side and importance of their radical or amino group is given below.

A. Neutral, aliphatic amino acids
1. Glycine (GLY) – important in detoxifying compounds in the liver *(No Cα)*
2. Alanine (ALA) – associated with gluconeogenesis and the recycling of glucose carbon
3. Serine (SER)
4. Threonine (THR) – abundant in mucins
5. Asparagine (ASN) – converted back into aspartic acid releases energy
6. Glutamine (GLN) – is related to the immune system

B. Branched chain amino acids
1. Valine (VAL) – is involved in tissue repair and maintenance of nitrogen balance in the body
2. Leucine (LEU) – may have a unique role in the regulation of muscle protein turnover
3. Isoleucine (ILE) – is part of the immune responses

Handwritten annotation: "R ⇒ what makes the 20 AA different"

C. Acidic amino acids
   1. Aspartic acid (ASP)
   2. Glutamic acid (GLU) – very important amino acid associated with gluconeogenesis and the recycling of glucose carbon
D. Basic amino acids
   1. Lysine (LYS) – acts as a precursor for carnitine but derived from protein catabolism
   2. Arginine (ARG) – part of the urea cycle which is important in disposing of toxic ammonia produced during protein and amino acid degradation
   3. Histidine (HIS) – capable of attaching to metallic ions
E. Sulfur amino acids (nonpolar amino acids)
   1. Cysteine (CYS) – constituent of glutathione and can be converted into taurine. These amino acids are important in disulfide linkages
   2. Methionine (MET) – plays a role in the synthesis of creatine and polyamines and it is a precursor of choline
F. Aromatic amino acids (important in hydrophobic interactions)
   1. Phenylalanine (PHE) – a precursor of hormones and neurotransmitters
   2. Tyrosine (TYR) – can be synthesized from phenylalanine
G. Heterocyclic
   1. Tryptophan (TRP) – precursor of serotonin, a neurotransmitter, and can be synthesized into niacin
   2. Proline (PRO) – is a major nitrogenous substrate for the synthesis of polyamines

AAs were traditionally classified as nutritionally essential (indispensable) or nonessential (indispensable) for fish.

4.2.3.1 Nonessential amino acids

By definition these are those AA which can be synthesized adequately by fish and crustaceans and hence need not be provided in the diet. Amino acids considered nonessential to fish and shrimp are alanine, asparagine, aspartate, cysteine, glutamate, glutamine, glycine, proline, serine, and tyrosine.

4.2.3.2 Essential (indispensable) amino acids

EAAs cannot be synthesized in the animal's body or are inadequately synthesized de novo by animals at a sufficient enough rate to maintain maximum performance by the animal. Thus they must be provided from the diet under conditions where rates of utilization are greater than rates of synthesis. Amino acids considered essential to fish and shrimp are methionine, arginine, threonine, tryptophan, histidine, isoleucine, leucine, lysine, valine, and phenylalanine.

4.2.4 Amino acid oxidation

As AA cannot be stored in the body for later use, any AA not required for immediate biosynthetic needs is deaminated and the carbon skeleton is used as metabolic fuel (10–20% in normal conditions) or converted into fatty acids (FAs) via acetyl-CoA.

Krebs Cycle ⇒ sequence of reactions where living cells generate energy during aerobic respiration in mitochondria
84   *– uses O_2, produces CO_2 + H_2O as waste*   Aquafeed Formulation
and converting ADP to ATP (energy rich)

The main products of the catabolism of the carbon skeleton of AA are pyruvate, oxalacetate, α-ketoglutarate, succinyl-CoA, fumarate, acetyl-CoA, and acetoacetyl-CoA. The fate of the amino group of oxidized AA can be used to synthesize nonessential AA by transamination (α-ketoglutaric acid + ALA = GLU + pyruvate) and can be converted to ammonia by deamination (GLU + NAD + H_2O = α-ketoglutaric acid + NH_4^+).

AAs can also be classified according to the metabolic fate of the carbon skeleton after removal of ammonia from ketogenic, glucogenic, or both glucogenic and ketogenic AAs.

Ketogenic AAs yield acetyl-CoA or acetoacetyl-CoA but they do not produce metabolites that can be converted into glucose (Cowey and Walton, 1989). These AAs increase blood acetoacetate (a ketone body) when fed in high amounts. Lysine and leucine are the only AA that are exclusively ketogenics. Glucogenic AAs, whose catabolism yields to the formation of pyruvate or Krebs cycle metabolites (α-ketoglurate, oxalacetate, enol pyruvate, etc.), can be converted into glucose through gluconeogenesis (Kaushik, 2001) which increase blood glucose when fed in high amounts. Glucogenic AAs are alanine, arginine, asparagine, aspartate, cysteine, glutamate, glycine, histidine, methionine, proline, serine, and valine (Cowey and Walton, 1989).

Some AAs can be both glucogenic and ketogenic because they can yield some precursors that can become glucose and FAs. AAs of this kind are isoleucine, phenylalanine, tryptophan, tyrosine, and threonine.

4.2.5   Amino acids requirements

Animals do not require dietary protein per se, but rather some individual AA forming part of them. As mentioned above, 10 AA are considered essential and this essentiality is determined solely by the need to be supplied to animals through feed as the body cannot synthesize them at an appropriate rate to meet their metabolic needs. Therefore the essentiality or nonessentiality of an AA refers to dietary requirements, bearing two important functions in the metabolism. The determination of the quantitative requirements of EAAs has been generally established on experiments where dose–response curves uses growth as a response criterion. For example, for lysine requirement (3.25%, 3.98%, 5.3%, 5.04%, 5.63% of protein) of Atlantic salmon up to date, five studies have been published by Berge et al. (1998), Anderson et al. (1993), Rollin et al. (2003), Espe et al. (2007), and Grisdale-Helland et al. (2011). These last researchers used a dose–response method to determine the minimum dietary lysine requirement of Atlantic salmon smolts during the last week in freshwater and the following 9 weeks in seawater. The increase in dietary lysine concentrations resulted in linear increases in final weights, feed intake, growth, and feed efficiency. Base on this the authors concluded that the minimum dietary lysine requirement was 2.88% (5.63% of protein).

Some studies in aquatic animals have shown antagonism between Lys and Arg (Kaushik and Fauconneau, 1984; Kim et al., 1992b). Recently, Feng et al. (2013)

found that the growth rate, FCR, body protein, body Lys and Arg content for shrimp *L. vannamei* were significantly affected by dietary Lys and Arg, improving when the ratio of these two was between 1:0.88 and 1:1.05. However, no antagonism has been found in other aquatic animals such as fingerling channel catfish (Robinson et al., 1981), *Dicentrarchus labrax* (Tibaldi et al., 1994), and *Marsupenaeus japonicus* (Hew and Cuzon, 1982). Further studies on the relationship between Lys and Arg are required to help to elucidate these contrasting reports. Contrary to *M. japonicus* and *P. monodon*, where the AAE requirements have been established, *L. vannamei* have five (three in low salinity water) AAEs despite its economic importance.

Another methodological approach has been based on determination of free AA concentrations in plasma and tissues, against a concentration gradient of dietary AA. Oxidation of the free AA in the tissue using radioactive labels is another method used, although its use has been questioned in fish (Kim et al., 1992c) and shrimp (Amouroux et al., 1997). It is supposed to increase when an EAA is in excess. A simple method that allows an approximation to the requirements of EAA, is the analysis based on EAA profile of a muscle or whole animal. Wilson and Poe (1985) demonstrated a good correlation between the estimated requirements in *Ictalurus punctatus* and EAA levels in body tissues. In shrimp, Millamena et al. (1997) also reported that the threonine requirement in *P. monodon* was consistent with the levels of this AA found in muscle. One way to know AA requirements in broodstocks would be determining the profile of these in your lipovitellin (Harrison, 1990).

Although the EAA profile of muscle is indicative of EAA requirements of the species, other factors may reduce or increase the effect of a diet with a certain balance of AA (i) there is a loss of AA for energy production, irrespective of the AA profile of the feed, and (ii) the energy of the diet may vary the requirement of EAA (Cowey, 1994). Among EAA required by fish and shrimp, methionine is known to be the most limiting EAA in ingredients used in aquafeeds so much effort has been directed in determining methionine requirement. However, according to Wilson and Halver (2002) fish have a total sulfur amino acid (TSAA) requirement rather than a specific methionine requirement because cystine can spare a portion of methionine when this is not adequate in the diet for maximum growth (Kim et al., 1992a; Goff and Gatlin, 2004). He et al. (2014) conducted a feeding experiment where they evaluated various ratios of DL-methionine (Met) to L-cystine (Cys) (53:47, 49:51, 34:66, and 31:69) at a constant TSAA level in practical diets of juvenile Nile tilapia (*O. niloticus*). Crystalline DL-methionine (Met) to L-cystine (Cys) were supplemented to fulfill based on an equimolar sulfur basis. The results showed that Cys could replace up to approximately 47% of the TSAA requirement without affecting growth performance and nutrient accretion in juvenile Nile tilapia. In an earlier study, Abidi and Khan (2011), using different ratios of L-cystine and L-methionine on equimolar sulfur, determined that optimum inclusion of TSAA is between 25.2 and 31.31 g/kg of protein of which 33–39% could be spared by cystine in diets for fingerling *Labeo rohita*.

EAA requirements have generally been established in fish (Table 4.4) but few studies have reported about EAA maintenance requirements (Richard et al., 2010; He et al., 2013). Hua (2013a) realized a study through a mathematical modeling

Table 4.4 **Quantitative requirements (percentage of dietary protein) for EAA have been established in following selected species**

Species	Atlantic salmon (*S. salar*)	Asian sea bass (*Lates calcarifer*)	Black sea bream (*Sparus macrocephalus*)	Channel catfish (*I. punctatus*)	Chinnok salmon (*O. tshawytscha*)	Coho salmon (*O. kisutch*)	Common carp (*C. carpio*)	European sea bass (*D. labrax*)	Hybrid catfish (*C. gariepinus* × *Clarias macrocephalus*)	Hybrid striped bass (*M. saxatilis* × *M. chrysops*)
Arginine	4.1–4.8 (a,b)	3.8(j)	7.7–8.1(l)	4,3(n)	6.0(u)	4.9–5.8 (w,x)	4,3(y)	3.9(z)	4.5–5.0 (dd)	4.4(ff)
Histidine				1,5(n)	1.8(u)	1,8(x)	2.1(y)			
Isoleucine				2,6(n)	2,6(v)		2.5(y)			
Leucine				3,5(n)	3,9(v)		3.3(y)			
Lysine	3.3–5.6 (c,d,e, f,g)	4,5(j)	8,6 (m)	5.0–5.1 (o,p)			5.7(y)	4.4 (aa)		4.0 (ff,gg)
Methionine	1,73 (h)	2,2(k)		2,3(q)			2(y)	1.8–1.9 (bb)	1.8–2.0 (ee)	
Phenylalanine				2.1 (0.6% Tyr) (r)	4.4 (0.4% Tyr) (v)		3.3 (2.9% Tyr) (y)			
Threonine	2,6(i)			2,2(s)			3.9(y)	2.3–2.6 (cc)		1.8–2.6 (hh)
Tryptophan				0,5(s)			0.8(y)			0.6–0.7 (ii)
Valine				3.0(t)			3.6(y)			

(a) Lall et al. (1994); (b) Berge et al. (1997); (c) Berge et al. (1998); (d) Anderson et al. (1993); (e) Rollin et al. (2003); (f) Espe et al. (2007); (g) Grisdale-Helland et al. (2011); (h) Espe et al. (2008); (i) Bodin et al. (2008); (j) Murillo-Gurrea et al. (2001); (k) Coloso et al. (1999); (l) Zhou et al. (2010a); (m) Zhou et al. (2010b); (n) Robinson et al. (1981); (o) Robinson et al. (1980a); (p) Wilson et al. (1977); (q) Harding et al. (1977); (r) Robinson et al. (1980b); (s) Wilson et al. (1978); (t) Wilson et al. (1980); (u) Klein and Halver (1970); (v) Chance et al. (1964); (w) Luzzana et al. (1998); (x) Klein and Halver (1970); (y) Nose (1979); (z) Tibaldi et al. (1994); (aa) Tibaldi and Lanari (1991); (bb) Tulli et al. (2010); (cc) Tibaldi and Tulli (1999); (dd) Singh and Khan (2007); (ee) Keembiyehetty and Garlin (1993); (ff) Griffin et al. (1992); (gg) Keembiyehetty and Gatlin (1992); (hh) Keembiyehetty and Gatlin (1997); (ii) Gaylord et al. (2005); (jj) Ahmed and Khan (2004a); (kk) Ahmed and Khan (2005a); (ll) Ahmed and Khan (2006); (mm) Ahmed and Khan (2004b); (nn) Ahmed et al. (2003); (oo) Ahmed (2009); (pp) Ahmed et al. (2004); (qq) Ahmed and Khan (2005b); (rr) Alam et al. (2002a,b); (ss) Forster and Ogata (1998); (tt) Alam et al. (2000); (uu) Furuya et al. (2004); (vv) Furuya et al. (2006); (ww) Furuya et al. (2001); (xx) Nguyen and Davis (2009); (yy) Yue et al. (2014); (zz) Cho et al. (1992); (ab) Santiago and Lovell (1988); (ac) Chiu et al. (1988); (ad) Fournier et al. (2003); (ae) Kim et al. (1992b); (af) Rodehutscord et al. (1997); (ag) Choo et al. (1991); (ah) Walton et al. (1984b); (ai) Wang et al. (2010a); (aj) Walton et al. (1986); (ak) Cheng et al. (2003); (al) Encamacao et al. (2004); (am) Cowey et al. (1992); (an) Kim et al. (1992a); (ao) Rodehutscord et al. (1995); (ap) Kim (1993); (aq) Bodin et al. (2008); (ar) Poston and Rumsey (1983); (as) Walton et al. (1984a); (at) Khan and Abidi (2007a); (au) Abidi and Khan (2007); (av) Khan and Abidi (2007b); (aw) Abidi and Khan (2008); (ax) Abidi and Khan (2004); (ay) Ren et al. (2014); (az) Zhou et al. (2007); (ba) Zhou et al. (2006); (bd) Peres and Oliva-Teles (2008); (be) Twibell and Brown (1997); (bf) Twibell et al. (2000); (bg) Millamena et al. (1998); (bh) Chen et al. (1992); (bi) Millamena et al. (1999); (bj) Richard et al. (2010); (bk) Millamena et al. (1996a); (bl) Huai et al. (2009); (bm) Millamena et al. (1997); (bn) Millamena et al. (1996b); (bo) Zhou et al. (2012); (bp) Liu et al. (2014a); (bq) Liu et al. (2014b); (br) Xie et al. (2012); (bs) Fox et al. (1995); (bt) Zhou et al. (2013); (bu) Teshima et al. (2002); (bv) Alam et al. (2004).

Mrigal carp (*C. mrigala*)	Japanese flounder (*Paralichthys olivaceus*)	Nile tilapia (*O. niloticus*)	Rainbow trout (*O. mykiss*)	Rohu carp (*L. rohita*)	Cobia (*Rachycentron canadum*)	Turbot (*Psetta maxima*)	Yellow perch (*Perca flavescens*)	Tiger shrimp (*P. monodon*)	White shrimp (*L. vannamei*)	Kuruma shrimp (*M. japonicus*)
4.6(jj)	4.1–4.9 (rr)	4.2 (ab)	3.5–4.2 (zz,ac, ad,ae, aj)		6.2 (ay)	3.0–5.4 (ad)	4.9 (be)	5.3–5.5 (bg, bh)	5.6 (bo)	3.2–5.3 (bu, bv)
2.1(kk)		1.7 (ab)	1.0–1.2 (af)					2.2(bi)		1.2 (bu)
3.2(ll)		3.1 (ab)	1.5–2.8 (af)	3.8–4.0 (at)				2.7(bi)	3,9 (bp)	2.6 (bu)
3.9(ll)		3.4 (ab)	2.3–9.2 (af,ag)	3.8–3.9 (au)				4.3(bi)	5.8 (bq)	3.8 (bu)
5.8 (mm)	3.3–4.6 (ss)	5.1–5.7 (ab, uu, vv)	3.0–8.4 (ae,af, ah,ai,aj, ak,al)		5.3 (az)	5.0 (bd)		5.2 (bg,bj)	3.9–4.7 (br, bs)	3.8 (bu)
2.0–3.0 (nn)	2.9–3.0 (tt)	2.1–2.8 (ab, ww, xx)	0.7–1.9 (am, an,ao)		2.6 (ba)		2.5–3.1 (bf)	2.4–2.9 (bj,bk)		1.4 (bu)
3.3 (0.1% Tyr) (oo)		3.8 (0.5% Tyr) (ab)	2.0(ap)	2.9–3.1 (2.5% Tyr) (av)				3.7(bi)		3.0 (bu)
4.5 (pp)		3.8–4.7 (ab, yy)	2.6(aq)	3.8–4.2 (aw)				3.5(bm)	3.5–3.8 (bl,bt)	2.6 (bu)
1.0 (qq)		1.0(ab)	0.3–0.9 (af,aj, ar,as)					0.5(bi)		0.8 (bu)
3.8(ll)		2.8(ab)	1.7–3.4 (af)	3.75 (ax)				3.4(bn)		2.8 (bu)

approach using 59 studies excluding those cases where fish had been fed in excess or there was no report on whether feeds were all consumed or how much feed was wasted. The analysis determined the following maintenance requirements (mg/kg $BW^{0.75}$/day) for arginine (7.7 ± 6.6), histidine (9.8 ± 2.4), isoleucine (6.4 ± 24.8), leucine (9.1 ± 13.7), lysine (15.6 ± 8.2), methionine (18.4 ± 1.9), phenylalanine (14.6 ± 3.2), threonine (5.4 ± 4.4), tryptophan (0.5 ± 1.3), and valine (9.3 ± 26.6). Due to large standard errors associated with a very limited number of studies it was not possible to establish a final conclusion for isoleucine, leucine, and valine.

The supplementation with crystalline amino acids (CAA) to the diet is a practice commonly adopted to meet dietary EAA requirements through a least costing exercise. However, in aquaculture industry there is still controversy about biological efficiency and gastrointestinal absorption rates in the literature. Several studies have demonstrated that CAA utilization is as efficient as protein-bound AA (Kim et al., 1991; Rollin et al., 2003; Chi et al., 2014), whereas others studies have found lower biological efficiency when compared to intact protein (Zarate et al., 1999; Dabrowski et al., 2010).

During digestion the digestive enzymes digest long-chain polypeptides into short peptides that are absorbed more efficiently. When CAAs are delivered in feed this process does not take place and hence free AA are removed from the gastrointestinal tract and its absorption is faster than protein-bound AAs, thus resulting in a metabolic dyssynchronic with AAs released from digestion and hence affecting protein synthesis. As the fate of absorbed AAs is to be used for protein synthesis the redundant AAs will be oxidized to yield energy. To slow down the absorption rate and water leaching of CAA, important in shrimps due to the slow consumption, encapsulation, coating, and polymerization have been developed to achieve its utilization. Gu et al. (2013) conducted an assay to evaluate the effects of dietary crystalline methionine or oligomethionine on performance of juvenile *L. vannamei*. Shrimps fed on low-fish-meal-based diets supplemented with 1 g/kg of synthesized oligomethionine showed significantly higher growth rates, FCR, and PER than those shrimps fed on crystalline methionine.

Methionine and lysine are the main AAs limited in low-fish-meal aquafeeds (Gatlin et al., 2007) that is overcome with supplementation of DL-methionine or Met hydroxy analogs and L-lysine. The L-lysine, produced by fermentation and established under the form HCl lysine (Tosaka et al., 1983), is the most popular CAA used as a nutritional supplement in animal feeds. A methionine hydroxyanalog also known as DL-2-hydroxy-4-methylthiobutanoic acid (DL-HMTBa) has started being supplemented to aquafeeds as a methionine source although its relative bioavailability has been questioned by NRC (2011). However, Nunes et al. (2014) argue that an efficacy of 75% to 80% reported by NRC was not based on all studies published in aquaculture and poultry that evaluated DL-HMTBa's relative bioavailability to DL-Met. These authors consider that relative bioavailability should be between 81% and 100% based on DL-HMTBa bioavailability estimates realized by Vedenov and Pesti (2010) and Vázquez-Añón et al. (2006).

The mode of expression of AA requirements of fish is a topic of disagreement between fish nutritionists. Kim et al. (1991) consider that EAA requirements are best expressed as a percentage of diet, while Cowey and Cho (1993) reported EAA requirements in relation to the dietary protein content (% protein or g/16 g N) and Rodehutscord et al. (1997) indicate that EAA requirements should be expressed in relation to the diet energy content (e.g., digestible energy, DE). Hua (2013b) elucidated this controversy after carrying out nonlinear mixed model analysis and multilevel analysis on a lysine requirement dataset. Results showed that expressing requirement as a percentage of dietary protein provides a better goodness-of-fit than a fixed concentration of diet as a ratio to DE content. However, to establish the quantity of AA required per kg fish per day is of most interest not only for scientific approach but also for calculation of ration size.

4.3   Lipids and fatty acids

Lipids are organic compounds that are insoluble in water but soluble in organic solvents (nonpolar solvents). Lipids are a concentrated source of energy (9.44 kcal/g or 39.5 kJ/g) compared to carbohydrates, containing approximately 2.25-fold more energy. Other functions are structural components of membranes and of some hormones and fat-soluble vitamins, such as vitamins A, D, E, and K. There are basically two classes of lipids – neutral and polar lipids. Neutral lipids (NL) include FAs, alcohols, triacylglycerols (TAG), waxes, and sterols which mostly serve as storage and transport of metabolic fuels while polar lipids include glycerophospholipids and glyceroglycolipids.

FAs are carboxylic acids that are components of lipids. Over 70 different FAs are known in nature. The properties of the FAs are most important to the properties of conjugated fat. A carbon chain ends with a carboxylic acid. The FA is named on the basis of the number of carbon atoms and the number (if any) of the unsaturations and their positions relative to the carboxyl groups. Thus, for example, 18:2ω-6 denotes an unsaturated FA (USFA) containing 18 carbons and two double bonds with the first bond at the number 6 carbon atom from the methyl end of the molecule. FAs are generally unbranched, and can be classified as:

- *Saturated FA (SFAs)* – every carbon has two hydrogens attached to it except for the terminal carbon that has three hydrogens attached to it. All carbon atoms are attached by single bonds. Basic formula – $C_nH_{2n}O_{2n}$ = number of carbons. These FAs are solid at room temperature and are more characteristic of animal fats, for instance palmitic acid ($C_{16}H_{32}O_2$) melts at 63°C. Longer chains have higher melting points.
- *USFAs* – Some of the carbons are attached by at least one double bond. These FAs may be liquid at room temperature as a result of their double bonds. Thus, monounsaturated FAs contain one carbon-to-carbon double bond, and polyunsaturated FAs (PUFAs) contain two or more carbon-to-carbon double bonds. For example, linoleic acid (LA) (18:2ω-6) which melts at −12°C is a PUFA with two double bonds. The position of the double bond is important to its biological activity. Because of the double bonds, there are two types of FAs: (i) *cis* and (ii) *trans*.

The presence of the double bond also makes the FA more susceptible to oxidation. Light, heat, and minerals may catalyze a free radical formation which reacts with

[Handwritten margin notes at top: "Esterified → general name for a chemical reaction where two reactan (typically an alchol + acid) form an ester as the reaction product"
"→ glycerides are fatty acid esters of glycerols."
"Ester → chemical compound derived from an acid in which at least one OH (Hydroxyl) group is replaced by O (alkyl) group"]

atmospheric O_2. Molecule breakup forming keto and hydroxyketo acids gives fat its rancid taste. Antioxidants are added to feeds that have a high fat content to scavenge free radicals that may form. The same reactions may occur in animal tissues where vitamin E and selenium-containing enzymes act as the free radical scavengers. PUFAs increase the requirement for vitamin E and possibly selenium.

[Handwritten margin note: "NL - Neutral Lipid"]

TAG, the main NL, are composed of three FA hydrocarbon chains esterified to a single glycerol molecule by an ester linkage and thus are called triglycerides. This is the most common form of fat in storage depots of plants and animal tissue. A triglyceride that is solid at room temperature is called a fat and one that is liquid is called an oil.

Phosphoglycerides, also called phospholipids (PL), are FA esters with a glycerol backbone but containing an additional phosphoric acid and nitrogen group. For example, lecithin uses phosphatidylcholine (PC) as its nitrogenous base and this is the most abundant PL in biological membranes.

Glycolipids are fats containing carbohydrates (generally glucose or galactose) and often a nitrogen base, for example, galactolipids with a galactose molecule attached to the end hydroxyl group of a glyceride. Waxes are a combination of a FA with a long-chain monohydroxy or dihydroxy alcohol. They are insoluble in water, with high melting points and are difficult to hydrolyze and not of much nutritional value. Examples are lanolin in sheep wool or waxes of leaves. Marine zooplankton are also rich in wax ester. Their presence in feed may overestimate the ether extract value of the feed.

Sterols are a particular type of lipid that consists of the same basic structure as the phenanthrene ring nucleus linked to a cyclopentane ring. They differ in side chains and positions of double bonds in the molecule. The prototypical sterol is cholesterol, which is an important component of tissues and hormones. Other sterols include bile acids, vitamin D, and molting hormone.

Terpenes are based on the condensation of isoprene units into long chains. The most well known is β-carotene, the precursor to vitamin A.

4.3.1 Lipid digestion and absorption

Relative to carbohydrates and proteins, fats are slowly discharged from the stomach. Digestion basically begins in the proximal part of the intestine and pyloric caeca if present, but lipolytic activity can extend into the lower parts of the intestine with the activity deceasing progressively. A higher activity can occur in the distal part of the intestine in certain fish with a short digestive tract with few pyloric caeca. The pancreas or hepatopancreas is generally assumed to be the major source of digestive lipase enzymes in fish and shrimp, as it is in mammals.

Two lipases are present for TAG digestion; activated by calcium ions in the intestinal lumen. TAG lipase – the most important lipase – and phospholipase fats are first finely emulsified with the detergent action of bile salts, FAs, and glycerides. Emulsification allows for better penetration of the lipase to the lipid. Emulsified lipid

Table 4.7 **Quantitative essential fatty acid (EFA) requirements of juvenile and subadult marine fish species**

Gilthead sea bream (*S. aurata*)	n-3 HUFA	0.9 (DHA:EPA = 1)	Kalogeropoulos et al. (1992)
	n-3 HUFA	1.9 (DHA:EPA = 0.5)	Ibeas et al. (1994)
	DHA:EPA	0.5	Ibeas et al. (1997)
Japanese flounder (*Paralicthys olivaceus*)	n-3 HUFA	1.4	Takeuchi (1997)
Korean rockfish (*Sebastes schlegeli*)	n-3 HUFA	0.9	Lee et al. (1993a,b)
	EPA or DHA	1.0	Lee et al. (1993a,b)
Red drum (*S. ocellatus*)	n-3 HUFA	0.5–1.0	Lochmann and Gatlin (1993)
	EPA + DHA	0.3–0.6	Lochmann and Gatlin (1993)
Red sea bream (*P. major*)	n-3 HUFA or EPA	0.5	Yone (1978)
	EPA	1.0	Takeuchi et al. (1990)
	DHA	0.5	Takeuchi et al. (1990)
Silver bream (*Rhabdosargus sarba*)	n-3 HUFA	1.3	Leu et al. (1994)
Starry flounder (*Platichthys stellatus*)	n-3 HUFA	0.9	Lee et al. (2003)
Striped jack (*P. dentex*)	DHA	1.7	Takeuchi et al. (1992a)
Turbot (*P. maxima*)	n-3 HUFA	0.8	Gatesoupe et al. (1977)
	ARA	≈0.3	Castell et al. (1994)
Yellowtail flounder (*Pleuronectes ferrugineus*)	n-3 HUFA	2.5	Whalen et al. (1999)
Sea bass (*D. labrax* L.)	n-3 HUFA	1.0	Coutteau et al. (1996a)

HUFA, highly unsaturated fatty acids; ARA, arachidonic acid; EPA, eicosapentaenoic acid; DHA, docosahexaenoic acid.
Source: Based on Tocher (2010). With kind permission of John Wiley & Sons, Inc.

a diet containing 4% PL (Ebrahimnezhadarabi et al., 2011). Phospholipase A2 secretion increased in sea bass larvae with increasing dietary PL (from 2.7% to 11.6%). Lipase activity was the same in larvae fed 13% and 17% NL (mainly TAG) but was four times higher in the groups fed either 20% or 23% NL (Cahu et al., 2003). These differences between lipase and phospholipase A2 expression strongly suggest that fish larvae are better prepared to digest PL rather than TAG. PLs were found to decrease the incidence of larvae malformations in larval Ayu and Atlantic cod (Kanazawa et al., 1981; Kjørsvik et al., 2009). Increasing dietary vegetable PL (soy lecithin) level in 40-day-old sea bass larvae diet from 2.7% to 11.6% decreased the malformation rate (Cahu et al., 2003). PLs have also been found to mitigate fish larvae stress caused by increasing water temperature and a reduction of dissolved oxygen and salinity

Table 4.8 **Essential fatty acid requirements of shrimps species reported in various studies**

Species	18:3ω-3	18:2ω-6	C20:5ω-3	C22:6ω-3	C20:4ω-6	ω-3 to ω-6 ratio	References
Blue shrimp (*Penaeus stylirostris*)						1.8:1	Fenucci et al. (1981)
Brown shrimp (*F. aztecus*)	1–2%						Shewbart and Mies (1973)
Fleshy prawn (*Fenneropenaeus chinensis*)	0.7–1.0%			1.0%			Xu et al. (1993) Xu et al. (1994)
Giant river prawn (*M. rosenbergii*)				0.075%	0.08%		D'Abramo and Sheen (1993) D'Abramo and Sheen (1993)
Kuruma prawn (*M. japonicus*)	–	–	1.1% or	1.1%			Kanazawa et al. (1977) Kanazawa et al. (1978, 1979a)
Tiger shrimp (*P. monodon*)	1.0% or 1.5% and	1.5% 1.0%	0.9% or	0.9%	NR		Glencross and Smith (1999) Glencross and Smith (1999) Glencross and Smith (2001a) Glencross and Smith (2001b) Glencross et al. (2002)
White shrimp (*L. vannamei*)	2.5		0.5% or	1.44 0.5%	0.5%	2.5:1	Merican and Shim (1997) González Félix et al. (2003) González Félix et al. (2003)

(Camara, 1994; Kanazawa, 1997). Zwingelstein et al. (1998) reported that euryhaline fish and crustaceans adapt to changes in salinity and/or temperature through the mechanism of PS decarboxylation and subsequent methylation of PE species.

PLs were found to be an important source of energy and EFA (Tocher, 1995), supply for dietary EPA and DHA (Gisbert et al., 2005), phosphorous (Lall, 2002; Uyan et al., 2007) choline, and inositol (Geurden et al., 1995a,c). PLs enable hydrophobic lipids to be transported in aqueous environments through forming lipid–water interfaces (biles salts; cholic acid and derivatives). Maintenance distribution of sources is between the extracellular and intracellular fluids (Kanazawa, 1985; Tocher, 2003). PLs have an important structural role in emulsifying dietary lipids along with bile salts that have hydrophilic and hydrophobic domains (Olsen and Ringø, 1997; Kasper and Brown, 2003). PLs have an important role in lipid transport (Fontagné et al., 1998) and accumulation of lipid vacuoles in the intestinal enterocytes of Arctic char fed deprived soybean lecithin has been reported (Olsen et al., 1999). PLs are important precursors for a range of highly biologically active mediators of metabolism, including eicosanoids, diacylglycerol, inositol phosphates, and platelet activating factors (Sargent et al., 2002; Tocher, 2003).

PL can serve as a source of energy for fish in certain circumstances such as embryonic and early larval development. During embryogenesis and early larval development, the absolute amount of PL decreased particularly in PL-rich marine fish eggs. A complete catabolism of PL for energy would be the loss of important PUFA.

Table 4.9 summarizes all studies published to date. Requirements were lower in the freshwater fish, carp (2%), and ayu (3%), and higher in the marine fish red sea bream (5%), knife jaw (3–7%), and Japanese flounder (7%), followed by the highest reported value for European sea bass at 12% of diet. In juveniles, the PL requirements ranged from 1.5% for striped jack, 2–3% for European sea bass, 4–6% for Atlantic salmon, and up to 7% for Japanese flounder.

The PL requirement generally decreases with age or developmental stage, as shown for sea bass where required levels decreased from 12% (Cahu et al., 2003) to 3% (Geurden et al., 1995b) of PL or rockbream/knife jaw from 5% to 3% of soybean lecithin (Kanazawa, 1993) in both species from larva to juvenile. In Atlantic salmon it was also reported that the PL requirement was 4–6% of diet in 180-mg fish, 4% in 1–1.7-g fish and no requirement was observed in 7.5-g fish (Poston, 1990, 1991). Some crustaceans are able to biosynthesize PLs from FAs and diglycerides as shown by Shieh (1969) in the lobster *Homarus americanus*. However, the rate of biosynthesis, especially during larval growth, seems to be lower than metabolic demand (D'Abramo et al., 1981; Kanazawa et al., 1985; Teshima et al., 1986a). There have been several reviews (Coutteau et al., 1997; Teshima, 1997; Gong et al., 2004b) about the action of the PLs in shrimp reporting the results obtained in various species. The requirements of PLs in diets for different species of penaeid shrimp are varied (Table 4.10). Gong et al. (2001) found no differences in growth and survival of juvenile *L. vannamei* to test with three types of soybean lecithin to the same levels of inclusion in feed: type I (97.6% FL), type II (71.4%), and type III (48.4%); they recommended levels of 3–5% in the diet. González Felix et al. (2002b) found that *L. vannamei* juveniles had higher final weights and feed conversion with diets containing 5% different oils supplemented with 3.1% lecithin compared to nonsupplemented with this compound.

Table 4.9 Responses of some fish species to dietary phospholipids and sources of phospholipids used in feeding experiments

Species	Developmental stage	Optimal requirement (%)	Reference
Soybean lecithin			
Atlantic salmon (*Salmo salar*)	Juvenile (0.18 g)	6	Poston (1991)
	Juvenile (1–1.7 g)	4[a]	Poston (1990a)
	Juvenile (7.5 g)	NR	Poston (1990a)
Ayu (*Plecoglossus altivelus*)	Larvae	3	Kanazawa et al. (1983a)
	Larvae	3[a]	Kanazawa et al. (1981)
	Juvenile	3[a]	Kanazawa et al. (1981)
Channel catfish (*Ictalurus punctatus*)	Juvenile	≤4.3	Sink and Lochmann (2014)
European sea bass (*Dicentrarchus labrax*)	Larvae	12[a]	Cahu et al. (2003)
	Juvenile	3[a]	Geurden et al. (1995b)
Gilthead sea bream (*Sparus aurata*)	Larvae	>9	Seiliez et al. (2006)
	Larvae	8	Saleh et al. (2013)
Japanese flounder (*Paralichthys olivaceus*)	Larvae	7[a]	Kanazawa (1993)
	Juvenile	7[a]	Kanazawa (1993)
Knife jaw (*Oplegnathus fasciatus*)	Larvae	7.4[a]	Kanazawa et al. (1983b)
	Larvae	5	Kanazawa (1993)
	Juvenile	3	Kanazawa (1993)
Pikeperch (*Sander lucioperca*)	Larvae	9[a]	Hamza et al. (2008)
	Larvae	9[a]	Hamza et al. (2012)
Rainbow trout (*Oncorhynchus mykiss*)	Juvenile	4	Poston (1990b)
	Juvenile	14[a]	Rinchard et al. (2007)
Red sea bream (*Pagrus major*)	Larvae	5[a]	Kanazawa et al. (1983b)
White sturgeon (*Acipenser trans montanus*)	Juvenile	NR	Hung and Lutes (1988)
Other phospholipid sources			
Rainbow trout (*Oncorhynchus mykiss*)	Fry	2 CEL	Azarm et al. (2013)
Common carp (*Cyprinus carpio*)	Larvae	2[a] SPC, SPI or CEL	Geurden et al. (1995a, 1997a)
European sea bass (*Dicentrarchus labrax*)	Juvenile	2[a] EPC or SPC	Geurden et al. (1995b)

(Continued)

Table 4.9 **(Continued)**

Species	Developmental stage	Optimal requirement (%)	Reference
Turbot (*Psetta maxima*)	Juvenile	2[a] CEL	Geurden et al. (1997b)
Stripedjack (*Pseudocaranx dentex*)	Juvenile	1.5[a] SPC	Takeuchi et al. (1992a)
	Juvenile	1.5[a] SPE	Takeuchi et al. (1992a)
Ayu (*Plecoglossus altivelus*)	Larvae	3[a] CEL or BPL	Kanazawa et al. (1983a)
	Juvenile	3 CEL	Kanazawa et al. (1981)

BPL, bonito egg polar lipid; CPL, corn polar lipid; CEL, chicken egg lecithin; SPC, purified soybean; PC,
SPI, purified soybean; PI SPE, purified soybean PE NR, not required.
[a]Maximum value of range tested.

Table 4.10 **Published quantitative and qualitative phospholipid requirements of shrimp**

Species	Stage	Optimal requirement (%)	References
Tiger shrimp (*P. monodon*)	Juvenile	1.0–1.5 Soybean lecithin	Paibulkichakui et al. (1998)
White shrimp (*L. vannamei*)	Juvenile	1.5 PC (from soybean)	Coutteau et al. (1996b)
	Juvenile	6.5 deoiled soybean lecithin	Coutteau et al. (1996b)
	Larvae	4.6 Soybean lecithin (60% PC)	Niu et al. (2012a)
Kuruma prawn (*M. japonicus*)	Juvenile	1.0 (PC + PE)	Kanazawa et al. (1979b)
	Juvenile	3.0 Soybean lecithin PE and PI	Teshima et al. (1986a,b)
	Larvae	3.0 Soybean lecithin	Kanazawa (1983)
	Larvae	0.5 to 1.0 (PC and PI)	Kanazawa et al. (1985)
Fleshy prawn (*F. chinensis*)	Juvenile	2 Soybean lecithin	Kanazawa (1993)
Redtail prawn (*Fenneropenaeus penicillatus*)	Juvenile	1.25 PC from soybean lecithin	Chen and Jenn (1991)
Banana shrimp (*Fenneropenaeus menguiensis*)	Juvenile	1–2 Soybean lecithin (40%PC + 15%PE)	Thongrod and Boonyaratpalin (1998)

In *M. japonicus* postlarvae, Tackaert et al. (1991) observed that semipurified diets without adding PLs but with cholesterol, promoted better growth and higher resistance to osmotic stress than obtained with diets supplemented only with PC. Gong et al. (2004) attributed the beneficial effect of PLs on the osmotic stress due to PLs facilitating lipid metabolism of shrimp and improving utilization of energy reserves. According to this study, the PLs are superior to NL to meet energy requirements.

Evidence on the importance of dietary PL in marine crustacean nutrition was gathered in several studies. D'Abramo et al. (1982) reported that the absence of PC in purified diets fed to lobsters caused a significant decrease in serum cholesterol levels which was attributed to a decrease in lipoproteins. In contrast, Hilton et al. (1984) and Kanazawa (1993) did not find any beneficial effect by supplementing lecithin to diets fed to postlarval and juvenile freshwater prawn *Macrobrachium rosenbergii*.

The few studies evaluating single PL demonstrate that PC and PI are the most efficient in most species. For most of the fish and crustacean species examined, the estimated PL requirement of larvae is in the range of 1–3% PC + PI of diet dry weight.

4.3.6 Sterols

Mammals can synthesize cholesterol from low-molecular-weight precursors such as acetate or mevalonate. Crustaceans and some molluscs require dietary sources of sterol for growth and survival because of the absence of de novo sterol-synthesizing ability (Kanazawa, 2001). Cholesterol is the major sterol in crustaceans because they do not possess the ability to synthesize de novo, steroid ring from acetate therefore is an essential component to be provided in the diet (Teshima and Kanazawa, 1971a; Teshima et al., 1976; Teshima, 1982). Besides its function in the membranes it is also a precursor of several bioactive molecules, such as steroid hormones in special molting hormone, bile acids, and vitamin D_3 (Teshima and Kanazawa, 1971a; Teshima, 1982). To facilitate solubilization and absorption of cholesterol a set of biological detergents are synthesized by the hepatopancreas, among them *N*-(*N*-dodecanoyl-sarcosyl)taurine (Lester et al., 1975), and secreted into the intestine lumen. Teshima et al. (1974) found that *M. japonicus* digests 82.6% of cholesterol and 77.3–98.3% of phytosterols as ergosterol, 24-methylene cholesterol, brassicasterol, β-sitosterol. The authors also suggest the existence of a relationship between cholesterol digestion and feed composition. In shrimp *Artemesia longinaris* maximum cholesterol digestibility is 87.5% (Martínez Romero et al., 1991), whereas in the *Pleoticus muelleri* it is 80.1% (Harán and Fenucci, 1996). In the last two studies, digestibility values of cholesterol were related to the dietary sterol level, reaching a plateau on an asymptotic curve. This indicates the probable existence of a system of saturable carriers in the epithelium where cholesterol is absorbed.

Likewise, the ability to bioconvert sterols de novo is generally low or absent, and varies among crustacean species. Crustaceans possess the ability to dealkylate some C_{28} and C_{29} sterols to cholesterol (Kanazawa et al., 1971b; Teshima, 1982). Several studies have indicated that dietary sterols such as β-sitosterol, ergosterol, and stigmasterol are inferior to cholesterol in promoting performance in *M. japonicus* (Kanazawa et al., 1971a), *M. rosenbergii* (D'Abramo and Daniels, 1994), *Homarus* sp. (D'Abramo et al., 1984), and *L. vannamei* (Castille et al., 2004). This implies that a dietary supply of sterol is necessary for growth. However, conversion of sterols to cholesterol has been demonstrated for brassicasterol in *Artemia salina* Leach (Teshima and Kanazawa, 1973a), β-sitosterol in *Portunus trituberculatus* (Teshima and Kanazawa, 1972), demosterol in *Palaemon serratus* (Teshima et al., 1975) and *M. japonicas* (Teshima and Kanazawa, 1973b) and ergosterol in *A. salina* (Teshima and Kanazawa, 1971b) and *P. trituberculatus* (Teshima, 1971). Morris et al. (2011) feeding shrimps, *L. vannamei*, on a diet supplemented at 0.1% of phytosterol extracted from soybeans (containing brassicasterol 2.5%, campesterol 26.5%, stigmasterol 18.2%, β-sitosterol 46.9%, β-sitostanol 2.2%, capestanol 0.6%, and other sterols 3.1%), found a growth response similar to those shrimps fed on cholesterol.

The optimum or required levels of dietary cholesterol for the different species of crustaceans reported to date are summarized to be approximately 0.1–2.0% of diet (Table 4.11), and may be age- and diet-dependent. Examples of the estimated

Table 4.11 **Cholesterol requirement of several crustacean**

Species	Optimum requirement (% in diets)	References
Kuruma prawn (*M. japonicus*)	0.5–1.0	Kanazawa et al. (1971a)
	0.2	Shudo et al. (1971)
	2.1	Deshimaru and Kuroki (1974a)
	1[a]	Teshima et al. (1982a)
	1[a]	Teshima et al. (1982b)
Tiger shrimp (*P. monodon*)	0.54	Kai and Kanazawa (1989)
	0.5	Chen (1993)
Redtail prawn (*Fenneropenaeus penicillatus*)	0.5	Chen and Jenn (1991)
Banana shrimp (*Fenneropenaeus menguiensis*)	NR in juvenile	Thongrod and Boonyaratpalin (1998)
Artemesia longinaris	0.5	Petriella et al. (1984)
Giant river prawn (*M. rosenbergii*)	0.12	Briggs et al. (1988)
	0.11–0.26	Teshima et al. (1997)
	0.6	D'Abramo (1998)
American lobster (*H. americanus*)	0.5	Castell et al. (1975)
	NR in adult	Castell and Covey (1976)
	0.12	D'Abramo et al. (1984)
	0.25	Kean et al. (1985)
Crayfish (*Pacifastacus leniusculus*)	0.4	D'Abramo et al. (1985)
C. maenas	1.4–2.1	Ponat and Adelung (1983)
Mud Crab (*Scylla serrata*)	0.61[a]	Suprayudi et al. (2012)
White shrimp (*L. vannamei*)	0.35	Gong et al. (2000)
	0.14 (1.5% deoiled soybean lecithin)	Gong et al. (2000)
	0.13 (3.0% deoiled soybean lecithin)	Gong et al. (2000)

[a]Larval stage; NR: not required.

cholesterol requirements range from 0.05 to 0.5% for *L. vannamei* (Duerr and Walsh, 1996; Gong et al., 2000). Morris et al. (2011), under the presence of natural productivity, stated that the cholesterol requirement for *L. vannamei* grow-out appears to be around 0.11%, which resulted in similar shrimp performance as those with higher levels, however, a regression analysis predicted the cholesterol requirement for maximum growth to be 0.15% of the diet. Variations reported between previous studies may be due to a variety of factors which include a lack of dietary PLs (Gong et al., 2000) and can vary with feed rates. Teshima (1998), using a factorial method, estimated optimum dietary cholesterol levels for *P. japonicus* to be 0.50%, 0.40%, and 0.29% at daily feeding rates of 3%, 5%, and 7% of body weight. Excess dietary cholesterol has been reported to have adverse effects on the growth performance of postlarval shrimp (Niu et al., 2012b).

4.4   Carbohydrates

Carbohydrates are polyhydroxylated aldehydes or ketones, that is, they yield these products on hydrolysis. They are comprised of carbon, hydrogen, and oxygen, with the last two elements being present in the same ratio as in water ($[CH_2O]_n$). Carbohydrates are only a small fraction of the body mass of animals but comprise the bulk of the mass of plants. In mammals and fish they serve as energy sources to run metabolic functions with little storage. In plants they serve a structural function as well as being an energy source to run the plant's metabolic functions. They form the bulk of the organic matter of plants: 50% in forages and 80% in grains.

Plants synthesize carbohydrates of various sizes (carbon chain length) and complexity from carbon dioxide and water by using solar energy in the process of photosynthesis.

Plants can use glucose as an energy source or convert it to larger carbohydrates such as starch or cellulose. Starch provides energy for later use, while cellulose is the structural material of plants.

The simplest carbohydrates are those that cannot be hydrolyzed to produce even smaller carbohydrates and are called **monosaccharides**. The naturally occurring monosaccharides contain three to seven carbon atoms per molecule. They give poor results because of immediate absorption. **Disaccharide** molecules, composed by two monosaccharide units, show a good performance because of steady digestion and gradual absorption. The **oligosaccharides** are formed by between three and ten monosaccharide units joined by glycoside bonds. Raffinose, a trisaccharide, acts as an antinutritional factor present in pulses. **Polysaccharides** are formed by polymerization of more than 10 monosaccharides with either straight or branched chains in which the monosaccharides are joined together by glycosidic linkages. They serve as structural elements in plant cell walls and exoskeletons. Polysaccharides differ from each other by the monosaccharides recurring in the structure, by the length and the degree of branching of chains, or by the type of links between units. Thus, polysaccharides may be homopolysaccharides or heteropolysaccharides. The following table is a list of biologically important polysaccharides and their functions (Table 4.12).

Table 4.12 **Importance and biological function of several polysaccharides**

Name of the polysaccharide	Composition of polymer	Occurrence	Functions
Starch	α-Glucose containing a straight chain of glucose molecules (amylose) and a branched chain of glucose molecules (amylopectin)	In several plant species	Reserve of energy
Glycogen	α-Glucose	Animals (equivalent of starch)	Energy source
Callose	β-Glucose	Different parts of plant	Formed often as a response to wounds
Cellulose	β-Glucose	Plant cell wall	Cell wall matrix
Chitin	β-Glucose	Body wall of crustacean. In some fungi also	Exoskeleton Impermeable to water
Lignin	β-Glucose	Plant cell wall	Cell wall matrix
Insulin	Fructose	In roots and tubers	Storage of reserve food
Pectin	Galactose and its derivatives	Plant cell wall	Cell wall matrix
Hemicellulose	Pentoses and sugar acids	Plant cell wall	Cell wall matrix
Murein	Polysaccharide cross linked with amino acids	Cell wall of prokaryotic cells	Structural protection
Hyaluronic acid	Sugar acids	Connective tissue matrix, Outer coat of mammalian eggs	Ground substance, protection
Chrondroitin sulfate	Sugar acids	Connective tissue matrix	Ground substance
Heparin	Closely related to chrondroitin	Connective tissue cells	Anticoagulant
Gums and mucilages	Sugars and sugar acids	Gums – bark or trees. Mucilages	Retain water in dry seasons

4.4.1 Classifications of carbohydrates

Carbohydrates are classified into three major classes on the basis of complexity and behavior on hydrolysis.

A. Monosaccharides (simple sugars)
  1. Pentoses ($C_5H_{10}O_5$) – five carbon sugars: D-xylose, L-arabinose, D-ribose
  2. Hexoses ($C_6H_{12}O_6$) – six carbon sugars
  3. Four and seven carbon sugars are found only in insignificant quantities. They may be found in metabolic cycles in the catabolism of carbohydrates

B. Olysaccharides: These are polymers of monosaccharides containing two to ten residues

  a. *Disaccharides* ($C_{12}H_{22}O_{11}$): lactose, maltose, sucrose. They yield two monosaccharides on hydrolysis

  b. *Trisaccharides* ($C_{18}H_{32}O_{16}$) – three simple sugars. Raffinose (glucose + fructose + galactose) found in cotton seed.

  c. *Tetrasaccharides*: Yield four monosaccharides on hydrolysis, for example, stachyose (glucose + fructose + galactose + galactose) that is the only tetrasaccharide known to exist in plants.

C. Polysaccharides

  1. Homopolysaccharides: on hydrolysis gives single monosaccharide units

    i. Pentosan: string of pentoses ($C_5H_{10}O_5$)

    ii. Hexosans: contains hexoses ($C_6H_{12}O_6$) subdivided into

      a. Glucosans: polymer of glucose (starch, glycogen, cellulose)

      b. Fructosans: polymer of fructose (inulin)

      c. Galactans: polymer of galactose (galactan)

      d. Mannans: polymer of mannose (mananas)

  2. Heteropolysaccharide: pectin, hemicellulose, hyaluronic acid, chondroitin sulfates

    a. Gum: consist of arabinose, rhamnose, galactose, and glucoronic acid

    b. Agar: the sulfuric acid esters of galactans consists of galactose, galactouronic acid

    c. Pectins: fundamental unit is pectic acid, consist of arabinose, galactose, galactouronic acid

    d. Chitin: polymer of *N*-acetyl-glucosamine.

The three most abundant polysaccharides are starch, glycogen, and cellulose. These three are referred to as *homopolysaccharides* because each yields only one type of monosaccharide (glucose) after complete hydrolysis. Starch found in plants is quantitatively the most important and abundant carbohydrate. It is a polymerized molecule composed by 100 to several hundred units. It is insoluble in cold water but swells when the water is heated. Starches are made up of two main polymers of glucose, a straight-chain glucose polymer called amylose and the branched-chain glucose polymer, amylopectin. Amylose, formed from linear chains of glucose units in -1,4 linkages, forms a 3D network when cooked. It presents a structure of a gel with a degree of polymerization of approximately 1,000 or less glucose units per chain. Starch rich in amylose is poorly digestible compared to starch rich in amylopectin. Amylopectin is a highly branched polymer of glucose. Branching takes place with α-(1→6) bonds occurring every 24–30 glucose units. It is the major component of starch and its content varies 91–100% in waxy starches, 70–80% in normal starches, and 30–50% in high-amylose starches. It plays a dominant role in the properties of starch, such as gelatinization and pasting properties. it thickens but does not form a gel.

Starch is the most important, abundant, digestible food polysaccharide. It occurs as discrete, partially crystalline granules whose size, shape, and gelatinization temperature depend on the individual plant species.

4.4.2 Non-starch polysaccharides

The term nonstarch polysaccharides (NSPs) covers a large variety of polysaccharides, other than starches which are composed of different kinds of monomers linked predominantly by β-glycosidic bonds. One of the first classifications was based on the differences in solubility which includes three categories of NSPs, namely crude fiber (CF), neutral detergent fiber (NDF), and acid detergent fiber (ADF) (Choct, 1997). The residue of plant material after a series of acid and alkali extractions was named CF while NDF refers to the neutral detergent-insoluble portion of the hemicellulose together with cellulose and lignin; and ADF is a portion of insoluble NSP formed mainly, but not only, of cellulose, lignin, and acid-insoluble hemicelullose. NSP are the major part of dietary fiber and comprised largely of cellulose, β-glucans, pectins, gums, mucilages. The NSP in wheat, maize, and rice are mainly insoluble (cellulose and mayorit hemicelullose) and have a laxative effect, while those in oats, barley, rye, and beans are mainly soluble (pectins, gums, and some hemicelullose).

Sinha et al. (2011) have written a comprehensive review on NSP and their role on fish nutrition. The authors summarized the following factors as causing antinutritional effects: changes in digesta viscosity and different kinds of alterations in native gut microflora, gut mucus layer, gut morphology, gut physiology, and in rate of gut transit and hence gastric emptying. All of these alterations cause: interference in the interaction of digestive enzyme with feed, reduce mixing of enzymes with feed bolus, impaired nutrient digestion, hampered absorption of water, cholesterol and lipids in intestine, reduce villi length and increase the depth of intestinal crypts and concentration of luminal mucin, and reduce fish performance among others. There are three types of polysaccharides that make up the plant cell wall: cellulose, hemicellulose, and pectin or pectic polysaccharides.

Cellulose is a linear chain of glucose molecules consisting of 7,000–10,000 subunits that comprise over 50% of the carbon in vegetation. It is resistant to most chemical reagents and cannot be hydrolyzed by any enzyme secreted by animal tissues, including aquatic species, because they do not have enzymes to cleave β-1,4 glycosidic linkage. Bacteria in gut contain cellulose which can cleave this linkage. The difference between cellulose and starch lies in the type of linkages between subunits. Cellulose has β-1,4 glycosidic linkages compared to α-1,4 glycosidic linkages in starch. Partial hydrolysis of cellulose releases cellobiose that completely hydrolyzed produces glucose. In contrast to cellulose that is crystalline, strong, and resistant to hydrolysis, hemicellulose has a random, amorphous structure with little strength. Hemicelluloses are low-molecular-weight polymers, however, cellulose has a very high degree of polymerization. This heteropolysaccharide is linear, flat, with a β-1,4 backbone and relatively short side chains. It is easily hydrolyzed by dilute acid or base. Two common types include xyloglucans and glucuronarabinoxylans. Other less common ones include glucomannans, galactoglucomannans, and galactomannans.

Pectins are the widest complex family of polysaccharides in nature. The structural classes of the pectic polysaccharides include homogalacturonan, xylogalacturonan, apiogalacturonan, rhamnogalacturonan I, and rhamnogalacturonan II (Caffall and Mohnen, 2009; Voragen et al., 2009). Homogalacturonan, a polymer of α-1,4-linked-D-galacturonic acid, is the major type of pectin in cell walls, that can account for greater than 60% of pectins. These polysaccharides contribute to the structural integrity of the wall, cell strength, cell adhesion, stomatal function, and are a defense mechanism and protect against wounding. Pectic polysaccharides are found in cell walls of cereals but not in endosperm, with the exception of rice (Shibuya and Nakane, 1984). Although **pentosans** occur as a minor carbohydrate fraction in cereals such as maize, wheat, and rye, they have important water-binding capacity and viscosity-enhancing properties. They are composed predominantly of two pentoses, arabinose and xylose, structured in a linear $(1\rightarrow4)$-β-xylan backbone. The β-glucans are constituents of the cell wall in most cereals, being particularly high in barley and oats at concentrations of about 7% and 5%, respectively. These polysaccharides consist of a linear chain of glucose units bound by both β-$(1\rightarrow3)$ and β-$(1\rightarrow4)$ linkages. Growing conditions and variety affect β-glucan content and linkage types. In barley, the β-glucan content ranges from 4% to 11% and they contain approximately 70% $(1\rightarrow4)$ linkages and 30% $(1\rightarrow3)$ linkages (Choct, 1997; Gatlin et al., 2007). The metabolic effects of glucans is not clear in fish but it has been demonstrated that these substances increase immune defense by an activating system (Sealey et al., 2008; Meena et al., 2013), so prolonged feeding may be detrimental.

The adverse effects include impaired nutrient digestion and absorption, lesser mixing of digestive enzymes and substrates, reduced rate of gastric emptying, villi length and animal performance among others that are attributed to changes in digesta viscosity, alteration in the gastric emptying and rate of passage, alteration of gut in its morphology, physiology, native microflora, and mucus layer. Feeding rainbow trout (*Oncorhynchus mykiss*) with dietary-soluble NSP (guar gum) at 2.5%, 5%, 10% inclusion levels led to an increase in digesta viscosity and depressed digestibility and growth (Storebakken, 1985), whereas in African catfish Leenhouwers et al. (2006) reported that dietary-soluble NSP (guar gum: 4% and 8%) increases digesta viscosity, reduces nutrient digestibility, and increases organ weight without affecting growth and FCR. On the contrary, Enes et al. (2013) also evaluating guar gum at three dietary levels (4%, 8%, 12%) on white sea bream, *Diplodus sargus*, found no effect on growth performance, feed efficiency, glucose utilization, liver glycogen content, and in lowering plasma cholesterol and triacylglyceride levels. Tilapia is very tolerant to NSPs of cereal grains (Leenhouwers et al., 2007). Amirkolaie et al. (2005) compared cellulose (insoluble NSP), guar gum (soluble NSP), and their combination at 8% dietary inclusion in tilapia diets. Cellulose inclusion did not influence digesta viscosity, growth, and digestibility of protein and starch, whereas guar gum inclusion increased digesta viscosity and reduced the growth and digestibility of protein, fat, and starch. The growth reduction with guar gum was, however, of practical insignificance and more a demonstration of the tolerance of tilapia.

The structure of NSP can be classified based on their physicochemical properties: viscosity, water-holding capacity, fermentability, and the capacity to bind organic and inorganic molecules and also based on the reaction with water as either soluble or insoluble. The NSP content varies (Table 4.13) in intra- and interingredients due to variety and the geographical location where it is grown.

Apart from NSP, other unavailable carbohydrates in food are inulin and lignin. Inulin (fructo-oligosacchirdes) consists of a straight chain of fructose units which are linked by $(1\rightarrow2)$-β-glycosidic bonds. The number of fructose units ranges from 2 to 60 and often terminates in a α-D-glucose (Green, 2001). Fructo-oligosaccharides

Table 4.13 Nonstarch polysaccharide (NSP) content (g/kg dry matter) present in ingredients used in fish feed

Ingredients	Starch	Total dietary fiber	Cellulose	Soluble NSP	Insoluble NSP	Total NSP
Alfalfa	68	457	139	77	113	329
Barley: hulled	587	221	43	56	88	186
Barley: hull-less	654	146	19	50	58	127
Corn whole grain	690	108	20	9	66	119
Corn gluten meal	282	383	75	242	34	351
Cottonseed cake	18	340	92	61	103	257
Cottonseed meal	19	375	90	66	127	283
Faba beans	407	210	81	50	59	665
Linseed meal	27	423	53	138	112	303
Lupins	14	416	131	144	139	405
Oats: hulled	468	298	82	40	110	232
Oats: hull-less	557	148	14	54	49	116
Palm cake	11	602	73	32	361	466
Peas	454	192	53	52	76	180
Rapeseed cake	15	295	59	43	103	205
Rapeseed meal	18	354	52	55	123	220
Rice polishing	180	131	112	5	213	218
Rice pearled	770	4	3	3	6	60
Rye bran	87	490	39	63	321	422
Rye whole grain	613	174	16	42	94	152
Sorghum[a]	712	93	34	99		67
Soybean meal	27	233	62	63	92	217
Sunflower cake	10	448	123	57	136	315
Tapioca	768	106	27	23	33	84
Wheat whole grain	651	138	20	25	74	119
Wheat bran	222	449	72	29	273	374
Wheat middlings	36	201	19	71	101	190

Based on Sinha et al. (2011); Bach Knudsen (1997); Choct (1997).
[a]Average of three varieties (Bach Knudsen et al., 1988).

(FOSs) have been shown to be nondigestible and beneficially affect the host by selectively stimulating the growth and/or the activity of the intestinal microflora. FOSs have important beneficial effects, such as a prebiotic effect, improving growth, feed conversion, gut microbiota, cell morphology, resistance against pathogenic bacteria, and immune response (Ortiz et al., 2012; Luna-González et al., 2012). Lignins are phenolic heteropolymers that confer mechanical support to plant cell walls. They contain more gross energy (GE) than other cell wall components under thermochemical processes such as direct combustion. In animal nutrition, lignin is considered an antinutritive component because it affects the digestibility of feeds.

Glycogen is to animals as starch is to plants. Glycogen is very similar to amylopectin, having a high molecular weight and branched chain structures made up of thousands of glucose molecules. The main difference between glycogen and amylopectin is that glycogen is much more highly branched, with branches at 8–12 residues, resulting in a more compact molecule with greater solubility and lower viscosity. Glycogen stored in the liver (1–12% of tissue weight) and in muscle (<1%) (Kaushik, 2001) functions as a readily available energy source and not long-term storage. Glycogen synthesis occurs through a different metabolic pathway. The first step in glycogen synthesis is glucose activation to glucose-6-phosphate, a reaction catalyzed by glucokinase, also called hexokinase-IV. Glucose-6-phosphate polymerizes forming glycogen plus phosphoric acid. Hepatic glycogen synthesis has been reported under defined nutritional conditions in different fish species, including *S. salar, Sparus aurata*, brown trout *Salmo trutta, D. labrax* (Soengas et al., 1996; Metón et al., 1999; Blasco et al., 2001; Perez-Jiménez et al., 2007). Felip et al. (2012) using stable isotopes ([^{13}C] starch) demonstrated the incorporation of ^{13}C from dietary starch to muscle glycogen in rainbow trout (*S. trutta*). In carnivorous species that poorly digest carbohydrate the liver can synthesize glycogen from gluconeogenic AA as has been reported by Enes et al. (2009) and Viegas et al. (2012).

Glucose does not require digestion and is the most rapidly absorbed monosaccharide across the intestines. In peneids, the glucose level of plasma increased rapidly after they were fed a diet containing glucose and remained at high levels for 24 h. Disaccharides and polysaccharides increased to a maximum at 3 h with excess glucose stored as glycogen or fat (main storage of glucose energy in large excesses). In fish, glucose makes a small contribution during swimming. According to West et al. (1993, 1994) the white muscle in fish was only responsible for 10–15% of the glucose utilization, while other smaller organs including spleen, kidney, and gills, used 60–90% of the glucose taken up (Blasco et al., 2001).

Metabolism is the biological utilization of absorbed nutrients for synthesis (e.g., growth) and energy expenditure. As mentioned above, for most aquatic species, the protein-sparing effect of carbohydrates is good. However, carbohydrate metabolism has a long lag time associated with it once carbohydrate is ingested/digested, blood glucose levels quickly rise, but require extended periods to decline. This lag response is considered similar in effect to that of diabetes, thus turnover of carbohydrate by aquatics is much slower than that of land animals because aquatic species often prefer to oxidize AA for energy. Carbohydrates have different metabolic roles, such as (i) immediate source of energy for certain tissues including brain, muscle, red blood cells; (ii) energy reserve (glycogen); (iii) converted to lipids (lipogenesis) when glucose is in large excesses; and (iv) synthesis of nonessential amino acids. The normal pathway of converting carbohydrate to energy is known as glycolysis, where one mole of glucose (6C) is converted into two moles of pyruvate (3C). In these reactions, glucose is first phosphorylated into glucose 6-phosphate. This irreversible reaction is catalyzed by hexokinase. The molecule thus formed is converted by means of phosphohexose isomerase to fructose-6-phosphate, which is again phosphorylated to yield fructose-1,6-bisphosphate by enzyme phosphofructokinase-1. The last step is the transfer of the phosphoryl group from phosphoenolpyruvate to ADP-producing pyruvate, catalyzed by pyruvate kinase (PK). Couto et al. (2008) reported that the PK activity increased with dietary gelatinized starch level.

Gluconeogenesis is a very important pathway in glucose homeostasis and takes place mainly in the liver. The glucose produced passes into the blood to supply other tissues. The key enzymes involved in the gluconeogenesis pathway are phosphoenolpyruvate carboxykinase, fructose-1,6-bisphosphatase (FBPase), and glucose-6-phosphatase that use nonglycosidic substrates, such as lactate, glycerol, or α-ketoacids.

4.4.3   Carbohydrate metabolic pathways

4.4.3.1   Fish insulin and glucagon

Insulin produced in the pancreas B cells is high during feeding because it is required for the transportation and utilization of glucose in tissues. Thus, it increases the uptake capacity from blood, decreasing blood glucose levels. On the contrary, glucagon is released from A cells in the pancreas during starvation (and insulin is low) and increases blood glucose. The glucagon:insulin ratio is under hormonal control (epinephrine), and norepinephrine has been identified in fish (Moon et al., 1999). Glucagon stimulates glycogen breakdown into glucose subunits and also stimulates the synthesis of glucose in the liver mainly and the kidneys to a lesser extent. Thus, the blood glucose concentration is maintained during the absence of food in certain species such as carp (and other warmwater species) and American eel (Cornish and Moon, 1986), while in other species, such as sea bass, brown trout, and dogfish, a lower glycemia can be found after a few days of fasting (Navarro and Gutiérrez, 1995). Blood glucose levels in all teleost, which account for 96% of extant fish, fed on a carbohydrate-rich meal reach 300 mg/100 ml (similar to chicken) and can be prolonged more than 40 h. Fish blood glucose levels fluctuate between intraspecies and interspecies, and under different physiological conditions and environmental challenges. It has been reported that after an oral glucose tolerance test, the blood glucose level in fish remains high for several hours, returning to the basal level at greater than 5 h for carp, red sea bream, and yellowtail (Furuichi and Yone, 1981), greater than 6 h in catfish (Wilson and Poe, 1987), greater than 24 h in European silver eel, tilapia (Lin et al., 2000), and sturgeon (Hung, 1991), and greater than 40 h in salmon (Mazur et al., 1992). These studies show that teleost fish are glucose-intolerant compared to mammals.

Another distinction from mammals is the prolonged tolerance of fish to hypoglycemia without any distinguishable symptoms of neural disorders (Mommsen and Plisetskaya, 1991), this is due to very low glucose requirements by tissues and their predominant dependence on protein and lipid metabolism. However, fish maintain homeostatic control of glucose by a feedback mechanism conformed by glucosensors located in different parts of the body and linked to insulin secretion. In teleost, there are several studies available regarding the existence of glucosensing-dependent mechanisms involved in insulin release in the pancreas but not in elasmobranch fish. There is a greater amount of studies in carnivorous species than omnivorous and herbivorous species that have reported evidence that suggest the presence of glucosensing components: glucose transporter type 2 (GLU-2), glucokinase (GK), glycolytic pathway, K_{ATP} channel, and calcium channel (Polakof et al., 2011).

4.4.4   Utilization of carbohydrates

The use of carbohydrate in formulated diets is justified because it is the cheapest dietary energy source in nature and its inclusion in feeds allows more protein to be used for growth rather than as an energy source, enhance feed conversion rates and by their binding properties (gelatinized starch, alginates, gums) reducing fines in pelleted and extruded feeds. Moreover, they are intermediate metabolites of biological components, e.g., chitin, nucleic acids, and mucopolysaccharides (Stone, 2003). The utilization of carbohydrate by aquatic species is lower than in terrestrial animals. Warmwater fish can use much greater amounts of dietary carbohydrate than coldwater and marine fish. Generally, complex carbohydrates are better used than simple carbohydrates (monosaccharides and disaccharides) (Deshimaru and Yone, 1978b; Arnesen et al., 1995; Shiau and Liang, 1995). Carbohydrate utilization depends on species and development stage, the complexity of the molecule and the amount thereof in the feed, interaction with other dietary components, technological process, and water temperature and salinity (Médale et al., 1999; Hemre et al., 2002; Stone, 2003; Gaxiola et al., 2005; Alexander et al., 2011). The fish and shrimp in general possess enzymes to hydrolyze both simple and more complex carbohydrates. Several carbohydrases have been identified in fish, namely α-amylase, disaccharidase, and α-glucosidase (Stone, 2003) and in crustaceans α- and β-amylase, maltase, sucrase, chitinase, and cellulose (Guillaume and Ceccaldi, 2001). However, it is believed that the latter two enzymes are produced by cellulite and chitinolytic bacteria. Soler (1996) reported that 60% of the bacteria present in the gut of *Litopenaeus setiferus* show rapid growth, tolerance to low pH, some enzyme activity, and 85–100% chitinolytic activity.

All species of fish have been shown to secrete α-amylase. α-Amylase is a carbohydrase glycoprotein that catalyzes the hydrolysis 1–4 bound of carbohydrates. The ability of fish and shrimps to utilize dietary carbohydrate varies considerably among species. The α-amylase in common carp is 80 times and 10–30 times greater than in yellowtail and rainbow trout, respectively (Shimeno et al., 1977; Wilson, 1994). Silver perch possess seven and 15 times higher activities than those reported for carnivorous species (barramundi and Atlantic salmon) and around half that found in tilapia (Anderson and Lipovsek, 1998). In general, the total amylase activity of the digestive tract is higher in the omnivorous species (carp, goldfish, and tench) than in the carnivores (rainbow trout, gilthead sea bream, European eel) (Hidalgo et al., 1999). Studies have indicated that common carp, channel catfish, red sea bream, and tilapia use higher levels of carbohydrates than yellowtail and salmonids. In Nile tilapia juveniles (8 g) dietary starch levels up to 10% as energy source had no effect on fish performance (Boscolo et al., 2008). When comparing dietary starch levels from gelatinized maize, growth performance of 50-g tilapia fed low-starch diets (10–12%) were significantly higher than that of fish fed high-starch diets (30–36%). There were no effects of starch levels on feed intake (Tran-Duy et al., 2008). In general, less than 20% digestible carbohydrate seems to be optimal for marine or coldwater fish (rainbow trout, plaice, and yellowtail, etc.) whereas fresh- or warmwater fish (channel catfish, common carp, eel, red drum, etc.) can use a higher percentage (Wilson, 1994). Pretreating (cooking, extrusion) carbohydrates improves their digestibility (Stone, 2003).

Three isoforms of amylase have been determined in *L. vannamei* (Van Wormhoudt et al., 1996) but no zymogen for this enzyme has been characterized in crustaceans (Carrillo and González, 1998). Depending on the degree of amylolytic activity in penaeid shrimp different proportions of carbohydrate compounds, usually starches, may be included in diets. Considering that starch is the cheapest component in diets for aquatic organisms, it is of importance to shrimp farming because it partially replaces protein (Cousin et al., 1996) as an energy source. Vega-villasante et al. (1993) determined that the amylase activity in the digestive tract of *Litopenaeus californiensis* is halotolerant and appears to be better activated in the presence of low ion concentrations, such as magnesium, calcium, and sodium. Note that the freshwater crustaceans have more capacity to digest especially complex carbohydrates, compared to sea crustaceans (Lee and Lawrence, 1997).

With respect to the use of sources of raw starch in diets for *L. vannamei*, it is reported that starch from wheat is more digestible than from maize and sorghum (Cousin et al., 1996). Davis and Arnold (1993) evaluated in *L. vannamei* the apparent digestibility of five carbohydrate sources: wheat starch, whole wheat, Nutribinder, sorghum, and corn when cooked. The results show that gelatinization improved digestibilities of corn flour and sorghum but not in whole wheat and rice flour. Catacutan (1991), who fed shrimp, *P. monodon*, on isoproteic diets, observed that by increasing the level of carbohydrates the digestibility of dry matter and lipids were affected positively and negatively, respectively. In shrimp, *P. orientalis*, Shen and Liu (1992) observed that the starch level in the diet is negatively correlated to its digestibility. A similar behavior was reported by Velurtas et al. (2011), who evaluated the effect of different starch/cellulose ratios (30/0, 20/10, 10/20, 0/30) on apparent digestibility in two species of penaeids: *A. longinaris* and *P. muelleri*. The apparent digestibility coefficients decreased from 83.7% to 51.2% for *A. longinaris* and from 71.9% to 7.6% for *P. muelleri* as the dietary starch levels increased, suggesting that the former has herbivorous behavior and the latter has omnivorous habits.

According to Niu et al. (2012a,b) wheat starch and sucrose are better utilized as carbohydrate sources for juvenile *P. monodon* due to the higher hepatopancreas amylase activity, nutrient digestibility, weight gain, and biomass gain obtained with shrimp fed on these carbohydrate sources as compared to other sources (potato starch, maize starch, dextrin, maltose, and glucose). Moreover, it was also reported that the phosphogluconate dehydrogenase (6PGDH) and hexokinase activities increased as dietary carbohydrates became more suitable to support the best growth performance.

4.4.5 Carbohydrate digestibility

Source (amylose/amylopectin content, granule size; Table 4.14), dietary level, and physical state (degree of gelatinization and molecular complexity) affect the

Table 4.14 **Amylose content and granule size of various starches**

Starch source	Amylose (g/100 g starch)	Granule size range (µm)
Corn	37.1	5–20
Rice	34.1	3–8
Sorghum	17.1	15–35
Wheat	28	2–30
Barley	22	2–25
Potato	21	5–80
Cassava	17	3–30

digestibility of carbohydrates in fish. The differences in digestibility among starch sources have been attributed to mylose/amylopectin content. In rainbow trout, Gaylord et al. (2009) reported that amylose digestibility is reduced compared to amylopectin. Carbohydrate digestibilities are also affected by the size of granules that is given by botanical origins. Molina-Poveda and Gómez (2002) reported that cassava starch resulted in a significantly higher carbohydrate digestibility due to their granules have a particle size of 4–24 microns which is less than the 30–70 microns of the starch granules of banana and plantain. This smaller particle size increases the contact surface for carbohydrase enzymes, making them more suitable as a carbohydrate source.

A negative effect of the carbohydrate inclusion level on starch digestibility has been reported in several species (Arnesen et al., 1995; Stone et al., 2003). Thus in Atlantic halibut the increase in wheat content from 8% to 17% resulted in a decrease in starch digestibility from 84% to 53% (Grisdale-Helland and Helland, 1998). Starch gelatinization enhances its digestibility by making it more susceptible to carbohydrase enzyme action and can thus make it a valuable energy source for carnivorous species deficient in amylase. During gelatinization starch granules swell and then broken. The moist-heat cooking produces not only the breakdown of starch granules, which are insoluble in coldwater, but it also generates an irreversible change in the molecule structure. Several works show that the digestibility of diets with gelatinized carbohydrate sources is higher than diets containing the same source of carbohydrates ungelatinized. In fish it has been observed that cooked starches are more digestible than raw for silver perch and Indian major carp (Erfanullah and Jafri, 1998; Stone et al., 2003). In shrimp the same effect has been reported, Cousin et al. (1996), who fed *L. vannamei* (18–25 g) with different carbohydrate sources at a level of inclusion of 35% in the diet, observed a higher ADC in diets containing gelatinized starches. In general, fish and shrimp digest better purified or processed carbohydrates sources than when raw (Table 4.15).

Table 4.15 Effect of source and dietary inclusion level on carbohydrate apparent digestibility (ADC) reported in several fish and shrimps species

Species	Starch source		Inclusion level (%)	ADC (%)	References
Channel catfish	Corn grain	Native	30	–	Cruz (1975)
	Wheat grain	Native	30	59	
	Corn	Native	30	66	
	Corn	Native	60	59	
	Corn	Cooked	30	78	
	Corn	Cooked	60	62	
	Corn starch	Native	12.5	72.8	Saad (1989)
			25	60.9	
			50	55.1	
	Corn starch	Cooked	12.5	83.1	
			25	78.3	
			50	66.5	
Common carp	Wheat starch	Native	10	77	Appleford and
	Wheat starch	Native	20	88	Anderson
	Wheat starch	Native	30	90	(1996)
	Wheat starch	Native	40	94	
	Potato starch	Native		55	Chiou and
		Cooked		85	Ogino (1975)
Silver perch	Wheat starch	Native	30	76	Stone et al.
	Wheat starch	Gelatinized	30	89	(2003)
	Wheat starch	Native	60	41	
	Wheat starch	Gelatinized	60	70	
Tilapia	Corn starch	Native	31	92	Shiau and Liang
			47	86	(1995)
Rainbow trout	Potato starch	Cooked	20	69	Singh and Nose
		Cooked	30	65	(1967)
		Cooked	40	53	
		Cooked	50	38	
		Cooked	60	26	
	Dextrin		20	77	
			30	74	
			40	60	
			50	50	
			60	46	
	Whole corn	Native	10	43	Ufodike and
	Whole corn	Native	20	43	Matty (1989)
	Whole corn	Native	30	54	
	Whole potato	Native	10	16	
	Whole potato	Native	20	18	
	Whole potato	Native	30	35	

(Continued)

Table 4.15 (**Continued**)

Species	Starch source		Inclusion level (%)	ADC (%)	References
Atlantic salmon	Wheat starch	Native	10	88	Grisdale-
	Wheat starch	Native	19	78	Helland and
	Wheat starch	Native	25	75	Helland
					(1997)
	Whole wheat	Native	15	50	Arnesen and
	Whole wheat	Native	30	73	Krogdahl
	Whole wheat	Native	45	34	(1993)
	Whole wheat	Extruded	15	48	
	Whole wheat	Extruded	30	41	
	Whole wheat	Extruded	45	28	
Cod	Potato starch	Dextrinized	3	40	Hemre et al.
	Potato starch	Dextrinized	7	33	(1989)
	Potato starch	Dextrinized	12	26	
White shrimp	Corn starch	Raw	35	85	Cousin et al.
(*L. vannamei*)	Corn starch	High amylose	35	63	(1996)
	Corn starch	Waxy	35	85	
	Corn starch	Gelatinized	35	94	
	Corn starch	Gelatinized waxy	35	96	
	Potato starch	Raw	35	72	
	Potato starch	Gelatinized	35	93	
	Wheat starch	Raw	35	92	
	Wheat flour		36	78	Rivas-Vega
	Cowpea	Raw	15	77	et al. (2006)
	Cowpea	Cooked	15	83	
	Cowpea	Extruded	15	82	
	Cassava	Native	35	75	Molina-Poveda
		Gelatinized	35	88	and Gómez
	Plantain	Native	35	69	(2002)
		Gelatinized	35	81	
	Banana	Native	35	45	
		Gelatinized	35	74	

4.5   Nutritional energetics

Bioenergetics or nutritional energetics studies energy flow through living systems and its regulation through the balance between energy intake in the form of food and energy utilization by animals for life-sustaining processes. All processes in the animal body involve the exchange and transformation of energy. Thus nutritional energetics involves the study of the sources and transformations of energy into new products, such as tissue deposition, being energy defined as the capacity to do work or utilized for chemical synthesis and anabolic processes in growth. Animal metabolism is primarily concerned with the utilization of chemical energy.

Energy is not a nutrient but a product of the absorption and metabolism of organic food components such as proteins, lipids, and carbohydrates. Fish and crustaceans preferably use proteins as an energy source due mainly to poor utilization of low-digestibility carbohydrate sources and moreover because in some species high lipid levels can cause abnormalities in animals, reduce growth and increase mortalities. Fish and shrimp require energy for growth, reproduction, and muscle activity which is taken from the oxidation of food. The amount of energy needed by an organism depends on the stage of life cycle, year season, and environmental conditions. An organism will need more energy per unit weight in its early stages than as an adult; likewise, the environmental temperature has a decisive effect on the metabolic rate of the organisms.

It is considered that aquatic organisms have lower energy requirements than terrestrial animals because they are poikilotherms, regulating their body temperature to medium, requiring less energy to maintain its position and move in the water compared to terrestrial organisms. Besides, nitrogenous wastes are excreted as ammonia, urea, or instead of uric acid, wasting less energy in protein catabolism and excretion of nitrogenous wastes.

4.5.1 Units of energy

Chemical energy is measured as heat. A calorie (cal) is the amount of heat required to raise one gram of water up one degree centigrade from 16.5°C to 17.5°C. A kilocalorie (kcal) equals 1,000 calories and a megacalorie (Mcal) equals 1,000 kilocalories. One cal is defined more precisely as 4.184 joules (J) and 1 J equal 0.238 cal. A Joule (James P. Joule, English physicist) is the amount of work converted into heat, equivalent to 10^7 ergs (1 erg being the energy quantity expended to accelerate a mass of 1 g by 1 cm/s). A kilojoule (kJ) equals 1,000 J and a megajoule (MJ) equals 1,000,000 J. According to the Le Systéme International d'Unites (SI; International System of Units) and the US National Bureau of Standards, the unit for expressing all forms of energy is the Joule (J) however; calorie is still used because it is the standard energy.

4.5.2 Partition of dietary energy

The energy flow can be expressed mathematically using the following equation in which all elements are expressed in Joules/unit weight/unit time.

$$IE = FE + R + W + SDA + UE + ZE + B + G$$

where:

IE: intake energy is GE consumed in food (carbohydrate, lipid, protein)
FE: fecal energy is GE of feces (undigested feed, metabolic products, gut epithelial cells, digestive enzymes, and excretory products)
R: maintenance functions
W: nonmaintenance activities
SDA: specific dynamic action

UE: urinary energy is total GE of urinary products of unused ingested compounds and metabolic products

ZE: gill excretion is GE of products excreted through gills

B: support organismal growth, and

G: gonad maturation.

Using this approach it is possible to obtain information on the different processes involved in the energy flow of organisms (Figure 4.2) from both an integral perspective or through the evaluation of some of its elements.

GE is the amount of energy released as heat of combustion when a substance is completely burnt to carbon dioxide and water in a bomb calorimeter under 25–30 atmospheres of oxygen. The energetic content of feedstuffs can also be calculated indirectly from the nutritional composition using this energy equivalent of carbohydrate, lipids, and protein (Table 4.16). Fats (triglycerides) have about twice the GE as carbohydrates because of the relative amounts of oxygen, hydrogen, and carbon in the compounds. Energy is derived from the heat of combustion of these elements: C = 8 kcal/g, H = 34.5 kcal/g.

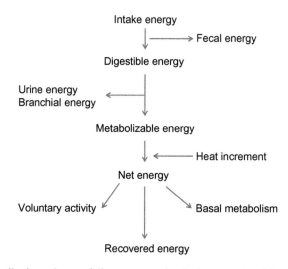

Figure 4.2 Distribution scheme of dietary energy intake in a growing fish.

Table 4.16 **Energetic values provided by each organic component of the diet**

	kcal/g	kJ/g
Carbohydrates	4.15	17.2
Fats	9.4	39.5
Proteins	5.65	23.7

Table 4.17 Gross energy values for some sources of carbohydrates, fats, and proteins determined by bomb calorimeter

Substrate	kcal/g
Glucose	3.74
Corn starch	4.21
Soybean oil	9.28
Casein	5.86
Corn	4.43
Oat straw	4.5

GE present in feedstuffs is not a measurement of its energy value to the consuming animal. The difference between GE and energy available to the animal varies greatly for different feedstuffs. For example, from Table 4.17 oat straw does not have the same feeding value as corn.

DE is GE consumed by fish in its food less GE of the feces. Fecal material is from undigested feed, body secretions as the product of metabolism, gut epithelial cells, and bacterial residues. As a result, DE is only an "apparent DE" and not a true DE. (DE = [E of food per unit dry weight × dry weight of food] – [E of feces per unit dry weight × dry weight of feces]). Metabolizable energy (ME) is defined as the amount of energy available from feed once the energy from feces, urine, and gills has been subtracted. Essentially, ME is the energy left for an animal's body to use once all processes of digestion and absorption are complete. In other words, ME is DE less GE in urine and gill products. The most common system used to express dietary energy for fish is the analog system utilized for poultry because urine and feces cannot be separated in this species. This is difficult to determine in aquatics due to the inherent inability to quantify losses. In addition, in fish confined in a small volume, excretion of urine and feces is increased by stress which can make it difficult to determine ME values.

ME is then calculated as:

$$ME = IE - (FE + UE + ZE)$$

where:

IE: intake energy
FE: fecal energy
UE: urinary energy
ZE: gill excretion energy.

The availability of energy varies according to feed ingredients and species. For finfish, DE approximates ME because most energy is used for digestion and is a reasonable indicator of quality of energy sources in feeds, although this depend on the trophic level (herbivore, carnivore, omnivore). Energy losses in fish through urine and gills do not vary much by feedstuff. Surface energy (SE) is energy lost to sloughing of mucus, scales, epithelial cells, and in crustacean exoskeleton. These losses are difficult to estimate and considered generally of little value in the energetic budget.

Net energy (NE) is ME minus the energy lost as the heat increment (HI): NE = ME − HI. Thus, for instance, the breakdown of glycogen to ATP during work causes heat. HI is also called SDA. NE is the energy used at different levels of efficiency for maintenance only or for maintenance as basal metabolism and production of tissue and eggs. There is no absolute NE value for each feedstuff. The magnitude of each type of energy depends on the quantity of intake plus the animal's ability to digest and utilize that energy. In a series of studies on nutritional energetics in carp, *Cyprinus carpio*, Ohta and Watanabe (1996) determined that the GE intake (100%) for maximum growth was partitioned as follows: 29.9% lost as fecal energy, 1.5% as nonfecal energy, 31.9% as SDA, and 36.7% as NE (including 12.6% for maintenance and activity and 24.1% as productive energy).

4.5.3   Energy utilization

Total daily energy expenditure can be subdivided into three components: basal metabolism, specific dynamic action, and voluntary physical activity.

4.5.3.1   Standard metabolic rate

The biological processes of energy use are defined as metabolism, while the rate at which energy is used is called the standard metabolic rate (SMR), which represents the minimum rate of necessary activity to support the structure and function of the body's tissues. The SMR is measured in organisms that are resting in a postabsorptive state and unstressed conditions in a thermally controlled environment for which the animals have had ample time to acclimate. In aquatic animals, SMR can be calculated in terms of oxygen consumption per unit time due to oxygen being the last electron acceptor in the respiratory chain. This consumption can be used to quantify the amount of energy available under a given nutritional status (Secor, 2011). Metabolic processes in fish and shrimp are influenced by factors such as species, age, or body size, genetic difference, physical activity, and body functions; other parameters such as water temperature, salinity, concentrations of ammonia, oxygen, or carbon dioxide, and pH also influence the metabolic rate. Relating the basal metabolic rate with body weight or volume unit, a specific weight metabolic rate is obtained with increasing body weight. The allometric relation that describes this is nonlinear, $y = aW^b$, for most physiological variables, b values usually range between 0.6 and 0.9, with being 0.86 more appropriate for many species (Jobling, 1994).

The metabolic rate has also been used to assess differences found between normal and genetically modified animals. Thus, Cook et al. (2000) found that transgenic Atlantic salmon had higher metabolic rates than normal salmon, inclusive of the SDA associated with feeding, but an oxygen consumption 42% less than normal salmon over the time to reach smolt size. Changes in salinity result in an increased cost of energy. Dalla Via et al. (1998) conducted a study on the metabolic rate of juvenile sea bass (*Discentrarchus labrax*) under stepwise changes in salinity (37→20→5→2→5→20→37 and 37→50 ppt). Metabolic rate increases above 80% of the routine level in each salinity change, but these increases were transient, returning to normal levels within 3–10 h after salinity changes. In the salinity ranges 37→20, 20→5 and 5→2 ppt, sea bass took about 3, 6–9, and 9–10 h, respectively, in returning to their levels of normal metabolism. For changes in salinity from 37 to 50 ppt, the rate of metabolism showed a temporal increase of 30% with a return to previous levels of exposure after 5 h.

It is known that in some crustaceans the metabolic rate is affected by the osmotic pressure of the environment. Thus, when *Astacus astacus* is transferred from freshwater to 15‰ seawater, oxygen utilization decreased by 40%. Furthermore, *Carcinus maenas* move from very high concentrations of seawater at 25‰ salinity, increases in 40% oxygen uptake (Waterman, 1960). An increase in SMR also results from heightened levels of waste, low oxygen, crowding, handling, pollution, etc. According to Alcaraz et al. (1999) postlarvae shrimp *L. setiferus* exposed to different levels of ionized ammonia (0.4–0.7 mg/l) and nitrite reduced oxygen consumption, which is further reduced as the oxygen concentration drops in the medium.

4.5.3.2 Specific dynamic action

The SDA of a food represents the oxygen or energy that the body has to use to ingest (prehension, chewing), digest (enzymatic hydrolysis, peristalsis), and assimilate a meal and protein turnover. SDA is largely calculated from the accumulated amount of oxygen consumed above the resting metabolic rate due to the cost of processing food. The heat produced by the food metabolism is unavoidable and mostly lost to the animal.

The amount of oxygen consumed can be expressed in energy units using the oxy-calorific coefficient which depends on the nature of the metabolic substrate used by the organisms being studied.

Respiratory quotient (RQ) or respiratory coefficient is a measurement of the ratio between oxygen an organism intakes and carbon dioxide the organism eliminates, expressed with the formula $RQ = CO_2$ eliminated/O_2 uptake. The RQ is a unitless number used in calculations.

The range of RQ for organisms in metabolic balance usually ranges from 1.0 representing the value expected for pure carbohydrate oxidation

Glucose
$$C_6H_{12}O_6 + 6O_2 = 6CO_2 + 6H_2O + 38ATP$$
$$6CO_2/6O_2 = 1$$

to approximately 0.7 representing the value expected for pure fat oxidation

Palmitic acid
$$C_{16}H_{32}O_2 + 23O_2 = 16CO_2 + 16H_2O + 129ATP$$
$$16CO_2/23O_2 = 0.7$$

Albumin
$$C_{72}H_{112}N_2O_{22}S + 77O_2 \rightarrow 63CO_2 + 38H_2O + SO_3 + 9CO(NH_2)_2$$
$$63CO_2/77O_2 = 0.82$$

Table 4.18 **Values of equivalent energy per unit of oxygen consumed necessary for degradation of different energy substrates in aquatic organisms**

Respiratory substrate	End products	Caloric value		RQ	Oxycalorific coefficient (j per mg O_2)
		O_2 (Cal/L)	CO_2 (Cal/L)		
Carbohydrate	CO_2, H_2O	5.047	5.047	1.0	14.76
Lipids	CO_2, H_2O	4.686	6.629	0.7	13.72
Protein	CO_2, H_2O, urea	4.485	5.599	0.84	13.6

RQ as measured also includes a contribution from the energy produced from protein, but due to each AA being able to be oxidized in different ways, no single RQ can be assigned (Table 4.18).

SDA has been quantified for more than 60 fish species and 30 crustacean species (Secor, 2009) but Table 4.19 summarizes only a group of selected species by their commercial importance. In this review, Secor (2009) standardized SDA of each study to kJ using the following conversion factors 1 cal = 4.184 J, 1 mg O_2 = 14 J, 1 mL O_2 = 19.5 J and 1 umol O_2 = 0.45 J. Among tabulated studies, body mass ranges from 0.0001 g for red drum, *Sciaenops ocellatus*, to 1,025 g for channel catfish, *I. punctatus* (Table 4.19). For most studies, temperatures were maintained between 10°C and 30°C. The food consumed was typically a natural prey item, such as rotifer, algae, brine shrimp, worms, mussel, squid, clam, and formulated feeds for fishes and shrimps. Although meal sizes range from 0.98% for *C. carpio* to 10% for penaeid shrimps' body mass, most meals for fish are less than 6% of body mass (Table 4.19). Factors affecting the magnitude and duration of the SDA response or the efficiency of utilization of the ME are size, type, and composition of meal, fish size, and salinity and temperature. For a greater feed intake there is greater heat loss per unit of ME (Houlihan et al., 1990) as a consequence of the increase in time and effort needed to digest and assimilate a larger meal. There are several studies that show a linear relationship between meal size and peak metabolism, duration, and SDA (Secor and Diamond, 1997; Fu et al., 2005), although Jobling and Davies (1980) suggested that SDA reaches a maximum level limited by the digestive process.

As illustrated in Table 4.19, a wide variety of natural prey (e.g., algae, shrimp, worms, clam, and various fish) has been fed to fish to assess their SDA response, however only a couple of studies have explored the effects of different natural food items on SDA. Nelson et al. (1977, 1985), feeding *Crangon franciscorum* and *M. rosenbergii* with tubificid worms, found a larger metabolic response than diets of fish, mysid shrimp, or algae.

Feed composition relative to percentages of protein, lipids, and carbohydrates affects the SDA response differently. Dietary fat has a lower SDA compared to carbohydrate, which means that fat is used with greater efficiency, and these two generate lower SDA than protein-based meals. At maintenance, SDA of fat and carbohydrate

Table 4.19 Resume of selected fish and crustacean studies on SDA response

Species	Body mass (g)	T (°C)	Meal type	Meal size (%)	Scope	Duration (h)	SDA (kJ)	SDA coefficient (%)	References
C. carpio	1.36	23	Formulated diet		1.51	12	0.038		Kaushik and Dabrowski (1983)
	72	28	Formulated diet	1	2.66	18	2.24	15.5	Chakraborty et al. (1992)
	150	20	Formulated diet		2.1				Yarzhombek et al. (1984)
	318	25	Formulated diet	0.94	2		2.57		Hamada and Maeda (1983)
O. niloticus	95	28	Formulated diet	1	1.7	15	1.05	6.35	Ross et al. (1992)
	58	27	Formulated diet	1.3	2.81		3.08	20.6	Mamun et al. (2007)
	6.3	30	Formulated diet	6	2.83		0.96	26.9	Xie et al. (1997)
Anguilla rostrata	1.5	20	Formulated diet		6.6	18	0.15		Gallagher and Matthews (1987)
Anguilla anguilla	72.8	25	Formulated diet	1.3	11		1.97	11.3	Owen (2001)
S. aurata	99.8	21	Formulated diet	1.5	3.18		0.95	19.8	Guinea and Fernandez (1997)
I. punctatus	1025	22	Formulated diet	2	2.28	18			Brown and Cameron (1991a,b)
D. labrax	42	25	Formulated diet	2	2.07	2.49	2.49	14.9	Peres and Oliva-Teles (2001)
Oncorhynchus rhodurus	50	10	Fish	2.4		30	0.68	16.6	Miura et al. (1976)
S. salar	3.7	18	Formulated diet	2.7	1.71	3.5	0.041	2.48	Smith et al. (1978)
G. morhua[a]	61	18	Fish	5.2	2.71		7.59	17.1	Soofiani and Hawkins (1982)
Tiliapia rendalli[a]	100	23.5	Algae	6.8			0.88	8.53	Caulton (1978)
O. tshawytscha	520	10	Formulated diet		2.28				Thorarensen and Farrell (2006)
Seriola dumerili	308	19	Formulated diet		1.52				De la Gándara et al. (2002)
Perca fluviatilis	2.45	20	Fish		4.19		0.65	13.5	Wieser and Medgyesy (1991)
S. ocellatus	0.0001	24	Rotifer		2.52				Torres et al. (1996)
Oreochromis mossambicus	1.18	25	Tubificid worms		5.01		0.52	40	Cui and Liu (1990)
C. gariepinus	0.002	28	Brine shrimp	3.33	1.63		0.24		Coneição et al. (1998)
C. maenas	20	15	Squid		3.08	20		10.5	Robertson et al. (2002)

(Continued)

Table 4.19 (Continued)

Species	Body mass (g)	T (°C)	Meal type	Meal size (%)	Scope	Duration (h)	SDA (kJ)	SDA coefficient (%)	References
C. maenas	49	15	Mussel	4.1	1.79				Legeay and Massabuau (1999)
Homaraus americanus	3.2		Formulated diet		1.51				Koshio et al. (1992)
L. vannamei	2.6	28	Formulated diet	2.3	2.08	96	0.031	3.9	Rosas et al. (2001)
M. rosenbergii	0.035[b]	28	Tubificid worm		1.41				Nelson et al. (1977)
M. rosenbergii	0.57	28	Formulated diet		1.73	17	0.024	7.1	Du and Niu (2002)
Penaeus duorarum	0.031	28	Formulated diet	10	2.89				Rosas et al. (1996)
Penaeus esculentus	0.27	30	Formulated diet		1.33				Hewitt and Irving (1990)
P. esculentus	17.7	25	Shrimp		1.39				Dall and Smith (1986)
P. monodon	5.58	28	Shrimp	2.86	2.25	5	0.061	9	Du Preez et al. (1992)
Penaeus notialis	0.027	28	Formulated diet	10	3.12				Rosas et al. (1996)
Penaeus schmitii	0.028	28	Formulated diet	10	2.84				Rosas et al. (1996)
Penaeus setiferus	0.023	28	Formulated diet	10	2.11				Rosas et al. (1996)

Source: Based on Secor (2009). With kind permission from Springer Science + Business Media.

Baseline: Metabolic rate of postabsorptive individuals. Scope: Postprandial peak divided by baseline. SDA coefficient: SDA divided by meal energy.

[a]Studies for scope, SDA, and/or SDA coefficient are calculated from published information.

[b]Body mass reported as dry mass.

are about the same (Beamish and Trippel, 1990). When nutritional requirements are covered, the magnitude of SDA minimizes and maximizes growth but deficiencies reduce the efficiency of metabolism and raise the SDA. Salinity has an effect on the SDA of crustaceans. For shrimp, *P. monodon* fed on a commercial diet, Du Preez et al. (1992) found greater SDA when maintained at 5 ppt compared to those exposed to the highest salinity. In crab, *Cancer gracilis*, McGaw (2006) also observed a decrease in oxygen uptake when animals were exposed to 21 ppt salinity but oxygen consumption was restored when placed back into seawater at 32 ppt. Several studies in aquatic species report that the peak of SDA increases with temperature, whereas the peak time and the duration of SDA decrease with temperature during digestion (Jobling, 1982; Robertson et al., 2002; Luo and Xie, 2008).

4.5.3.3   Voluntary physical activity

Energy expenditure rises above resting energy expenditure when physical activity is performed. Differences in duration, frequency, and intensity of physical activities caused by too high a water current in tanks, seeking food, stress, maintaining position, may create considerable variations in total energy expenditure depending on the activity that is performed. The physical activity levels have been shown to result in increases in total energy expenditure.

4.5.4   Protein to energy ratio

The knowledge of the optimal levels of protein and economization of protein to energy ratio, using nonprotein DE sources as carbohydrates and lipids, are necessary to reduce feed costs and produce maximum growth (Bautista, 1986). In addition to economic benefits, this allows a more environmentally friendly production system, avoiding excess protein in the diet, consequently decreasing the amount of excreted ammonium (Shiau and Chou, 1991). One way of reducing the content of protein in the diets is through a suitable balance between protein and lipid. Lipids are the energy source more concentrated from nutrients food, having 2.25 times more energy per unit weight of proteins and carbohydrates. In catfish, weight gain and protein deposition level increase when fish oil was as high as 15% in the dry feed, but with levels of 20% decreases weight gain (Piper et al., 1982). In rainbow trout, ranging lipid levels between 5% and 25% with intervals of 5% in diets with different protein levels (16–48%), an optimum protein:lipid ratio of 35:18% in diets was found (Takeuchi et al., 1978). Most recently, Liu et al. (2014c) found that turbot (*Scophthalmus maximus*) fed diets with 55% protein and 12% lipid with protein:energy of 110.9 mg protein/kcal reached the maximum growth, FER, PER, and energy retention of fish weighing 5–300 g after a 9-week feeding period.

The fish and shrimp have the ability to better utilize carbohydrate-type polysaccharides than di- and monosaccharides (Shiau and Peng, 1992, 1993). In shrimp, *P. monodon*, Shiau and Peng (1992) found that when the protein level decreased from 40% to 30% and the starch level increased from 20% to 30% this did not reduce weight gain, survival, or feed efficiency. Shiau and Peng (1993) also carried out a study to evaluate the possible protein-sparing effects of carbohydrates in juvenile tilapia (*O. niloticus* × *Oreochromis aureus*). These authors reported that a decrease in the dietary protein level from 28% to 24% by increasing the starch or dextrin from 37% to 41% in diets did not reduce weight gain and feed efficiency. These two studies suggest that starch is efficiently used as an energy source instead of the protein.

The optimum level of P/E balance in fish and crustaceans is dependent on the species, age, and life cycle. Studies of P/E ratio have shown that:

1. When the total energy rate is greater than the protein, feed consumption and hence dietary protein can be restricted and thus growth retardation. Also, an excess in dietary energy may result in an increased deposition of body fat.
2. Supplementation of a low-energy diet would result in AA catabolism to produce sufficient energy for the normal metabolism of the animal, resulting in a low efficiency of protein and a lower growth.

Some investigations to determine an appropriate protein/energy balance have been made in several species of fish and shrimps with various results reported in Table 4.20.

Determining the energy requirement of fish has been a difficult task because energy needs calculated in some studies are valid only for those reared under specific conditions (feed composition, feeding rate, temperature, genetic, life stages, rearing periods). In addition, another major problem associated with the determination of dietary energy requirements is that fish have different nutrient depositions directly linked to growth rates and hence different nutrient and energy requirements. There are several models on bioenergetics developed to predict growth, feed ration, FCR, and waste output of several fish species (Rosas et al., 1998; Bureau et al., 2002, 2003; Glencross et al., 2011). However, these factorial models do not work well under the wide husbandry practices, sanitary conditions, and environmental variables encountered in fish culture (Dumas et al., 2008).

4.6 Vitamins

Vitamins are defined as organic compounds required in small amounts for normal growth, reproduction, health, and general maintenance of animals. Vitamins are responsible for 5–8% of the total cost of the diet, but they are responsible for 100% of the cellular metabolism of other nutrients. Vitamins can be divided into two major groups: fat- (A, D, E, K) and water-soluble (B complex, choline, and vitamin C) compounds. Fat-soluble vitamins are not rapidly metabolized when storage depots become full showing more problems with toxicity compared to the water-soluble vitamins. Most of the 15 shown are essential for fish, but not for all species (Tables 4.21–4.25).

4.6.1 Water-soluble vitamins

Water-soluble enzymes are all parts of coenzymes and hence these combine with proteins to form active enzymes and catalyze many reactions involving the catabolism and anabolism of carbohydrates, proteins, and lipids. Vitamins are part of the enzyme

Table 4.20 Optimum ratio of dietary protein to energy in various species of fish and shrimps

Species	Protein/energy ratio	Body weight	References
S. salar	97 mg/kcal	Fingerlings	Storebakken (2002)
	84 mg/kcal	Smoltification	Storebakken (2002)
	80 mg/kcal	1.0–2.5 kg	Einen and Roem (1997)
	67–71 mg/kcal	2.5–5.0 kg	Einen and Roem (1997)
S. gairdneri	92 mg/kcal digestible energy (DE)	90 g	Cho and Kaushik (1985)
	105 mg/kcal DE	94 g	Cho and Woodward (1989)
C. carpio	97–116 mg/kcal DE	4.3	Takeuchi et al. (1979b)
O. niloticus	110 mg/kcal gross energy (GE)	0.012	El-Sayed and Teshima (1992)
O. niloticus × O. aureus[a]	68–104 mg/kcal metabolizable energy (ME)	1.6	Shiau and Huang (1990)
O. aureus	123 mg/kcal DE	2.5	Winfree and Stickney (1981)
O. niloticus[b]	75 mg/kcal ME	2.9	Fineman-Kalio and Camacho (1987)
O. mosambicus	100 mg/kcal DE	5.2	El-Dahhar and Lovell (1995)
O. aureus	106 mg/kcal DE	7.5	Winfree and Stickney (1981)
Tilapia zilli	103 mg/kcal DE	50	El-Sayed (1987)
I. punctatus	97 mg/kcal DE	10	Mangalik (1986)
	94 mg/kcal DE	34	Garling Jr. and Wilson (1976)
	86 mg/kcal DE	266	Mangalik (1986)
	95 mg/kcal DE	526	Page and Andrews (1973)
	81 mg/kcal DE	600	Li and Lovell (1992)
M. saxatilis × M. chrysops	112 mg/kcal DE	35	Nematipour et al. (1992)
S. ocellatus	98 mg/kcal DE	43	Daniels and Robinson (1986)
Fenneropenaeus indicus	90–211 mg/kcal	0.95 g	Colvin (1976)
Litopenaeus merguiensis	112 mg/kcal	0.3 g	Sedgwick (1979)
P. monodon	120 mg/kcal	0.6 g	Bautista (1986)
	112.2 mg/kcal	0.5 g	Hajra et al. (1988)
	109.1 mg/kcal	0.8 g	Shiau and Chou (1991)
Litopenaeus schimitti	119–147 mg/kcal	0.25 g	Fraga et al. (1992)
L. vannamei	80–120	1 g	Cousin et al. (1993)

[a]Reared at 32–34 ppt.
[b]Reared at 18–50 ppt.

Table 4.21 Responses of various fish species to dietary vitamin C and sources used in feeding experiments

Species	Vitamin C source	Requirement (mg/kg diet)	References
African catfish (*Clarius gariepinus*)	AA	46	Eya (1996)
Angelfish (*Pterophylum scalare*)	C2MP-Mg	360	Blom et al. (2000)
Asian sea bass (*L. calcarifer*)	C2MP-Mg	30	Phromkunthong et al. (1997)
Atlantic salmon (*S. salar*)	AA	50	Lall et al. (1991)
	C2MP-Ca	10	Sandnes et al. (1992)
		20	Sandnes et al. (1992)
Ayu (*Plecoglossus altivelis*)	C2PP	116	Xie and Niu (2006)
		47	Xie and Niu (2006)
Blue tilapia (*O. aureus*)	AA	50	Stickney et al. (1984)
Carp (*C. mrigala*)	C2PP	102–120	Zehra and Khan (2012)
Channel catfish (*I. punctatus*)	AA	50	Andrews and Murai (1975)
	AA	60	Lim and Lovell (1978)
	AA	45	Robinson (1990)
	AA	25	Murai et al. (1978)
		50	Murai et al. (1978)
	C2S	2,000	Murai et al. (1978)
	AA	30	Durve and Lovell (1982)
	C2MP	11	El Naggar and Lovell (1991)
	C2MP-Na	15	Mustin and Lovell (1992)
	C2MP	50	Li et al. (1998)
		150	Li et al. (1998)
Clarias hybrid catfish (*Ctarias gariepinus* × *Clarias inacrocephaluss*)	C2D	42	Khajarern and Khajarem (1997)
	C2MP-Ca	12.6	Boonyaratpalin and Phromkunthong (2001)
Cobia (*Raehycentwn canadum*)	C2PP	44.7–53.9	Xiao et al. (2009)
Coho salmon (*Oncorhyncluts kisutch*)	AA	50–100	Halver et al. (1969)
Common carp (*C. carpio*)	C2PP	45	Gouillou-Coustans et al. (1998)
		354	Gouillou-Coustans et al. (1998)
European sea bass (*Dicemrarchus labrax*)	AA	200	Saroglia and Scarano (1992)
	C2PP	20	Merchie et al. (1996)
	C2PP	5	Foumier et al. (2000)
		5–31	Foumier et al. (2000)
		121	Foumier et al. (2000)
Gilthead sea bream (*Spams auratus*)	AA	63	Alexis et al. (1997)

(Continued)

Table 4.21 **(Continued)**

Species	Vitamin C source	Requirement (mg/kg diet)	References
Grouper (*Epinephelus malabaricus*)	AA	45.3	Lin and Shiau (2005a)
	C2MP-Mg	17.9	Lin and Shiau (2004)
	C2MP-Na	8.3	Lin and Shiau (2004)
	C2PP	17.8	Lin and Shiau (2005b)
	C2S	46.2	Lin and Shiau (2005b)
Hybrid striped bass (*M. chrysops* female × *M. saxatilis*)	C2PP	22	Sealey and Gatlin (1999)
Hybrid tilapia	AA	79	Shiau and Jan (1992a)
(*O. niloticus* ×	C2MP	17–20	Shiau and Hsu (1995)
O. aureus)	C2S	19–23	Shiau and Hsu (1995)
	C2MP-Mg	18.82	Shiau and Hsu (1999a)
	C2MP-Na	15.98	Shiau and Hsu (1999a)
Indian catfish (*Heteropneustes*	AA	69	Mishra and Mukhopadhyay (1996)
fossilis)	AA	82.2	Ibiyo et al. (2007)
Japanese sea bass	C2PP	53.5	Ai et al. (2004)
(*Lateolabrax*		93.4	Ai et al. (2004)
japonicits)		207.2	Ai et al. (2004)
Korean rockfish	AA	144	Lee et al. (1998)
(*Sebastes schhgeli*)	AA	100–102	Bai (2001)
	C2MP-Ca	112	Wang et al. (2003b)
	C2MP-Na/Ca	101	Wang et al. (2003b)
	C2D	50	Wang et al. (2003c)
Mexican cichlid (*Cichlasoma urophthalmus*)	AA	40	Chavez de Martinez (1990)
Mrigal (*C. mrigala*)	C2PP	36–42	Zehra and Khan (2012)
Nile tilapia	AA	420	Soliman et al. (1994)
(*O. niloticus*)	C2PP	50	Abdelghany (1996)
	C2S	50	Abdelghany (1996)
Olive flounder	C2MP-Mg	28–47	Teshima et al. (1991)
(*Paralichrhys olivaceus*)	C2PP	91–93	Wang et al. (2002)
Pacu (*Piaractus mesopotamicus*)	Ascorbyl-6-palmitate	139	Martins (1995)
Parrot fish (*O. fasciatus*)	AA	250	Ishibashi et al. (1992)
		500	Ishibashi et al. (1992)
	C2MP	118	Wang et al. (2003a)
Plaice (*Pleumnectes platessa*)	AA	200	Rosenlund et al. (1990)

(*Continued*)

Table 4.21 **(Continued)**

Species	Vitamin C source	Requirement (mg/kg diet)	References
Rainbow trout	AA	250–500	McLaren et al. (1947)
(*O. mykiss*)	AA	100	Halver et al. (1969)
	AA	40	Hilton et al. (1978)
	AA	20	Sato et al. (1982)
		5–100	Sato et al. (1982)
		500	Sato et al. (1982)
	C2PP	20	Grant et al. (1989)
Red drum (*S. ocellatus*)	C2PP	15	Aguirre and Gatlin (1999)
Rohu carp (*L. rohita*)	AA	670–750	Mahajan and Agrawal (1980)
	AA	200	Misra et al. (2007)
Sabaki tilapia	C2S	75	Al-Amoudi et al. (1992)
(*Oreochromis spilurus*)		400	Al-Amoudi et al. (1992)
Spotted rose snapper	C2PP	29	Chávez-Sánchez et al. (2014)
(*Lutjanus guttatus*)		>250	
Tiger puffer (*Takijugu rubripes*)	C2MP	29	Eo and Lee (2008)
Turbot (*Scophthahmis maximus*)	C2PP	20	Merchie et al. (1996)
Yellow croaker	C2PP	28.2	Ai et al. (2006)
(*Pseudosciaena crocea*)		87	Ai et al. (2006)
Yellowtail (*Seriola lalandi*)	AA	122	Shimeno (1991)
	C2MP-Mg	14–28	Kanazawa et al. (1992)
	C2MP-Mg	52	Ren et al. (2008)
	C2MP-Na/Ca	43	Ren et al. (2008)

Note: AA, L-ascorbic acid; C2D, L-ascorbyl-2-glucose; C2MP, ascorbyl-2-monophosphate; C2PP, L-ascorbyl-2-polyphosphate; C2S, L-ascorbyl-2-sulfate.

complex which cannot be synthesized by the animal. Microorganisms can synthesize B vitamins. Thus they are not required by ruminant species but, for fish where microbial synthesis is not a prominent feature in the GI tract. They still must be included in the diet and hardly induce toxicity.

4.6.1.1 Vitamin C

Ascorbic acid also known as vitamin C is an essential dietary ingredient for teleosts (Moreau and Dabrowski, 2001) due to the absence of L-gulonolactone oxidase, an enzyme required for biosynthesis of ascorbic acid from glucose. Most animals are known to be extremely sensitive to ascorbic acid deficiencies and several studies have shown the essential role of ascorbic acid in penaeid shrimp nutrition despite having limited ability to synthesize ascorbic acid (Deshimaru and Kuroki, 1976;

Table 4.22 **Responses of various shrimps species to dietary vitamin C and sources used in feeding experiments**

Species	Vitamin C source	Requirement (mg/kg diet)	References
Kuruma shrimp	AA	3,000	Deshimaru and Kuroki (1976)
(*M. japonicus*)	AA	5.000–10.000	Guary et al. (1976)
	C2MP-Mg	215–430	Shigueno (1988)
	C2MP-Mg	71	Moe et al. (2004)
	C2MP-Na/Ca	43	Moe et al. (2004)
	C2MP-Na/Ca	91.8	Moe et al. (2005)
Pacific white shrimp	C2PP	90–120	He and Lawrence (1993a)
(*L. vannamei*)	C2MP-Mg	100	Montoya and Molina (1995)
	C2MP	130	Lavens et al. (1999)
	C2PP	150	Zhou et al. (2004a)
	C2PP	191	Niu et al. (2009)
Fleshy prawn	C2PP	600	Qin et al. (2007)
(*F. chinensis*)			
Brown shrimp	AA	2,000	Lightner et al. (1979)
(*Farfantepenaeus californiensis*)			
Indian white prawn	AA	4,000–8,000	Boonyaratpalin (1998)
(*F. indicus*)			
Giant river prawn	C2MP-Ca	104	D'Abramo et al. (1994)
(*M. rosenbergii*)	Ascorbyl-6-palmitate	104	D'Abramo et al. (1994)
	C2PP	135	Hari and Kurup (2002)
Tiger shrimp	AA	2	Shiau and Jan (1992b)
(*P. monodon*)	C2PP	210	Chen and Chang (1994)
	C2MP-Mg	100–200	Catacutan and Lavilla-Pitogo (1994)
	C2MP-Mg	40	Shiau and Hsu (1994)
	C2S	157	Hsu and Shiau (1997)
	C2MP-Na	106	Hsu and Shiau (1998)

Note: AA, ʟ-ascorbic acid; C2MP, ascorbyl-2-monophosphate; C2PP, ʟ-ascorbyl-2-polyphosphate; C2S, ʟ-ascorbyl-2-sulfate.

Lightner et al., 1977; He and Lawrence, 1993b; Chen and Chang, 1994; Montoya and Molina, 1995). Ascorbic acid is related to the formation of collagen, an essential component of capillaries and connective tissue (Hunter et al., 1979). It is also involved in other physiological processes, such as protection of cells from oxidative damage and the regeneration of vitamin E (Frischknecht et al., 1994). Ascorbic acid affects immune functions in different ways: stimulation of proliferative response, chemotaxis, interferon and antibody production, and protection against free-radical-mediated protein inactivation associated with the oxidative burst of macrophages in fish (Lim et al., 2001) and influences the dietary requirement of vitamins such as folate (Duncan

Table 4.23 **Water-soluble vitamins requirement estimates for a range of fish species in a controlled environment**

Vitamin and fish	Requirement (units/kg diet)	References
Thiamin (vitamin B₁)		
Channel catfish (*I. punctams*)	1 mg	Murai and Andrews (1978a)
Common carp (*C. carpio*)	0.5 mg	Aoe et al. (1969)
Grass carp (*C. idella*)	1.3–5.0 mg	Jiang et al. (2014a)
Nile tilapia (*O. niloticus*)	3.5 mg	Lim et al. (2011)
Pacific salmon (*Oncorhynchus* spp.)	10–15 mg	Halver (1972)
Rainbow trout (*O. mykiss*)	1–10 mg	McLaren et al. (1947)
	1 mg	Morito et al. (1986)
Yellowtail (*S. lalandi*)	11.2 mg	Shimeno (1991)
Riboflavin (vitamin B₂)		
Blue tilapia (*O. aureus*)	6 mg	Soliman and Wilson (1992a)
Channel catfish (*I. punctams*)	9 mg	Murai and Andrews (1978b)
	6 mg	Serrini et al. (1996)
Common carp (*C. carpio*)	4 mg	Aoe et al. (1967a)
	6.2 mg	Aoe et al. (1967a)
	7 mg	Takeuchi et al. (1980b)
Hybrid striped bass (*M. chrysops* female × *M. saxatilis*)	4.1–5.0 mg	Deng and Wilson (2003)
Jian carp (*C. carpio* var. Jian)	5.0 mg	Li et al. (2010a)
Pacific salmon (*Oncorhynchus* spp.)	20–25 mg	Halver (1972)
Rainbow trout (*O. mykiss*)	5–15 mg	McLaren et al. (1947)
	6 mg	Takeuchi et al. (1980b)
	3 mg	Hughes et al. (1981)
	2.7 mg	Amezaga and Knox (1990)
Red hybrid tilapia (*O. mossambicus* × *O. niloticus*)	5 mg	Lim et al. (1993)
Yellowtail (*S. lalandi*)	11 mg	Shimeno (1991)
Pantothenic acid		
Blue tilapia (*O. aureus*)	10 mg	Soliman and Wilson (1992b)
Channel catfish (*I. punctatus*)	10 mg	Murai and Andrews (1979)
	15 mg	Wilson et al. (1983)
Common carp (*C. carpio*)	30–50 mg	Ogino (1967)
Grass carp (*C. idella*)	25 mg	Liu et al. (2007)
Grouper (*E. malabaricus*)	11 mg	Lin et al. (2012)
Jian carp (*C. carpio* var. Jian)	23 mg	Wen et al. (2009)
Pacific salmon (*Oncorhynchus* spp.)	40–50 mg	Halver (1972)
Rainbow trout (*O. mykiss*)	10–20 mg	McLaren et al. (1947)
	20 mg	Cho and Woodward (1990)
Yellowtail (*S. lalandi*)	35.9 mg	Shimeno (1991)

(Continued)

Table 4.23 (Continued)

Vitamin and fish	Requirement (units/kg diet)	References
Pyridoxine (vitamin B$_6$)		
Atlantic salmon (*S. salar*)	5 mg	Lall and Weerakoon (1990)
Channel catfish (*I. punctatus*)	3 mg	Andrews and Murai (1979)
Common carp (*C. carpio*)	5–6 mg	Ogino (1965)
Gibel carp (*C. auratus gibelio*)	7.62–11.36 mg	Wang et al. (2011a)
Hybrid tilapia (*O. niloticus* × *O. aureus*)	15–16.5 mg	Shiau and Hsieh (1997)
Indian catfish (*H. fossilis*)	3.21 mg	Shaik Mohamed (2001a)
Jian carp (*C. carpio* var. Jian)	6.07 mg	He et al. (2009)
Pacific salmon (*Oncorhynchus* spp.)	10–20 mg	Halver (1972)
Rainbow trout (*O. mykiss*)	1–10 mg	McLaren et al. (1947)
	2 mg	Woodward (1990)
	3–6 mg	Woodward (1990)
Red hybrid tilapia (*O. mossambicus* × *O. niloticus*)	3 mg	Lim et al. (1995)
Yellowtail (*S. lalandi*)	11.7 mg	Shimeno (1991)
Biotin		
Asian catfish (*Ciarías batrachus*)	2.49 mg	Shaik Mohamed et al. (2000)
Channel catfish (*I. punctatus*)	R	Robinson and Lovell (1978)
Common carp (*C. carpio*)	1 mg	Ogino et al. (1970b)
Goldspot mullet (*Liza parsia*)	1.6–3.2 mg	Chavan et al. (2003)
Hybrid tilapia (*O. nilolicus* × *O. aureus*)	0.06 mg	Shiau and Chin (1999)
Indian catfish (*H. fossilis*)	0.25 mg	Shaik Mohamed (2001b)
Japanese sea bass (*Lateolabrax japonicus*)	0.046 mg	Li et al. (2010b)
Lake trout (*Salvelinus namaycush*)	0.1 mg	Poston (1976b)
	0.5–1 mg	Poston (1976b)
Pacific salmon (*Oncorhynchus* spp.)	1–1.5 mg	Halver (1972)
Rainbow trout (*O. mykiss*)	0.05–0.25 mg	McLaren et al. (1947)
	0.08 mg	Woodward and Frigg (1989)
	0.14 mg	Woodward and Frigg (1989)
Yellowtail (*S. lalandi*)	0.67 mg	Shimeno (1991)
Zebrafish (*Danio rerio*)	0.51 mg	Yossa et al. (2014)
Folic acid		
Common carp (*C. carpio*)	NR	Aoe et al. (1967b)
Channel catfish (*I. punctatus*)	1.5 mg	Duncan and Lovell (1991)
	1.0 mg	Duncan et al. (1993)
Hybrid tilapia (*O. nilolicus* × *O. aureus*)	0.82	Shiau and Huang (2001a)
Grass carp (*C. idella*)	3.6–4.3 mg	Zhao et al. (2008)
Grouper (*E. malabaricus*)	0.8 mg	Lin et al. (2011)
Pacific salmon (*Oncorhynchus* spp.)	6–10 mg	Halver (1972)
Rainbow trout (*O. mykiss*)	1.0 mg	Cowey and Woodward (1993)
Yellowtail (*S. lalandi*)	1.2 mg	Shimeno (1991)

(Continued)

Table 4.23 **(Continued)**

Vitamin and fish	Requirement (units/kg diet)	References
Cyanocobalamine (vitamin B$_{12}$)		
Common carp (*C. carpio*)	NR	Kashiwada et al. (1970)
Channel catfish (*I. punctatus*)	R	Limsuwan and Lovell (1981)
Grass carp (*C. idella*)	0.094 mg	Wu et al. (2007b)
Hybrid tilapia (*O. nilolicus* × *O. aureus*)	NR	Shiau and Lung (1993a)
Nile tilapia (*O. nilolicus*)	NR	Lovell and Limsuwan (1982)
Pacific salmon (*Oncorhynchus* spp.)	0.015–0.02 mg	Halver (1972)
Yellowtail (*S. lalandi*)	0.053 mg	Shimeno (1991)
Niacin		
African catfish (*Heterobranchus longifilis*)	33.1 mg	Morris et al. (1998)
Channel catfish (*I. puncratus*)	14 mg 7.4 mg	Andrews and Murai (1978) Ng et al. (1997)
Common carp (*C. carpio*)	28 mg	Aoe et al. (1967a)
GIFT Tilapia (*O. niloticus*)	20–85 mg	Jiang et al. (2014b)
Grass carp (*C. idella*)	25.5 mg	Wu et al. (2007a)
Hybrid tilapia (*Oreochromis nilolicus* × *O. aureus*)	26 mg	Shiau and Suen (1992)
Indian catfish (*Heteropneustes fossilis*)	20 mg	Shaik Mohamed and Ibrahim (2001)
Pacific salmon (*Oncorhynchus* spp.)	150–200 mg	Halver(1972)
Rainbow trout (*O. mykiss*)	1–5 mg 10 mg	McLaren et al. (1947) Poston and Wolfe (1985)
Yellowtail (*S. lalandi*)	12 mg	Shimeno (1991)
Choline		
Channel catfish (*I. punctatus*)	400 mg	Wilson and Poe (1988)
Cobia (*R. canadum*)	696 mg	Mai et al. (2009)
Common carp (*C. carpio*)	1,500 mg	Ogino et al. (1970b)
Gibel carp (*C. auratus* gibelio)	2,500 mg	Duan et al. (2012)
Grass carp (*C. idella*)	3.000 mg	Wang et al. (1995)
Hybrid tilapia (*O. nilolicus* × *O. aureus*)	1,000 mg	Shiau and Lo (2000)
Hybrid striped bass (*M. chrysops* female × *M. saxatilis*)	500 mg	Griffin et al. (1994)
Lake trout (*S. namaycush*)	1.000 mg	Ketola (1976)
Pacific salmon (*Oncorhynchus* spp.)	600–800 mg	Halver (1972)
Rainbow trout (*O. mykiss*)	50–100 mg 714–813 mg	McLaren et al. (1947) Rumsey (1991)
Red drum (*S. ocellatus*)	588 mg	Craig and Gatlin (1996)
Yellowtail (*S. lalandi*)	2,920 mg	Shimeno (1991)
Yellow perch (*Perca fiavescens*)	598–634 mg	Twibell and Brown (2000)

(Continued)

Table 4.23 **(Continued)**

Vitamin and fish	Requirement (units/kg diet)	References
Myoinositol		
Channel catfish (*I. punctatus*)	NR	Burtle and Lovell (1989)
Common carp (*C. carpio*)	440 mg	Aoe and Masuda (1967)
Gibel carp (*C. auratus* gibelio)	165.3 mg	Gong et al. (2014)
Grass carp (*C. idella*)	166–214 mg	Wen et al. (2007)
Grouper (*E. malabaricus*)	335–365 mg	Su and Shiau (2004)
Hybrid striped bass (*M. chrysops* female × *M. saxatilis*)	NR	Deng et al. (2002)
Hybrid tilapia (*O. nilolicus* × *O. aureus*)	400 mg	Shiau and Su (2005)
Nile tilapia (*O. niloticus*)	NR	Peres et al. (2004)
Olive flounder (*P. olivaceus*)	617 mg	Lee et al. (2009)
Pacific salmon (*Oncorhynchus* spp.)	300–400 mg	Halver (1972)
Parrot fish (*O. fasciatus*)	94–121	Khosravi et al. (2015)
Rainbow trout (*O. mykiss*)	250–500 mg	McLaren et al. (1947)
Yellowtail (*S. lalandi*)	423 mg	Shimeno (1991)

Note: NR, no requirement determined; R, required but no value determined.

Table 4.24 **Fat-soluble vitamins requirement estimates for a range of fish species in a controlled environment**

Vitamin and fish	Requirement (units/kg diet)	References
Vitamin A[a]		
Atlantic halibut (*Hippoglossus hippoglossus*)	2.5 mg	Moren et al. (2004)
Channel catfish (*Icialunis punctatus*)	0.3–0.6 mg	Dupree(1970)
Common carp (*C. carpio*)	1.2–6 mg	Aoe et al. (1968)
European sea bass (*Dicenlrarchus labrax*)	31 mg	Villeneuve et al. (2005a,b)
Grouper (*Epinephelus* spp.)	0.93 mg	Shaik Mohamed et al. (2003)
Hybrid striped bass (*M. chrysops* female × *M. saxatilis*)	0.51–40.52 mg	Hemre et al. (2004)
Hybrid tilapia (*O. niloticus* × *O. aureus*)	1.76–2.09 mg	Hu et al. (2006)
Japanese flounder (*P. olivaceus*)	2.7 mg	Hernandez et al. (2005)
Pacific salmon (*Oncorhynchus* spp.)	R	Halver (1972)
Rainbow trout (*O. mykiss*)	0.75 mg	Kitamura et al. (1967a)
Yellowtail (*Seiiola lalandi*)	5.68 mg	Shimeno (1991)

(*Continued*)

Table 4.24 (**Continued**)

Vitamin and fish	Requirement (units/kg diet)	References
Vitamin D[b]		
Channel catfish (*I. punctatus*)	12.5 ug	Lovell and Lim (1978)
	25 ug	Andrews et al. (1980)
Hybrid tilapia (*O. niloticus* × *O. aureus*)	9.35 ug	Shiau and Hwang (1993)
Pacific salmon (*Oncorhynchus* spp.)	NR	Halver (1972)
Rainbow trout (*O. mykiss*)	40–60 ug	Barnett et al. (1982)
Yellowtail (*S. lalandi*)	NR	Shimeno (1991)
Vitamin E		
Atlantic salmon (*S. salar*)	35 mg	Lall et al. (1988)
	60 mg	Hamre and Lie (1995)
Blue tilapia (*O. aureus*)	25 mg	Roem et al. (1990)
Channel catfish (*I. punctatus*)	25 mg	Murai and Andrews (1974)
	50 mg	Wilson et al. (1984)
Common carp (*C. carpio*)	100 mg	Watanabe et al. (1970)
Eel (*Anguilla japonica*)	21.2–21.6 mg	Bae et al. (2013)
Grass carp (*C. idella*)	200 mg	Takeuchi et al. (1992b)
	62.9 mg	Wu et al. (1990)
Grouper (*Epinephelus* spp.)	104–115 mg	Lin and Shiau (2005c)
Hybrid striped bass (*M. chrysops* female × *M. saxatilis*)	28 mg	Kocabas and Gatlin (1999)
Hybrid tilapia (*O. niloticus* × *O. aureus*)	60–67 mg	Shiau and Shiau (2001)
Korean rockfish (*S. schlegeli*)	45 mg	Bai and Lee (1998)
Mrigal (*C. mrigala*)	99 mg	Paul et al. (2004a)
Nile tilapia (*O. niloticus*)	50–100 mg	Satoh et al. (1987)
Pacific salmon (*Oncorhynchus* spp.)	30 mg	Woodall et al. (1964)
	40–50 mg	Halver (1972)
Rainbow trout (*O. mykiss*)	30 mg	Woodall et al. (1964)
	25 mg	Hung et al. (1980)
	100 mg	Watanabe et al. (1981)
	50 mg	Cowey et al. (1983)
Red drum (*S. ocellatus*)	31 mg	Peng and Gadin (2009)
Rohu (*L. rohita*)	131.91 mg	Sau et al. (2004)
Spotted murrel (*Channa punctatus*)	140–169 mg	Abdel-Hameid et al. (2012)
Yellowtail (*S. lalandi*)	119 mg	Shimeno (1991)
Vitamin K		
Atlantic cod (*G. morhua*)	0.2 mg	Grahl-Madsen and Lie (1997)
Atlantic salmon (*S. salar*)	<10 mg	Krossøy et al. (2009)
Channel catfish (*I. punctatus*)	R	Dupree (1966)
	NR	Murai and Andrews (1977)
Grass carp (*C. idella*)	1.9 mg	Jiang et al. (2007)
Lake trout (*S. namaycush*)	0.5–1 mg	Poston (1976a)
Pacific salmon (*Oncorhynchus* spp.)	R	Halver (1972)
Yellowtail (*S. lalandi*)	NR	Shimeno (1991)

Note: NR, no requirement determined; and R, required but no value determined.
[a] 10.000 IU = 3.000 μg vitamin A (retinol).
[b] 1 IU = 0.025 μg cholecalciferol.

Table 4.25 Vitamins requirement estimates for a range of shrimps species in a controlled environment

Vitamin and shrimp	Requirement (units/kg diet)	References
Vitamin A[a]		
Tiger shrimp (*P. monodon*)	2.51 mg	Shiau and Chen (2000)
White shrimp (*L. vannamei*)	1.44 mg	He et al. (1992)
Fleshy prawn (*F. chinensis*)	36–54 mg	Chen and Li (1994)
Vitamin D[b]		
Tiger shrimp (*P. monodon*)	100 µg	Shiau and Hwang (1994)
Vitamin E		
Tiger shrimp (*P. monodon*)	85–89 mg	Lee and Shiau (2004)
White shrimp (*L. vannamei*)	99 mg	He and Lawrence (1993b)
Vitamin K		
Tiger shrimp (*P. monodon*)	30–40 mg	Shiau and Liu (1994a)
Fleshy prawn (*F. chinensis*)	185 mg	Shiau and Liu (1994b)
Thiamin (vitamin B$_1$)		
Tiger shrimp (*P. monodon*)	14 mg	Chen et al. (1991)
Kuruma shrimp (*M. japonicus*)	60–120 mg	Deshimaru and Kuroki (1979)
Indian white prawn (*F. indicus*)	100 mg	Boonyaratpalin (1998)
Riboflavin (vitamin B$_2$)		
Tiger shrimp (*P. monodon*)	22.5 mg	Chen and Hwang (1992)
Kuruma shrimp (*M. japonicus*)	80 mg	NRC (1983)
Pyridoxine (vitamin B$_6$)		
Tiger shrimp (*P. monodon*)	72–89 mg	Shiau and Wu (2003)
Kuruma shrimp (*M. japonicus*)	120 mg	Deshimaru and Kuroki (1979)
White shrimp (*L. vannamei*)	80–100 mg	He and Lawrence (1991)
White shrimp (*L. vannamei*)	107–152 mg[c]	Li et al. (2010c)
Indian white prawn (*F. indicus*)	100–200 mg	Boonyaratpalin (1998)
Pantothenic acid		
Tiger shrimp (*P. monodon*)	101–139 mg	Shiau and Hsu (1999b)
Fleshy prawn (*F. chinensis*)	100 mg	Liu et al. (1995)
Indian white prawn (*F. indicus*)	750 mg	Boonyaratpalin (1998)
Niacin		
Tiger shrimp (*P. monodon*)	7.2 mg	Shiau and Suen (1994)
Kuruma shrimp (*M. japonicus*)	400 mg	NRC (1983)
Indian white prawn (*F. indicus*)	250 mg	Boonyaratpalin (1998)
White shrimp (*L. vannamei*)	109.6 mg	Xia et al. (2014)

(*Continued*)

Table 4.25 **(Continued)**

Vitamin and shrimp	Requirement (units/kg diet)	References
Biotin		
Tiger shrimp (*P. monodon*) Fleshy prawn (*F. chinensis*)	2.0–2.4 mg 0.4 mg	Shiau and Chin (1998) Liu et al. (1995)
Vitamin B$_{12}$		
Tiger shrimp (*P. monodon*) Fleshy prawn (*F. chinensis*)	0.2 mg 0.01 mg	Shiau and Lung (1993b) Liu et al. (1995)
Folate		
Tiger shrimp (*P. monodon*) Fleshy prawn (*F. chinensis*)	1.9–2.1 mg 5 mg	Shiau and Huang (2001b) Liu et al. (1995)
Choline		
Tiger shrimp (*P. monodon*) Kuruma shrimp (*M. japonicus*)	6,200 mg 600 mg NR	Shiau and Lo (2001) Kanazawa et al. (1976) Deshimaru and Kuroki (1979)
Myoinositol		
Tiger shrimp (*P. monodon*) Kuruma shrimp (*M. japonicus*) Fleshy prawn (*F. chinensis*)	3,400 mg 2,000 mg 4,000 mg	Shiau and Su (2004) Kanazawa et al. (1976) Liu et al. (1993)

Note: NR, no requirement determined; R, required but no value determined.
[a]10.000 IU = 3.000 µg vitamin A (retinol).
[b]1 IU = 0.025 µg cholecalciferol.
[c]at 3‰ salinity.

and Lovell, 1994) and iron (Lim et al., 2000). Several authors have reported the impact of ascorbic acid on disease resistance in fish and shrimps. Mortality rates from 1% to 5% per day due to ascorbic acid deficiency were observed in tank-reared juveniles of *Penaeus californiensis*, *Penaeus stylirostris*, *Penaeus japonicus*, and *L. vannamei*, with cumulative mortalities reaching more than 70% (Lightner et al., 1979; Magarelli et al., 1979; Shigueno and Itoh, 1988; He and Lawrence, 1993a). Nutrional deficiencies has been described in *P. californiensis* and *P. aztecus*, and nominated as "black death" syndrome (Lightner et al., 1977). It is characterized by melanized hemocytic accumulations in collagenous tissue. Such a syndrome could indicate perturbation of the immune response since melanin formation is the final step in the prophenoloxidase activation cascade (proPo), which is a potent defense mechanism associated with phagocytosis, encapsulation, hemocytic nodule formation, and wound repair (Unestam and Beskow, 1976; Lightner and Redman, 1977; Lightner et al., 1979; Söderhäll, 1982; Johansson and Söderhäll, 1989). Channel catfish were fed different concentrations of ascorbic acid during challenge. Beyond the signs of deficiency, a decrease in susceptibility caused by stressors such as temperature, salinity, handling, ammonia, pesticides, and heavy metals has also been reported in several fish species (Norrgren et al., 2001).

4.6.1.2 Thiamin (vitamin B₁)

The active form of vitamin B_1 is the phosphorylated form (thiamine pyrophosphate) and food sources are dried peas, unmilled cereal grains, and brewer's yeast. Most practical diets are not deficient. Thiamin is part of a coenzyme needed for proper utilization of carbohydrates, particularly in oxidative decarboxylation reactions to convert pyruvic acid (3C) to acetyl-CoA (2C) and α-ketoglutarate (5C) to succinyl-CoA (4C) in the TCA cycle. Thiamine pyrophosphate plays a role in the direct oxidation of glucose in the cytoplasm. It is also involved in appetite, fertility, and nervous transmission.

4.6.1.3 Riboflavin (vitamin B₂)

Riboflavin was the second water-soluble discovered of the B vitamins. Vitamin B_2 is commonly found in germinated grains and soybeans and can be synthesized by microorganisms (like all B vitamins). It is heat-stable but is decomposed by light, hence feeds must be protected from sunlight. It is poorly absorbed from the intestines but once in the plasma it is associated with plasma proteins for transportation. Riboflavin functions as a coenzyme in the intermediary transfer of electrons such as AA oxidases, aldehyde oxidase, cytochrome c reductase, glucose oxidase, and dehydrogenases. It also plays a role within cells in the form of flavin mononucleotide, which should really be called riboflavin-5′-phosphate, and flavin adenine dinucleotide which is used to generate ATP in the electron transport chain. Riboflavin is involved in the conversion of tryptophan to nicotinic acid.

4.6.1.4 Niacin (nicotinic acid – nicotinamide)

Niacin (β-pyridine carboxylic acid) comes in several forms: (i) nicotinic acid (pyridine-3-carboxylic acid), (ii) nicotinamide (nicotinic acid amide), and (iii) other derivatives (e.g., inositol hexanicotinate). Nicotinamide, or niacinamide, performs all of the essential biochemical functions of niacin and prevents its deficiency. Niacin can be synthesized from tryptophan. Yeast is a good source, while cereal grains are low in niacin. Corn is low in both tryptophan and niacin. Niacin can be synthesized by microorganisms. It is part of the structure of NAD and nicotinamide adenine dinucleotide phosphate that are involved in the synthesis of high-energy phosphate bonds which are essential for lipids, carbohydrates, and protein metabolism, and in photosynthesis. NADH is also used to generate ATP in the electron transport chain.

4.6.1.5 Vitamin B₆ (pyridoxine, pyridoxamine, pyridoxal)

Vitamin B_6 represents a group of vitamer derivatives of pyridine which has different functions: pyridoxine or pyridoxol (alcohol form), pyridoxal (aldehyde form), and pyridoxamine (amine form). Pyridoxine hydrochloride (82.3% active) commonly used in aquatic feeds is relatively unstable in the presence of moisture and trace minerals present in premixes. Pyridoxine as an ester of phosphate acts like a coenzyme in many catalytic reactions, such as deamination, decarboxylation, desulfhydration, and nonoxidative deaminations. It is required in the process of transamination, in particular glutamate–oxaloacetate transaminase and glutamate–pyruvate transaminase. It is also involved in the synthesis of messenger RNA and the formation of biogenic amines that act as neuromediators.

4.6.1.6 Pantothenic acid

Pantothenic acid is usually supplemented in fish diet as calcium *d*-pantothenate, which is 92% of *d*-pantothenic acid. Calcium *d*-pantothenate is stable during pelleting and storage but with losses less than 20% after extrusion and storage.

It is a constituent of coenzyme-A and acyl-carrier protein which function as transporters of acyl radicals. The acetyl coenzyme-A system is involved in the acetylation of aromatic amines. Pantothenic acid is needed for intermediate metabolism of fat, carbohydrate, and AA.

4.6.1.7 Biotin

Biotin is a monocarboxylic acid whose biological active form is *d*-biotin. It is added to feeds as a dry dilution having 2% activity. Biotin is part of the coenzymes that participate in the reactions of carbon dioxide fixation and decarboxylation. Among the main enzymes are pyruvate carboxylase, which is involved in glycolysis and neoglucogenesis, acetyl-CoA carboxylase involved in lipogenesis, and propionyl-CoA carboxylase involved in the conversion of propionic acid to succinic acid into methyl-malonyl-CoA.

4.6.1.8 Folacin or folic acid

Pteroylglutamic acid is the chemical name of folic acid and consists of pteridine linked to p-aminobenzoic acid giving pteroic acid. This acid is conjugated to glutamic acid forming polyglutamates or folates. These molecules are activated into tetrahydrofolic acid or tetrahydrofolate (THF), which is a folic acid derivative. Prior to absorption of food folate in the proximal part of the intestines, the excess glutamate is split off. THF is transported in the blood in association with a protein. Folic acid as a crystalline or spray-dried form is added to vitamin premixes. Stability of folic acid in feeds after extrusion and storage is relatively low, ranging from 40% to 80%. Folic acid plays an important role in normal blood cell formation, blood glucose regulation, and improving cell membrane function and hatchability of eggs. It acts as a donor of a group with one carbon atom in glycine–serine interconversion, homocysteine to methionine synthesis, and in histidine and glycine metabolism. It is also a cofactor in the synthesis of AA and nucleic acids, and in the catabolism of histidine to glutamic acid.

4.6.1.9 Vitamin B_{12} (cyanocobalamin)

Vitamin B_{12}, also known as cobalamin, comprises a number of forms including cyano-, methyl-, deoxyadenosyl- (coenzyme B_{12}), and hydroxy-cobalamin. Note that the mineral cobalt is an essential part of the B_{12}. The cyano- form produced by fermentation is supplemented in feed as a dry dilution having 1% activity. The methyl- or 5-deoxyadenosyl forms are required as cofactors for methionine synthase, essential for the synthesis of nucleic acids and L-methyl-malonyl-CoA mutase to convert methylmalonyl-CoA to succinyl-CoA. Interrelated to folic acid, the cobalamines help package folic acid into cells, hence a decrease in input can lead to a deficiency of folic acids. It is also involved in myelin synthesis in nerve cells.

Choline and inositol are compounds classified differently because they are required in larger quantities than the normal B vitamins. Generally they do not fulfill a coenzyme function but contribute to structural components.

4.6.1.10   Choline

Choline refers to quaternary ammonium salts containing the trimethylammonium hydroxide and has three labile methyl groups which can be passed on to other compounds. It is not generally considered a vitamin but is an essential micronutrient. Choline is produced for feed use as a chloride salt which is available as a dry dilution, having a concentration of 50% or 60% or liquid product with 70% activity on a weight basis. Choline is an essential dietary nutrient that plays a critical role as a component of PC, thus functioning in membrane lipids, lipoproteins, and bile lipid, in the neurotransmitter acetylcholine, and as a precursor to betaine, which functions as a source of labile methyl groups. Choline also serves an important function as a donor of labile methyl groups. Choline allows methionine sparing. Choline aids in fat transport and by its lipotropic effect prevents the accumulation of liver fat.

4.6.1.11   Inositol

Inositol is a carbohydrate similar in structure to glucose that can be synthesized from phytic acid from plants. It exists in nine possible stereoisomers, of which the isomer with biological activity is the myoinositol form, which can be utilized by fish. Inositol acts as a quasi-vitamin and is not a coenzyme. Inositol has lipotropic action that prevents the accumulation of liver fat. It is part of PL, PI and a reserve carbohydrate in muscle.

4.6.2   Fat-soluble vitamins

Fat-soluble vitamins A, D, E, and K are essential for normal health and life functions, such as growth, development, maintenance, and reproduction in most animals.

4.6.2.1   Vitamin A

Active forms of vitamin A are an unsaturated monohydric alcohol. It occurs in two forms, retinol 1 (most common in marine fish) and retinol 2 (found in freshwater fish). Carotenoids are the precursors of vitamin A in plants. β-Carotene is the most widely distributed and is essentially two vitamin As hooked together. β-Carotene is hydrolyzed in the intestinal wall or liver to vitamin A. Absorbed vitamin A is transported by chylomicrons in the lymphatic system like other fat-soluble compounds and deposited in the livers of animals and to a lesser extent in fatty tissues. Fish liver generally stores the highest amounts, and in excess for winter which has great practical significance. Vitamin A is stored primarily in an ester form which protects it from oxidation and destruction. A liver ester hydrolase cleaves the FA ester liberating the active alcohol form of the vitamin. The main functions of vitamin A are related to normal vision, bone development, and maintenance of epithelium and other tissue. Vitamin A participates in several reproductive processes in crustaceans, such as spermatogenesis, oogenesis, and embryonic growth and differentiation (Harrison, 1990). Vitamin A interacts with vitamin E in stabilizing cell membranes and with vitamin D. Additionally, liver vitamin A is utilized for protein intake and metabolism in tissues. Alava et al. (1993), working on *M. japonicus*, found that dietary nonsupplementation of vitamins A, E, and C resulted in significantly lower weight gain and higher content of vitamins A and E in ovaries, hepatopancreas, and tissue, than those of shrimp fed diets deficient in these vitamins. Hypervitaminosis A is not generally a problem in farm animals. There is no possible toxicity from the β-carotenes, because they cannot be converted to vitamin A at a sufficiently high enough rate. Direct sources of vitamin A and β-carotene are marine fish oil and crustaceans or detritus. Phytoplankton is also a good source of β-carotene (chlorophyll).

4.6.2.2   Vitamin D

The main compounds with vitamin D activity are vitamin D_2 (ergocalciferol precursor in plants) and vitamin D_3 (cholecalciferol precursor in animals). The vitamin D requirement can be met through dietary intake of these precursors that are activated by ultraviolet irradiation. Absorption of vitamin D is like the absorption of any other fat which uses chylomicrons for transport in the lymphatic system. Absorbed or endogenously synthesized vitamin D is brought to the liver first by the lipoprotein system. Liver then hyroxylates cholecalciferol to 25-hydroxycholecalciferol which is the main circulating form of vitamin D. 25-Hydroxycholecalciferol can form 24,25-, 25,26-, or 1,25-dihydroxycholecalciferol, each of these metabolites with unique biological activities, and the 1,25 form being the most potent and its transformation is carried out in the kidneys. Supplementation of diets with ergocalciferol is less effective than cholecalciferol. Thus, it has been shown in rainbow trout, where D_3 is roughly 3.3 times more potent than dietary vitamin D_2 to improve growth (Bamett et al., 1982). Lock et al. (2010) discussed the physiological function of vitamin D in maintaining calcium and phosphate homeostasis. Vitamin D is also responsible for regulating intestinal absorption (via a luminal calcium channel), mobilization that involves the vitamin-D-binding protein and deposition of calcium for normal development and maintenance of fish bone and crustacean exoskeleton. In addition, vitamin D modulates the expression of Na^+-dependent P_i transporters, the main transport route of intestinal phosphate uptake and reabsorption through kidney (Taketani et al., 1998). Lack of vitamin D_3 reduces calcium metabolism, matrix ossification, and bone density. Although plankton aquatic web food produces and accumulates vitamin D which is available as a dietary source for fish. However, it is convenient a vitamin D supplementation through feed using either marine feed ingredients or in vitamin premix to cover the requirement. One IU = 0.025 micrograms of crystalline vitamin D_3.

4.6.2.3   Vitamin E

α-Tocopherol is the most active compound designated as vitamin E. Seven other compounds exist with various activities. Free α-tocopherol is similar to vitamin A, another alcohol vitamin that is also easily oxidized. α-Tocopheryl acetate that is commercially available is less oxidized than free α-tocopherol but has no antioxidant properties. Esterases in tissues cleave off the acetate forming the free α-tocopherol. Vitamin E is absorbed like other lipids. The liver and adipose tissue are likely the chief depositories but other tissues store vitamin E as well. A wide distribution in tissues is expected due to the antioxidant function. However, there is no long-term storage of vitamin E compared to vitamins A and D. Chemical antioxidants like butylated hydroxy toluene can prevent some vitamin E deficiency diseases.

As a fat-soluble antioxidant, vitamin E has a specific role in protecting the cell from reactive oxygen species and from free radicals, which are initiators of unsaturated PL oxidation in cellular and subcellular membranes of almost all types of cell. Vitamin E is not only important in providing protection against oxidative damage to tissues in vivo, but it has also been reported to reduce oxidation and improve the storage quality of fishery products. Vitamin E has been demonstrated to be of major importance for the health status of various species of fish in which it appears to influence both humoral and cellular immune responses. Supplementation with α-tocopherol enhanced antibody production, phagocytosis of peritoneal macrophages and resistance against bacterial infections (Blazer, 1992; Lall and Olivier, 1993; Waagbø, 1994). Vitamin E and other antioxidant nutrients (AA, β-carotene, and selenium) are required for the optimum function of the immune system. Vitamin E or α-tocopherol is a fat-soluble vitamin whose requirement depends on level and oxidation state of PUFA in the diet, presence of antioxidants and selenium. Furthermore, the vitamin E requirement for an optimal functioning of the immune system and a subsequent enhancement of the resistance to diseases has been reported to be higher than the levels recommended for normal growth (Hamre, 2011). Thus, an improvement of sperm count and the rate of ovarian maturation have also been reported in *L. setiferus* (Chamberlain, 1988). One IU = 1 mg D,L-α-tocopherol acetate.

4.6.2.4   Vitamin K

There are three major forms that act as vitamin K, derived from phylloquinone (vitamin K_1) found in plants, menaquinone (vitamin K_2) obtained by bacterial flora of the intestinal tract, and menadione (vitamin K_3) which is a synthetic product. Vitamin K absorption is similar to fats and toxicity reports caused by it in fish are contradictory as summarized by Krossøy et al. (2011). Vitamin K is needed for the synthesis of prothrombin and other proteins necessary for normal blood clotting. It is also involved, together with vitamin D, in calcium transportation. Its deficiencies increase blood clotting time and reduce the level of prothrombin in the blood (Krossøy et al., 2011). Most animals have microorganisms that can synthesize menaquinones in the alimentary tract but bacterial synthesis has not been established in fish or crustaceans (Tan and Mai, 2001). Sulfa drugs and moldy feeds increase the requirement for vitamin K. Requirements vary with species, size, growth rate, environment (temperature, presence of toxins, etc.), and metabolic function (growth, stress response, disease resistance) (Krossøy et al., 2011). Many species can utilize intestinal bacteria synthesis for meeting vitamin requirements.

Currently, there is information available on vitamin requirements of certain species of shrimps, especially those of high commercial value. Table 4.25 provides the amounts of vitamins required by various species of shrimps.

4.7 Minerals

Although there is relatively little attention on mineral requirements of fish and shrimps compared with other nutrients, it is likely that all of them require the same 13 minerals because the animals cannot synthesize. The minerals are considered a coenzyme that helps cellular activities as given in Table 4.26, provides protection to the body's tissues, blood clotting, and is involved in muscle contraction, and functioning of nerves. Unlike vitamins that are easily destroyed by different agents (heat,

Table 4.26 **Essential metalloenzymes in aquatic animals**

Trace elements	Enzyme	Function
Iron	Succinate dehydrogenase	Aerobic oxidation of carbohydrates
	Cytochromes (a,b,c)	Electron transfer
	Catalase	Protection against H_2O_2
Copper	Cytochromes oxidase	Terminal oxidase
	Lysyl oxidase	Lysine oxidation
	Ceruloplasmin (ferroxidase)	Iron utilization, copper transport
	Superoxide dismutase	Dimutation of the superoxidase free radical (O_2^-)
Zinc	Carbonic anhydrase	CO_2 formation
	Alcohol dehydrogenase	Alcohol metabolism
	Carboxypeptidases	Protein digestion
	Alkaline phosphatase	Hydrolisis phosphate esters
	Polymerases	Synthesis of RNA and DNA chains
	Collegenase	Wound healing
Manganese	Pyruvate carboxylase	Pyruvate metabolism
	Superoxide dismutase	Dimutation of the superoxidase free radical (O_2^-)
	Glycosylaminotransferases	Proteoglycan synthesis
Molybdenum	Xanthine dehydrogenase	Purine metabolism
	Sulfite oxidase	Sulfite oxidation
	Aldehide oxidase	Purine metabolism
Selenium	Glutathione peroxidase	Removal of H_2O_2
	Type I and III deiodinases	Conversion of thyroxide to the active form

Source: Lall (2002). Reprinted by permission of Elsevier Science.

pH, moisture, etc.), minerals always maintain their chemical structure. Aquatic species can absorb minerals from the surrounding water but require a dietary source of certain minerals to cover their metabolic needs since the ionic composition of water is fluctuant and site-dependent.

Based on the amount needed, mineral elements are usually classified into two groups:

A. *Macroelements* – Required in large quantities as % of diet. These macrominerals are divided into two groups
    1. Cationic – electron-deficient
        a. Calcium, phosphorous, sodium, potassium, and magnesium
    2. Anionic – electron surplus
        a. Chlorine and sulfur
B. *Micro- or trace elements* – Iron, copper, iodine, zinc, cobalt, molybdenum, selenium, and fluoride are required in small quantities and are added in parts per million or less to the diet. They have been shown to be required or utilized by fish.

4.7.1 Mineral requirement

Aquatic species can absorb dissolved minerals from water through the gills, oral epithelium (and other membranes), and through water intake. Fish and crustaceans can absorb minerals by routes other than the digestion of food, through the ingestion of seawater and exchange from their aquatic environment across skin and gill membranes. Dietary mineral requirements are influenced by water chemistry. Most required calcium comes from water. For marine species, water provides the majority of iron, magnesium, cobalt, potassium, sodium, and zinc.

Dietary requirements for approximately 12 minerals (calcium, phosphorus, copper, iron, iodine, magnesium, manganese, potassium, sodium/chloride, selenium, and zinc) have been identified for fish, and seven minerals (calcium, copper, phosphorus, potassium, magnesium, selenium, and zinc) have been recommended for inclusion in penaeid shrimp and lobster feeds.

4.7.2 Calcium and phosphorus

Calcium and phosphorus are two of the major inorganic constituents of feed and both are structural components of tissue.

Calcium is essential for blood clotting, muscle function, nerve transmission, membrane permeability, and correct osmoregulation. These processes are dependent on an effective control of the ionic calcium levels in physiological fluids, which are affected by the continuous exchange of calcium between fish and the aquatic environment, mainly via the gills. In the case of dietary calcium, this may be absorbed from the intestine by active transport, and relies on the solubility of the mineral in intestine pH. Minerals generally have a greater availability if they are soluble in water and their availability increases in an acidic stomach (Hossain and Yoshimatsu, 2014).

A calcium requirement has been studied for species of fish such as *Onchorhynchus mykiss* (Ogino and Takeda, 1978), *Onchorhynchus keta* (Watanabe et al., 1980),

I. punctatus (Andrews et al., 1973; Robinson et al., 1986), *C. carpio* (Ogino and Takeda, 1976), *O. aureus* (Robinson et al., 1984, 1987), *O. niloticus* × *Oreochromis aureas* (Shiau and Tseng, 2007), *Morone saxatilis* (Dougall et al., 1996), *Crysophrys major* (Sakamoto and Yone, 1976a), and *Ctenopharyngodon idella* (Liang et al., 2012b). Reported calcium requirements vary from about 0.17% to 1.5% of the diet (Table 4.27).

Phosphorus is used in the pH control system and is a component of: high-energy molecules (ADP, ATP, phosphocreatine), PLs of cell membranes, genetic material (DNA, RNA), and coenzymes. In fish and shrimp, the source and level of phosphorus in the diet

Table 4.27 Calcium (Ca) requirement reported on several fish species using chemically defined diets

Species	Ca source	Protein source (%)	Ca requirement (g kg/DM)	References
Channel catfish	$CaSO_4$	CS	4.5	Robinson et al. (1986)
Atlantic cod	$CaSO_4$	FM	R	Kousoulaki et al. (2010)
Barbell chub			R	Zheng et al. (2007)
Black sea bream	$C_6H_{10} CaO_6$, $Ca_3(PO_4)_2$	CS	NR	Hossain and Furuichi (1999)
Blue tilapia	$CaSO_4$	CS	7	Robinson et al. (1984)
Channel catfish		SBM	15	Andrews et al. (1973)
Common carp	$C_6H_{10} CaO_6$	EA	NR	Ogino and Takeda (1976)
Grass carp	$C6H_{10} CaO_6$	CS	10	Liang et al. (2012b)
Grouper	$C_6H_{10} CaO_6$	CS	6	Ye et al. (2006)
Guppy	$CaCO_3$	CS	NR	Shim and Ho (1989)
Hybrid tilapia	$C_6H_{10} CaO_6$	CS	3.5–4.3	Shiau and Tseng (2007)
Japanese flounder	$C_6H_{10} CaO_6$, $Ca_3(PO_4)_2$	CS	R	Hossain and Furuichi (2000c)
Mrigal		CS	1.9	Paul et al. (2004b)
Rainbow trout	$C_6H_{10} CaO_6$	EA	NR	Ogino and Takeda (1978)
Rainbow trout	$CaCO_3$	CS	NR	Fontagné et al. (2009)
Redlip mullet	$C6H_{10} CaO_6$, $Ca_3(PO_4)_2$	CS	R	Hossain and Furuichi (2000b)
Scorpian fish	$C_6H_{10} CaO_6$, $Ca_3(PO_4)_2$	CS	R	Hossain and Furuichi (2000a)
Tiger puffer	$C_6H_{10} CaO_6$	CS	1.3–2.4	Hossain and Furuichi (1998)

Note: EA, egg albumin; CS, casein; FM, fish meal; SBM, soybean meal; R – required but quantitative requirement not determined; NR – not required.

had a greater effect than the amount of phosphorus in the culture water (Wiesmann et al., 1988). Phosphorus is the main cost for mineral supplementation. Absorbed calcium is rapidly deposited as calcium salts in the skeleton but absorbed phosphorus is distributed to all the major tissues: viscera, skeleton, skin, and muscle. Since levels of phosphorus are low in most natural waters, there is a dietary requirement. Hence, phosphorus is a major mineral that must be supplied in the feed. However, much of the phosphorus in commercial fish diets may be released into the environment (Wiesmann et al., 1988) and is regulated in many countries. A phosphorus requirement has been determined for species of fish including *C. major* (Sakamoto and Yone, 1978), *S. salar* (Ketola, 1975), *C. carpio* (Ogino and Takeda, 1976), *I. punctatus* (Andrews et al., 1973; Lovell, 1978; Wilson et al., 1982), *O. niloticus* (Furuya et al., 2008), *O. aureus* (Robinson et al., 1987), *Poecilia reticulata* (Shim and Ho, 1989), *O. keta* (Watanabe et al., 1980), *Onchorynchus mykiss* (Ogino and Takeda, 1978; Rodehutscord, 1996), and *Odontesthes bonariensis* (Rocha et al., 2014). Reported phosphorus requirements for different fish species vary widely between about 0.33 and 1.00 g/kg of the diet, although this rather wide range may be growth-related. In addition, the range of phosphorus requirements for salmonids is quite wide at 0.5% to 0.8% of the diet (Table 4.28).

A meta-analysis of available data on dose–response to dietary phosphorus (P) in fish from over 70 feeding trials covering over 40 species of fish was performed by Prabhu et al. (2013). Using broken-line regression to model the data sets showed that the estimated minimal dietary P level varies with the response criterion selected. The authors found that WG was the most reliable, expressed in terms of g P intake per kg $BW^{0.8}$ per day. In general, vegetal feed materials (soybean, cereals, etc.) have a low P digestibility because most of the P is bound to phytic acid (also known as phytate or phytin), rendering it unavailable for absorption fish and crustaceans. Phytate phosphorous is an organic source of phosphorous but is not available (Satoh et al., 1989a). Fish meals vary both in total and digestible P content, mainly depending on the origin and the processing of the fish meal. The bone matrix is formed by hydroxylated polymers of calcium phosphate: commonly called hydroxyapatite (HA) $[Ca_{10}(PO_4)_6(OH)_2]$ or $[Ca_5(PO_4)_3(OH)]$, a product of the mineralization process of collagen (Simkiss and Wilbur, 1989). The complex HA has been reported to be less available to some fish species, such as chum salmon, rainbow trout, and channel catfish (Watanabe et al., 1980; Satoh et al., 1987a,c, 1989a). Thus, a high level of fish bone lowers the P digestibility. Among phosphorous sources available, monobasic sources like sodium phosphate are highly digestible (90–95%), the availability of di- and tribasic phosphorus sources varies with species, but is generally around 40–65%. Monobasic sources are more expensive.

4.7.2.1 Absorption of calcium and phosphorous

Since levels of P are low in most natural waters, there is a dietary requirement. Under practical farming conditions, mineral deficiency signs often arise from a dietary imbalance of calcium due to the antagonistic effect of excess dietary calcium on the absorption of phosphorus (Nakamura, 1982). When there is an excess of supplementation of dietary calcium in comparison with the phosphorus, the phosphorus is not

Table 4.28 Phosphorus (P) requirement reported on several fish species using chemically defined diets

Species	P source	Protein source	P requirement (mg/kg DM)	References
African catfish	$CaHPO_4$	SBM	12.3 TP	Nwanna et al. (2009)
Asian sea bass	Na_2HPO_4	FM	R TP	Chaimongkol and Boonyaratpalin (2001)
Atlantic cod	NaH_2PO_4: KH_2PO_4 (1:1)	FMM, SPC	R TP	Kousoulaki et al. (2010)
Atlantic salmon	KH_2PO_4		9 TP	Lall and Bishop (1977)
Atlantic salmon	Na_2HPO_4	SBM	6 AP	Ketola (1975)
Atlantic salmon	$Ca(H_2PO4)_2$	CS	9 AP	Åsgård and Shearer (1997)
Atlantic salmon	$Ca(H_2PO_4)_2$	CS, WGM	5.6 AP	Vielma and Lall (1998)
Black sea bream	Na_2HPO_4	CS	5.5–8.8 AP	Shao et al. (2008)
Blue tilapia	Na_2HPO_4	CS	3.0–5.0 TP	Robinson et al. (1987)
Catla	KH_2PO_4	EA	6.4–7.1 TP	Sukumaran et al. (2009)
Channel catfish	$Ca_2(H_2PO_4)_2$	SMB, CGM, YC	8 AP	Andrews et al. (1973)
Channel catfish	Na_2HPO_4	CS	4.2 TP	Lovell (1978)
Chum salmon	NaH_2PO_4	EA	4.6 AP	Watanabe et al. (1980)
Common carp	NaH_2PO_4: KH_2PO_4 (1:3)	EA	6.0–7.0 TP	Ogino and Takeda (1976)
Common carp	Na_2HPO_4	EA, PP, WG	5.55–13.2 AP	Nwanna et al. (2010)
Common carp (Jian)	Na_2HPO_4	Rice GM, FM	5.2 AP	Xie et al. (2011)
Common carp (Mirror)	$Ca(H_2PO_4)_2$	SMB	6 AP	Schaefer et al. (1995)
European sea bass	$CaHPO_4$	CS	6.5 TP	Oliva Teles and Pimentel Rodrigues (2004)
European white fish	KH_2PO_4	CS	6.2–6.5 AP	Vielma et al. (2002)
Gilthead sea bream	$CaHPO_4$	CS, GEL	7.5 TP	Pimentel Rodrigues and Oliva Teles (2001)

Grass carp	Na_2HPO_4	CS	8.5 AP	Liang et al. (2012b)
Grouper	$Ca(H_2PO_4)_2$	FM, CS, WG	8.6 TP	Zhou et al. (2004b)
Haddock	$Ca(H_2PO_4)_2$	CS, WGM, FM, CGM	7.2 AP	Roy and Lall (2003)
Hybrid sunshine bass	KH_2PO_4	Red drum muscle	4.1 AP	Brown et al. (1993)
Japanese flounder	Na_2HPO_4	CS, SqM, Flounder muscle	4.5–5.1 TP	Wang et al. (2005)
Japanese sea bass	$Ca(H_2PO_4)_2$	SBM, FM	6.8 AP	Zhang et al. (2006)
Milkfish	KH_2PO_4	CS	8.5 TP	Borlongan and Satoh (2001)
Mrigal carp	Na_2HPO_4	CS	7.5 TP	Paul et al. (2004b)
Nile tilapia	$Ca_2(H_2PO_4)_2$	SBM	5.5–6.4 AP	Furuya et al. (2008)
Nile tilapia	$Ca_2(H_2PO_4)_2$	SBM	6.1 AP	Schamber et al. (2014)
Olive flounder	Na_2HPO_4		4.5–5.7 TP	Choi et al. (2005)
Pejerrey	Na_2HPO_4	EA	6.3 AP	Rocha et al. (2014)
Rainbow trout	NaH_2PO_4: KH_2PO_4 (1:3)	EA	7.0–8.0 TP	Ogino and Takeda (1978)
Rainbow trout	Na_2HPO_4	CS, BM	4.1 AP	Ketola and Richmond (1994)
Rainbow trout	Na_2HPO_4	CG	3.4–5.4 AP	Ketola and Richmond (1994)
Rainbow trout	Na_2HPO_4	WG	3.7–5.6 AP	Rodehutscord (1996)
Rainbow trout	Na_2HPO_4	WG, BM	4.5–5.4 AP	Sugiura et al. (2007)
Rainbow trout	Na_2HPO_4	WG, CS	7.0–9.0 AP	Skonberg et al. (1997)
Rainbow trout	KH_2PO_4	WG	7.3–7.5 AP	Sugiura et al. (2000b)
Rainbow trout	KH_2PO_4	WG	5.2 AP	Sugiura et al. (2000a)
Rainbow trout	KH_2PO_4	WG	6.1 AP	Sugiura et al. (2000a)
Rainbow trout	KH_2PO_4	EA	5.2 TP	Sugiura et al. (2000b)
Rainbow trout	KH_2PO_4	WG	5.3 AP	Sugiura et al. (2000a)
Rainbow trout	Na_2HPO_4	CG, SBM	R AP	Coloso et al. (2003)
Rainbow trout	NaH_2PO_4: KH_2PO_4 (1:1)	CS	R AP	Fontagné et al. (2009)

(Continued)

Table 4.28 (Continued)

Species	P source	Protein source	P requirement (mg/kg DM)	References
Rainbow trout	Na_2HPO_4	WG	R AP	Rodehutscord et al. (2000)
Red drum			8.6 TP	Davis and Robinson (1987)
Red sea bream			6.8 TP	Sakamoto and Yone (1978)
Red tilapia	$Ca_2(H_2PO_4)_2$	SBM	7.6–7.9 AP	Phromkunthong and Udom (2008)
Rohu carp	Na_2HPO_4	CS	7.5 TP	Paul et al. (2006)
Siberian sturgeon	Na_2HPO_4	EA, WGM	4.9–8.7 TP	Xu et al. (2011)
Silver perch	Na_2HPO_4	CS	5.6–7.1 TP	Yang et al. (2005)
Striped bass	KH_2PO_4	EA, SBM	2.9–4 TP	Dougall et al. (1996)
Striped bass	KH_2PO_4	EA, SBM	2.9–5.8 TP	Dougall et al. (1996)
Striped bass	KH_2PO_4	EA	2.9–5.4 TP	Dougall et al. (1996)
Tiger barb	KH_2PO_4	CS	5.2 TP	Elangovan and Shim (1998)
walking catfish	$Ca(H_2PO_4)_2$	SBM, FM, CGM	5.8–7.5 AP	Yu et al. (2013)
Yellow catfish	Na_2HPO_4	CS	7.6–8.5 TP	Luo et al. (2010)
Yellow catfish			17 TP	Li et al. (2008)
Yellow croaker	$Ca(H_2PO_4)_2$	SBM, FM	7–9.1 TP	Mai et al. (2006)
Yellow tail	Na_2HPO_4	EA	4.4 AP	Sarker et al. (2009)

Note: TP, total phosphorus; AP, available phosphorus; EA, egg albumin; CS, casein; CGM, corn gluten meal; CG, corn gluten; FM, fish meal; GM, gluten meal; SBM, soybean meal; WGM, wheat gluten meal; WG, wheat gluten; R, required but quantitative requirement not determined.

absorbed by the intestine because it is combined with the calcium to form calcium phosphates that are not biologically available (Andrews et al, 1973; Cowey and Sargent, 1979). Most fish maintain a constant ratio of Ca:P in bone and hence this ratio needs to be kept in feeds. Thus, optimum dietary ratios Ca:P of 1:1 and less have been reported to achieve an optimum performance of several species of fish (Phillips, 1959; Sakamoto and Yone, 1973; Nose and Arai, 1979; Nakamura and Yamada, 1980; Ye et al., 2006).

In vertebrates, Ca and P levels in blood are regulated by PTH and cholecalciferol. When there is a low level of blood Ca, PTH is released which facilitates the production of 1,25 (OH) D_3, whereas when P falls production of 1,25 (OH) D_3 is stimulated. An excessively high or excessively low Ca:P ratio interferes with the absorption of both. A diet supplemented with 1% calcium resulted in a higher available phosphorous requirement (1.22%) for optimum growth of L. vannamei (Cheng et al., 2006). A similar result in P. inidicus was reported by Ambasankar and Ali (2006), who found that optimal weight gain was reached with 1% total phosphorous in the presence of 1.25% Ca, and body composition maintained Ca:P ratio at 2:1 regardless of the P level in the diet. Calcium-binding protein aids in calcium absorption and its synthesis is controlled by vitamin D, so calcium absorption is not dietary-dependent but regulated via vitamin D.

Absorption of phosphorous is not as highly regulated as calcium and thus high dietary phosphorous will result in high P absorption. Sodium ions constitute 93% of the ions (bases) found in the bloodstream.

4.7.3 Magnesium (Mg)

Magnesium plays a vital role in carrying out several biochemical processes in animals. It is required for maintaining proper bone density, formation of cartilage, and is a component of the crustacean exoskeleton. Magnesium as an enzyme cofactor is extensively involved in the metabolism of fats, carbohydrates, and proteins. Through its role in enzyme activation, this mineral stimulates muscular integrity, respiration, thyroid metabolism, and in energy metabolism is associated with ATP transformations. Thus, ATP must be bound to a magnesium ion in order to be biologically active, this is often called Mg-ATP. Additionally, this mineral is particularly involved in the regulation of intracellular acid–base balance and in algae is necessary for the synthesis of chlorophyll and photosynthesis.

Magnesium occurs typically as the Mg^{2+} ion and is a relatively abundant ion in oceanic waters. Based on the high levels present in salt water, except for low salinity, a dietary requirement for marine species is less expected (Table 4.29) and can be dispensable in the diet.

Fish and crustaceans can readily obtain magnesium from surrounding water through the gastrointestinal tract, gills, skin, and fins, as well as from feed ingredients. Although, magnesium is present in plant foodstuffs this may be in part under the form of Mg salt of phytic acid.

Table 4.29 **Magnesium (Mg) requirement reported on several fish species using chemically defined diets**

Species	Mg source	Protein source		Mg requirement (mg/kg DM)	References
Atlantic salmon	$MgSO_4$	CS		0.33	El-Mowafi and Maage (1998)
Blue tilapia	$MgSO_4$	CS		0.5	Reigh et al. (1991)
Channel catfish	$MgSO_4$	CS		0.4	Gatlin et al. (1982)
Common carp	$Mg (C_2H_3O_2)_2$	CS		0.6	Dabrowska et al. (1991)
Common carp	$MgSO_4$	CS		0.4–0.5	Ogino and Chiou (1976)
Gibel carp	$MgSO_4$	CS	7, 8	0.75	Han et al. (2012)
Grass carp	$MgSO_4$	CS		0.5–0.7	Wang et al. (2011b)
Grass carp	$MgSO_4$	CS	7, 8	0.6–0.7	Liang et al. (2012c)
Grouper	$MgSO_4$	CS		0.24	Ye et al. (2010)
Guppy	$MgSO_4$	CS		0.54	Shim and Ng (1988)
Hybrid tilapia	$MgSO_4$	CS	7, 8	0.2	Lin et al. (2013)
Hybrid tilapia	$MgSO_4$	CS	7, 8	0.02	Lin et al. (2013)
Milk fish	$MgSO_4$	CS		R	Minoso et al. (1999)
Nile tilapia	$MgSO_4$			0.6–0.77	Dabrowska et al. (1989)
Rainbow trout	$MgSO_4$			0.5	Knox et al. (1981)
Rainbow trout	$MgSO_4$	CS		0.6; 1.35	Shearer (1989)
Rainbow trout	$MgSO_4$	FM		0.5	Satoh et al. (1991)
Rainbow trout				0.6–0.7	Ogino et al. (1978)
Rainbow trout	$MgSO_4$	CS		0.33	Shearer and Åsgård (1992)
Rainbow trout	$MgSO_4$	CS		0.01	Shearer and Åsgård (1992)
Redlip mullet	$MgSO_4$	CS		NR	El-Zibdeh et al. (1996b)
Yellow croaker	$MgSO_4$	CS		NR	El-Zibdeh et al. (1996a)

Note: FM, fish meal; CS, casein; R, required but quantitative requirement not determined; NR, not required.

4.7.4   Sodium (Na), potassium (K), chlorine (Cl)

Chlorine is the main monovalent anion constituting about 65% of the total anions of blood plasma and other extracellular fluids within the body. Sodium and chlorine are found mainly in body fluids and potassium occurs mainly in the cells. Sodium ions constitute 93% of the ions (bases) found in the bloodstream. Sodium, potassium, and chloride are recognized as essential for a number of physiological processes, they serve to maintain homeostasis, the acid–base balance, and in controlling the water metabolism. Although the major role of sodium and potassium in aquatic species is connected with the regulation of osmotic pressure, sodium and potassium also have an effect on muscle irritability. Potassium is also related to protein synthesis, and

Table 4.30 Potassium (K) requirement reported on several fish species using chemically defined diets

Species	K source	Protein source	K requirement (mg/kg DM)	References
Channel catfish	KHCO₃	CS	2.6	Wilson and Naggar (1992)
Chinook salmon	KCl	CS	8	Shearer (1988)
Grass carp	KCl	CS	8.0–10.0	Liang et al. (2014)
Hybrid tilapia	KCl	CS	2.0–3.0	Shiau and Hsieh (2001a)
Redlip mullet	KHCO₃	CS	NR	El-Zibdeh et al. (1996b)

Note: CS, casein; NR, not required.

glycogen and glucose catabolism. The Na–K balance is also involved in the passage of nutrients through membranes and the transmission of nerve impulses.

Because fish and shrimp are able to obtain enough sodium and chlorine from most feedstuffs in practical diets and seawater; therefore, dietary supplements are not needed. However, in the red drum, supplementation of sodium chloride to the diet at low salinities may satisfy other metabolic needs, which results in increased growth (Holsapple, 1990). In shrimp culture under low salinity dietary sodium chloride may release chloride to maintain the haemolymph osmolality of shrimp (Gong et al., 2004a). However, an inclusion of 1–2% of sodium chloride in the diet did not make any improvement in *L. vannamei* grown at salinity of 4 ppt. Keshavanath et al. (2003) also established that higher growth was obtained with sodium chloride levels of 1% in rohu (*L. rohita*), 1.5% in mrigal (*C. mrigaia*) and common carp (*C. carpio*), and 2% in prawn (*M. rosenbergii*). Thus these results suggest that dietary inclusion of salt can be beneficial and is related to fish species (carp). Very few studies have been conducted on the dietary potassium requirements of fish and shrimps. In red sea bream appears that potassium supplementation in the diet is not essential (Sakamoto and Yone, 1978b). Tables 4.30 and 4.36 (see later in the chapter) present potassium requirement data on fish and shrimp species reported.

4.7.5 Iodine (I)

Iodine is an essential mineral for a variety of animals. Iodine is necessary for the biosynthesis of the hormone thyroxine and other thyroid-active compounds which regulate the level of metabolic rate in fish. In teleosts thyroxine is necessary for gonadal maturation (Sage, 1973). The hormones have a wide role in cellular oxidation, neuromuscular and circulation functioning, nutrient metabolism, reproduction, and growth. Iodine is absorbed in an inorganic form. Iodine is concentrated in the thyroid gland by the action of thyroid-stimulating hormone, which is secreted by the anterior pituitary gland. Iodine attaches to the hydroxyl groups on the hormone. Triiodothyronine (T) is the major hormone of the thyroid gland and is supposed to be the active precursor for thyroxine CT (Power et al., 2001). Deficiencies are normally observed in wild populations. Dietary requirements are not well defined.

4.7.6 Iron (Fe)

Iron has an active part in oxidation/reduction reactions and is electron transport associated. It is utilized in various enzyme systems including the cytochromes, catalases, peroxidases, and the enzymes xanthine and aldehyde oxidase, and succinic dehydrogenase. Iron is found in complexes bound to proteins forming compounds such as transferrin, ferritin, and flavin iron enzymes. It is an essential component of the respiratory pigments hemoglobin and myoglobin. The former occurs in red blood cells while the latter is found in the muscle tissue of vertebrates. Red blood cells are regenerated periodically and most of the iron is recycled. Transferrin found in plasma is the principal carrier of iron in blood.

Supplemented iron sources in the form of chloride, sulfate, and citrate can cover nutritional needs. An excess of iron can produce lipid peroxidation, mainly HUFA and carotenoids in fatty or pigmented diets. High levels of phosphates, calcium, phytates, copper, and zinc can reduce iron absorption. Blood meals, dried distillers' solubles, crab meal, fish solubles, and fish meal are considered iron sources. Requirements for the level of iron for fish are 30–330 mg/kg and are not required for shrimps (Table 4.31).

Table 4.31 **Iron (Fe) requirement reported on several fish species using chemically defined diets**

Species	Fe source	Protein source	Fe requirement (mg/kg DM)	References
Atlantic salmon	$FeSO_4 \cdot 7H_2O$	CS	60	Naser (2000)
Atlantic salmon	$FeSO_4 \cdot 7H_2O$	CS	60–100	Andersen et al. (1996)
Channel catfish	$FeSO_4 \cdot 7H_2O$	EA	30	Gatlin and Wilson (1986a)
Common carp			150	Takeuchi et al. (2002)
Common carp	$C4H_2 FeO_4$	CS	147.4	Ling et al. (2010)
Gibel carp	$FeSO_4 \cdot 7H_2O$	CS	202	Pan et al. (2009)
Gilthead sea bream	Org-Fe	FM	R	Rigos et al. (2010)
Grass carp	$FeSO_4 \cdot 7H_2O$		300	Su et al. (2007)
Orange grouper	$FeSO_4 \cdot 7H_2O$	CS	245	Ye et al. (2007)
Rainbow trout	$FeSO_4 \cdot 7H_2O$		R	Carriquiriborde et al. (2004)
Rainbow trout	$FeSO_4 \cdot 7H_2O$	CS		Desjardins et al. (1987)
Red drum	$FeSO_4 \cdot 7H_2O$	CS	330	Zhou et al. (2009a)
Red sea bream			150	Sakamoto and Yone (1976b)
Tilapia	$FeSO_4 \cdot 7H_2O$	CS	85	Shiau and Su (2003)
Tilapia	$C_4H_5 \cdot FeO_7$	CS	150–160	Shiau and Su (2003)
Yellow tail	Fe-Protonate	FM	101	Ukawa et al. (1994)

Note: EA, egg albumin; FM, fish meal; CS, casein; R, required but quantitative requirement not determined.

4.7.7   Copper (Cu)

Copper is an essential component of numerous oxidation–reduction enzyme systems such as cytochrome oxidase, superoxide dismutase, caeruloplasmin (ferroxidase) among others. Copper is involved in carrying oxygen as part of hemocyanin as the oxygen-carrying pigment of the blood in crustaceans and a component of the enzyme caeruloplasmin involved with iron metabolism, and therefore hemoglobin synthesis and red blood cell production.

Although copper can be readily absorbed from the gut tract, it can be reduced substantially in the presence of phytates, and high dietary levels of zinc, iron, inorganic sulfates, and calcium carbonate. Sources of copper include dried distiller solubles, fish solubles, krill meal, and yeast. Copper as inorganic salt may be supplemented as cupric sulfate, cupric oxide, and cupric chloride being first source the poorer utilization (Miller, 1980). The high solubility of sulfates is also a disadvantage because dissociated copper ions are very reactive. They initiate and speed up oxidation and therefore promote degradation and damage to sensitive feed nutrients such as vitamins or fat, as well as forming harmful free radicals and peroxides that reduce feed palatability (Miles et al., 1998; Lu et al., 2010; Shurson et al., 2011). Requirement levels for copper in fish and shrimp feeds are of the order of 1.5–5 ppm and 16–53 mg/kg (Table 4.32). Copper toxicity in freshwater is 0.8 mg/l and 730 mg Cu/kg in feed (Lanno et al., 1985).

Phytic acid (myoinositol hexaphosphate) is considered to be an antinutritional factor for animals as it may chelate nutritionally important cations, such as Cu^{2+},

Table **4.32** **Copper (Cu) requirement reported on several fish species using chemically defined diets**

Species	Cu source	Protein source	Cu requirement (mg/kg DM)	References
Atlantic salmon	$CuSO_4 \cdot 5H_2O$		8.5–13.7	Lorentzen et al. (1998)
Atlantic salmon	$CuSO_4 \cdot 5H_2O$	FM	37	Berntssen et al. (1999)
Blunt snout bream	$CuCl_2$	FM	12–18	Shao et al. (2012)
Channel catfish	$CuSO_4 \cdot 5H_2O$	EA	5	Gatlin and Wilson (1986b)
Channel catfish	$CuSO_4 \cdot 5H_2O$	EA	R	Gatlin et al. (1989)
Common carp	$CuCl_2$		3	Ogino and Yang (1980)
Crusian carp	Cu-AA	FM	6.5–12.5	Shao et al. (2010)
Crusian carp	$CuCl_2$	FM	6.5–12.5	Shao et al. (2010)
Crusian carp	$CuSO_4 \cdot 5H_2O$	FM	6.5–12.5	Shao et al. (2010)
Malabar grouper	$CuSO_4 \cdot 5H_2O$	CS	4.0–6.0	Lin et al. (2008a)
Malabar grouper	CuPep	CS	2.0–3.0	Lin et al. (2010)
Rainbow trout	$CuSO_4 \cdot 5H_2O$		R	Gundogdu et al. (2009)
Rainbow trout	$CuSO_4 \cdot 5H_2O$	EA	3	Ogino and Yang (1980)
Rockfish	$CuSO_4 \cdot 5H_2O$		R	Kim and Kang (2004)
Yellow catfish	$CuSO_4 \cdot 5H_2O$	CS	3.1; 4.2	Tan et al. (2011b)

Note: EA, egg albumin; FM, fish meal; CS, casein; R, required but quantitative requirement not determined.

Zn^{2+}, Co^{2+}, Cd^{2+}, Mg^{2+}, Mn^{2+}, Fe^{2+}, Fe^{3+}, Ni^{2+}, and Ca^{2+} (Persson et al., 1998; Maenz et al., 1999). The apparent absorption of chelated minerals, such as copper proteinate, iron proteinate, selenium, zinc proteinate, and manganese proteinate have been shown to be higher than corresponding inorganic sources, as are copper sulfate, ferrous sulfate, selenium selenite, zinc sulfate, and manganese sulfate, respectively (Paripatananont and Lovell, 1997). Santos et al. (2015) found that the proteinate sources were consistently and significantly less inhibitory than the majority of the other sources (sulfates, chelated organic forms, glycinates, polysaccharide complexes, and AA chelates) of Cu, Zn, Fe, and Mn. This study suggested that enzyme inhibition can be a possible indication of chelation stability.

4.7.8 Manganese (Mn)

Manganese activates specific enzymes such as pyruvate carboxylase, lipase, super-oxide dismutase, and other enzyme systems important to the synthesis of polysaccharides, glycoprotein, FA, and urea from ammonia, and in fish cholesterol synthesis. Manganese is also required for regeneration of red blood cells, the reproductive cycle, and chondroitin sulfate synthesis that is a mucopolysaccharide of the bone organic matrix. Manganese is poorly absorbed in the Mn^{2+} form and is then oxidized to the Mn^{3+} form for transport. Excess dietary calcium and phosphorous decrease Mn absorption, but in contrast iron deficiency increases Mn absorption.

A feeding experiment in 4-g carp was carried out by Satoh et al. (1989b) to examine the availability of Mn contained in various types of fish meals (white fish meal, brown fish meal, and sardine meal with or without solubles) supplemented or not with Mn. The results showed that growth and feed efficiency were poor in fish without supplementary Mn in each fish-meal diet group. Thus, Mn derived from fish meals was not sufficient to keep Mn content in vertebrae and reduce body dwarfism despite high Mn bioavailability in the different fish meals tested. Mn bioavailability was not affected by the content of tricalcium phosphate in diet. Manganese may be supplemented to feeds as manganese sulfate or chloride. In carp, it has been found that these Mn salts are more suitable than manganese oxide and manganese carbonate as Mn sources (Satoh et al., 1987b). Manganese deficiencies can lead to poor growth, skeletal abnormalities, high embryonic mortality, and a low hatching rate. Requirement levels for manganese in feeds are between 2.4 and 25 mg/kg (Table 4.33).

4.7.9 Selenium (Se)

Selenium is a component of the enzyme glutathione peroxidase, which serves to protect cellular tissues and membranes against oxidative damage by destroying hydrogen peroxide and lipid hydroperoxides. Glutathione peroxidase reduces H_2O_2 into water and lipid hydroperoxides into lipid alcohols. Hence, a deprivation of Se affects normal enzyme activity. Selenium and vitamin E function synergistically. Tocopherol inhibits the formation of peroxides, whereas selenium degrades those peroxides already present. Sources of selenium include blood meal, corn gluten meal, fish solubles, fish meal, rapeseed meal, and yeast.

Table 4.33 Manganese (Mn) requirement reported on several fish species using chemically defined diets

Species	Mn source	Protein source	Mn requirement (mg/kg DM)	References
Atlantic salmon	MnSO$_4$·H$_2$O	CS	7.5–10.5	Maage et al. (2000)
Atlantic salmon	MnSO$_4$·H$_2$O	FM	15	Lorentzen et al. (1996)
Channel catfish	MnSO$_4$·H$_2$O	CS	2.4	Gatlin and Wilson (1984b)
Cobia	MnSO$_4$·H$_2$O	CS	22–25	Liu et al. (2013)
Common carp	MnSO$_4$·H$_2$O	EA	12.0–13.0	Ogino and Yang (1980)
Common carp	MnSO$_4$·H$_2$O	FM	12.0–13.0	Satoh et al. (1987b)
Common carp	MnSO$_4$·H$_2$O	EA	13–15	Satoh et al. (1992)
Gibel carp	MnSO$_4$·H$_2$O	CS	12.6, 13.7	Pan et al. (2008)
Grouper	MnSO$_4$·H$_2$O	CS	19	Ye et al (2009).
Hybrid tilapia	MnSO$_4$·H$_2$O	CS	7	Lin et al. (2008b)
Milk fish	MnSO$_4$·H$_2$O	CS	R	Minoso et al. (1999)
Rainbow trout	MnSO$_4$·H$_2$O	CS	R	Knox et al. (1981)
Rainbow trout	MnSO$_4$·H$_2$O	FM	19	Satoh et al. (1991)
Rainbow trout	MnSO$_4$·H$_2$O	EA	12.0–13.0	Ogino and Yang (1980)
Rainbow trout	MnSO$_4$·H$_2$O	FM	R	Yamamoto et al. (1983)
Tilapia	MnSO$_4$·H$_2$O	CS	R	Ishac and Dollar (1968)
Yellow catfish	MnSO$_4$·H$_2$O	CS	5.5–6.4	Tan et al. (2012)

Note: EA, egg albumin; FM, fish meal; CS, casein; R, required but quantitative requirement not determined; NR, not required.

Selenium may be supplemented as sodium selenite and selenate, which are inorganic selenium compounds or as selenomethionine, selenium-methylselenomethionine, selenocystine, and selenocysteine, which are organic complexes. Bell and Cowey (1989) reported that among dietary selenium sources from fish meal were selenite, selenomethionine, and selenocystine in Atlantic salmon. Selenomethionine was the most digestible (92%) and fish meal (47%) the least digestible source of selenium. Selenium is available from various feed components and other selenium-containing compounds. The results of Sritunyalucksana et al. (2011) indicated that organic Se supplementation in shrimp feed could improve shrimp growth and survival after an experimental challenge with TSV. Se supplementation and particularly supplementation with organic Se can have a dramatic effect on shrimp growth, hemocyte number, and survival after TSV challenge. A greater potency was also found in organic selenium sources compared to inorganic sources in channel catfish challenged with *Edwardsiella ictaluri* (Wang et al., 1997).

Supplementation with Se improved the growth and antioxidant status of fish and the effects of selenomethionine were relatively greater than sodium selenite in the crowded groups. Results suggest that crowding conditions cause significant detrimental effects in rainbow trout indicated by increased oxidative stress, reduced feed intake, and body weight gain. They also indicate that dietary Se supplementation offers a feasible way of reducing the losses in performance of rainbow trout reared under crowding conditions. Selenomethionine seems to be more effective than sodium selenite and the higher dose in the present study also seems to be more effective than the lower dose (Küçükbay et al., 2009). In mammals, selenium protects against mercury toxicosis complexed as mercuric selenium. Trout fed on a selenite-supplemented ($\sim 10\,\mu g\,Se/g$) diet enhanced the elimination of organic mercury from muscles, liver, kidney, bile, and erythrocytes (Bjerregaard et al., 1999). The quantitative requirement levels of selenium for fish and shrimps are 0.15 and 11.6 mg/kg (Table 4.34).

Table 4.34 Selenium (Se) requirement reported on several fish species using chemically defined diets

Species	Se source	Protein source	Se requirement (mg/kg DM)	References
Atlantic salmon	Na SeO	FM	<1.2	Lorentzen et al. (1994)
Atlantic salmon	Se-Met	FM	<1.2	Lorentzen et al. (1994)
Atlantic salmon	Na_2SeO_3	CS	0.14	Poston et al. (1976)
Atlantic salmon	Na_2SeO_3	Yeast	R	Bell et al. (1987)
Beluga sturgeon	Se-Met	FM	11.6	Arshad et al. (2011)
Channel catfish	Na_2SeO_3	CS	0.1–0.5	Gatlin and Wilson (1984a)
Channel catfish	Na_2SeO_3	CS	0.25	Gatlin and Wilson (1984a)
Channel catfish	Na_2SeO_3	CS	0.2–0.3	Wang and Lovell (1997)
Channel catfish	Se-Met	CS	0.12	Wang and Lovell (1997)
Channel catfish	Se-Yeast	CS	0.12	Wang and Lovell (1997)
Channel catfish	Na_2SeO_3	CS	0.4	Wang et al. (1997)
Channel catfish	Se-Met	CS	0.2	Wang et al. (1997)
Channel catfish	Se-Yeast	CS	0.2–0.4	Wang et al. (1997)
Cobia	Se-Met	CS	0.8	Liu et al. (2010)
Coho salmon	Na_2SeO_3	FM	8.6 (Max)	Felton et al. (1996)
Coho salmon	Na_2SeO_3	FM	8.6 (Max)	Felton et al. (1996)
Crucian carp	Se-nan	CS	R	Zhou et al. (2009c)
Crucian carp	Se-Met	CS	R	Zhou et al. (2009c)
Crucian carp	Na_2SeO_3	CS	R	Wang et al. (2007)
Crucian carp	Se-Met	CS	R	Wang et al. (2007)
Cut throat trout	Se-Met	FM	11.2 (Max)	Hardy et al. (2010)
Gibel carp	Se-Met	CS	1.18	Han et al. (2011)
Grouper	Se-Met	CS	0.77	Lin and Shiau (2005)
Hyb. striped bass	Se-Yeast	FM	0.2	Cotter et al. (2008)
Hyb. striped bass	Na_2SeO_3	FM	0.2	Cotter et al. (2008)
Hyb. striped bass	Na_2SeO_3	CS	1.2	Jaramillo et al. (2009)
Hyb. striped bass	Se-Met	CS	0.9	Jaramillo et al. (2009)
Largemouth bass	Na_2SeO_3	FM	1.6–1.85	Zhu et al. (2012)
Rainbow trout	Na_2SeO_3	Yeast	0.15–0.38	Hilton et al. (1980)
Rainbow trout	Na_2SeO_3	FM	R	Küçükbay et al. (2009)
Rainbow trout	Se-Met	FM	R	Küçükbay et al. (2009)
Rainbow trout	Na_2SeO_3	FM	R	Rider et al. (2009)
Rainbow trout	Se-Yeast	FM	R	Rider et al. (2009)

Note: CS, casein; FM, fish meal; GOC, groundnut oil cake; Se-Met, selenomethionine; Se-nan, selenium nanoparticle; R – required but quantitative requirement not determined.

4.7.10   Zinc (Zn)

Zinc is a component in more than 80 metalloenzymes including carbonic anhy-drase, carboxypeptidases A and B, alcohol dehydrogenase, glutamic dehydrogenase, D-glyceraldehyde-3-phosphate dehydrogenase, lactate dehydrogenase, malic dehy-drogenase, alkaline phosphatase, aldolase, superoxide dismutase, ribonuclease, and DNA polymerase. As an active component or cofactor for many important enzyme systems zinc plays a vital role in lipid, protein, and carbohydrate metabolism; being particularly active in the synthesis and metabolism of nucleic acids (RNA) and proteins. Another biological function of zinc is believed to play a positive role for normal growth, development in wound healing. Sources of zinc include dried distillers' solubles, corn gluten meal, fish solubles, fish meal, krill meal, rice mill by-products, wheat bran, and yeast. The bioavailability of zinc in various fish meals has been found to be inversely related to the tricalcium phosphate content of the meal.

An excess of zinc will make copper transport through the enterocyte be blocked due to the formation of intestinal metallothionein. Bioavailability of zinc in the inorganic form is affected by the presence of P from phytate (plant ingredients) or HA (fish meal). The results of Satoh et al. (1987c) have also shown that trical-cium phosphate is one inhibitor against Zn bioavailability to rainbow trout. Zinc is currently supplemented as zinc sulfate. The requirement level of zinc in feeds is 15–240 mg/kg (Table 4.35).

Table 4.35   **Zinc (Zn) requirement reported on several fish species using chemically defined diets**

Ref. species	Zn source	Protein source	Zn requirement (mg/kg DM)	References
Asian sea bass	$ZnCl_2$	FM; GOC	45	Sapkale and Singh (2011)
Atlantic salmon	$ZnSO_4 \cdot H_2O$		37–67	Maage and Julshamn (1993)
Blue tilapia		EA	20	McClain and Gatlin (1988)
Channel catfish	$ZnSO_4 \cdot H_2O$	SBM	50	Li and Robinson (1996)
Channel catfish	Zn-Met	SBM	50	Li and Robinson (1996)
Channel catfish	$ZnSO_4 \cdot H_2O$	EA	22	Paripatananont and Lovell (1995)
Channel catfish	$ZnSO_4 \cdot H_2O$	EA	R	Gatlin et al. (1989)
Channel catfish	$ZnSO_4 \cdot H_2O$	EA	R	Gatlin and Wilson (1983)
Channel catfish	$ZnSO_4 \cdot H_2O$	EA	20	Gatlin and Wilson (1983)
Channel catfish	$ZnSO_4 \cdot H_2O$	EA	20	Scarpa and Gatlin (1992)
Channel catfish	$ZnSO_4 \cdot H_2O$	EA	80	Scarpa and Gatlin (1992)
Channel catfish	ZnO	SBM	150–200	Gatlin and Wilson (1984c)

(*Continued*)

Table 4.35 **(Continued)**

Ref. species	Zn source	Protein source	Zn requirement (mg/kg DM)	References
Cobia	$ZnSO_4 \cdot H_2O$	EA	42.9	Xu et al. (2007)
Common carp	$ZnSO_4 \cdot H_2O$	EA	15	Ogino and Yang (1979)
Common carp	$ZnSO_4 \cdot H_2O$	EA	15–30	Satoh et al. (1992)
European sea bass	Zn-Met	FM	240	Fountoulaki et al. (2010)
gilthead sea bream	$ZnSO_4 \cdot HO$	FM	R	Serra et al. (1996)
Grass carp	$ZnSO_4 \cdot H_2O$	CS	55	Liang et al. (2012d)
Hy stripped bass	$ZnSO_4 \cdot H_2O$		17	Buentello et al. (2009)
Japanese sea bass	$ZnSO_4 \cdot H_2O$		103	Zhou et al. (2009b)
Jian carp	$Zn(C_3H_5O_3)_2 \cdot 2H_2O$	EA	43–49	Tan et al. (2011a)
Magur catfish	$ZnCl_2$	FM; GOC	30	Sapkale and Singh (2011)
Milk fish	$ZnSO_4 \cdot H_2O$	CS	R	Minoso et al. (1999)
Nile tilapia	$ZnSO_4 \cdot H_2O$	EA	30	Eid and Ghonim (1994)
Nile tilapia	$ZnSO_4 \cdot H_2O$	SBM	45; 80	Sa et al. (2004)
Rainbow trout	$ZnSO_4 \cdot H_2O$	EA	15	Ogino and Yang (1978)
Rainbow trout	$ZnSO_4 \cdot H_2O$	EA	20	Satoh et al. (1987d)
Rainbow trout	$ZnSO_4 \cdot H_2O$	EA	40	Satoh et al. (1987c)
Rainbow trout	$ZnSO_4 \cdot H_2O$	EA	80[*]	Satoh et al. (1987c)
Rainbow trout	$ZnSO_4 \cdot H_2O$	FM	80	Satoh et al. (1987a,c)
Red drum	$ZnSO_4 \cdot H_2O$	FM	20	Gatlin et al. (1991)
Yellow catfish	$ZnSO_4 \cdot H_2O$	CS	17–21	Luo et al. (2011)

Note: CS, casein; EA, egg albumin; FM, fish meal; SBM, soya bean meal; GOC, groundnut oil cake; R, required but quantitative requirement not determined.
[*]in presence of 7% tricalium phosphate.

Finally, in Table 4.36 is shown the available information on the quantities of minerals required by various shrimp species. Several studies have demonstrated that certain minerals need to be supplied in shrimp diet.

Table 4.36 Minerals requirement reported on several shrimp species using chemically defined diets

Species	Mineral source	Protein source	Requirement (mg/kg DM)	References
Calcium				
Kuruma prawn (*M. japonicus*)	$CaCO_3$	SBP	1–2%	Kanazawa et al. (1984)
Kuruma prawn (*M. japonicus*)	$45Ca–CaCl_2$	CS, EA	NR	Deshimaru and Yone (1978a)
White shrimp (*L. vannamei*)	$CaCl_2$	CS, GEL	NR	Davis et al. (1993a)
Tiger shrimp (*P. monodon*)	$CaCl_2$	CS, GEL	NR	Penaflorida (1999)
Phosphorous				
Kuruma prawn (*M. japonicus*)	$NaH_2PO_4 \cdot 2H_2O$	CS, EA	2.00%	Deshimaru and Yone (1978a)
Kuruma prawn (*M. japonicus*)		SBP	1–2%	Kanazawa et al. (1984)
White shrimp (*L. vannamei*)	Na_2HPO_4	CS, GEL	0.93%P, 0.5%Ca	Cheng et al. (2006)
White shrimp (*L. vannamei*)	Na_2HPO_4; $Ca(H_2PO_4)_2$	CS, GEL	0.34% P, 0.03% Ca	Davis et al. (1993a)
White shrimp (*L. vannamei*)	Na_2HPO_4; $Ca(H_2PO_4)_2$	CS, GEL	0.5–1.0%P, 1%Ca	Davis et al. (1993a)
White shrimp (*L. vannamei*)	Na_2HPO_4; $Ca(H_2PO_4)_2$	CS, GEL	1–2% P, 2%Ca	Davis et al. (1993a)
White shrimp (*L. vannamei*)	$Ca(H_2PO_4)_2$	Pr	>1.33%	Pan et al. (2005)
Tiger shrimp (*P. monodon*)	KH_2PO_4	CS, GEL	0.7%P at low Ca	Penaflorida (1999)
Tiger shrimp (*P. monodon*)	Na_2HPO_4	CS, EA	1.0–1.5%P, 1.25%Ca	Ambasankar et al. (2006)
Potassium				
Kuruma prawn (*M. japonicus*)	KCl	CS, EA	1.0%	Deshimaru and Yone (1978)
Kuruma prawn (*M. japonicus*)	K_2HPO_4	SBP	0.9%	Kanazawa et al. (1984)
Tiger shrimp (*P. monodon*)	KCl	CS	1.2	Shiau and Hsieh (2001b)
White shrimp (*L. vannamei*)	KCl	CS	1.09	Zhu et al. (2006)
White shrimp (*L. vannamei*)	K_2SO_4	CS, GEL	Unclear	Davis (1990)

(Continued)

Table 4.36 (Continued)

Species	Mineral source	Protein source	Requirement (mg/kg DM)	References
Magnesium				
White shrimp (*L. vannamei*)	$MgSO_4 \cdot 7H_2O$	CS, GEL	0.26–0.35%	Cheng et al. (2005)
Kuruma prawn (*M. japonicus*)	$MgSO_4 \cdot 7H_2O$	SBP	0.30%	Kanazawa et al. (1984)
Kuruma prawn (*M. japonicus*)	$MgSO_4 \cdot 7H_2O$	CS, EA	NR	Deshimaru and Yone (1978a)
Banana prawn (*Fenneropenaeus merguiensis*)	$MgSO_4$	CS	0.30%	Aquacop (1978)
Copper				
Kuruma prawn (*M. japonicus*)	$CuSO_4$	CS, EA	NR	Deshimaru and Yone (1978a)
Kuruma prawn (*M. japonicus*)		SBP	Dipensable	Kanazawa et al. (1984)
Fleshy prawn (*P. orientalis*)		FM, PM	53 mg/kg	Liu et al. (1990)
White shrimp (*L. vannamei*)	$CuSO_4 \cdot 7H_2O$	CS, GEL	16–32 mg/kg	Davis et al. (1993b)
Tiger shrimp (*P. monodon*)	$CuCl_2$	CS	10–30 mg/kg	Lee and Shiau (2002)
Zinc				
Tiger shrimp (*P. monodon*)	$ZnSO_4 \cdot 7H_2O$	CS	32–34 for growth 35–48 for immunity	Shiau and Jiang (2006)
White shrimp (*L. vannamei*)	$ZnCO_3$	CS, GEL	15 (33 total) 200 (218 total) in the presence of phytate	Davis et al. (1993c)

Selenium			
White shrimp (*L. vannamei*)	Na₂SeO₄	CS, GEL	0.2–0.4	Davis (1990)
Manganese			
Kuruma prawn (*M. japonicus*)	MnSO₄·5H₂O	CS	Dipensable	Kanazawa et al. (1984)
White shrimp (*L. vannamei*)	MnSO₄	CS, GEL	R	Davis (1990)
Iron			
Kuruma prawn (*M. japonicus*)	FeSO₄·7H₂O	CS	Dipensable	Kanazawa et al. (1984)
White shrimp (*L. vannamei*)	FeSO₄·7H₂O	CS, GEL	Dipensable	Davis et al. (1992)

Note: Pr, fish meal, squid meal, shrimp meal; FM, fishmeal; PM, peanut meal; SBP, purified soy-bean protein; NR, no dietary requirement demonstrated; R, required but quantitative requirement not determined.

References

Abbas, G., Jamil, K., Akhtar, R., Hong, L., 2005. Effects of dietary protein level on growth and utilization of protein and energy by Juvenile Mangrove Red Snapper (*Lutjanus argentimaculatus*). J. Ocean Univ. China 4, 49–55.

Abdelghany, A.E., 1996. Growth response of Nile tilapia *Oreochromis niloticus* to dietary L-ascorbic acid, L-ascorbyl-2-sulfate, and L-ascorbyl 2-polyphosphate. J. World Aquacult. Soc. 27, 449–455.

Abdel-Hameid, N.-A.H., Abidi, S.F., Khan, M.A., 2012. Dietary vitamin E requirement for maximizing the growth, conversion efficiency, biochemical composition and haematological status of fingerling *Channa punctatus*. Aquacult. Res. 43, 226–238.

Abdel-Tawwab, M., Ahmad, M.H., Khattab, Y.A.E., Shalaby, A.M.E., 2010. Effect of dietary protein level, initial body weight, and their interaction on the growth, feed utilization, and physiological alterations of Nile tilapia, *Oreochromis niloticus* (L.). Aquaculture 298, 267–274.

Abidi, S.F., Khan, M.A., 2004. Dietary valine requirement of Indian major carp, *Labeo rohita* (Hamilton) fry. J. Appl. Ichthyol. 20, 118–122.

Abidi, S.F., Khan, M.A., 2007. Dietary leucine requirement of fingerling Indian major carp, *Labeo rohita* (Hamilton). Aquacult. Res. 38, 478–486.

Abidi, S.F., Khan, M.A., 2008. Dietary threonine requirement of fingerling Indian major carp, *Labeo rohita* (Hamilton). Aquacult. Res. 39, 1498–1505.

Abidi, S.F., Khan, M.A., 2011. Total sulphur amino acid requirement and cystine replacement value for fingerling rohu, *Labeo rohita*: effects on growth, nutrient retention and body composition. Aquacult. Nutr. 17, e583–e594.

Aguirre, P., Gatlin III, D.M., 1999. Dietary vitamin C requirement of red drum *Sciaenops ocellams*. Aquacult. Nutr. 5, 247–249.

Ahmed, I., 2009. Dietary total aromatic amino acid requirement and tyrosine replacement value for phenylalanine in Indian major carp: *Cirrhinus mrigala* (Hamilton) fingerlings. J. Appl. Ichthyol. 25, 719–727.

Ahmed, I., Khan, M.A., 2004a. Dietary arginine requirement of fingerling Indian major carp, *Cirrhinus mrigala* (Hamilton). Aquacult. Nutr. 10, 217–225.

Ahmed, I., Khan, M.A., 2004b. Dietary lysine requirement of fingerling Indian major carp, *Cirrhinus mrigala* (Hamilton). Aquaculture 235, 499–511.

Ahmed, I., Khan, M.A., 2005a. Dietary histidine requirements of fingerling Indian major carp, *Cirrhinus mrigala* (Hamilton). Aquacult. Nutr. 11, 359–366.

Ahmed, I., Khan, M.A., 2005b. Dietary tryptophan requirements of fingerling Indian major carp, *Cirrhinus mrigala* (Hamilton). Aquacult. Res. 36, 687–695.

Ahmed, I., Khan, M.A., 2006. Dietary branched-chain amino acid valine, isoleucine and leucine requirements of fingerling Indian major carp, *Cirrhinus mrigala* (Hamilton). Br. J. Nutr. 96, 450–460.

Ahmed, I., Khan, M.A., Jafri, A.K., 2003. Dietary methionine requirement of fingerling Indian major carp, *Cirrhinus mrigala* (Hamilton). Aquacult. Int. 11, 449–462.

Ahmed, I., Khan, M.A., Jafri, A.K., 2004. Dietary threonine requirement of fingerling Indian major carp. *Cirrhinus mrigala* (Hamilton). Aquacult. Res. 35, 162–170.

Ai, Q., Mai, K., Zhang, C., Xu, W., Duan, Q., Tan, B., et al., 2004. Effects of dietary vitamin C on growth and immune response of Japanese seabass. *Lateolabrax japonicus*. Aquaculture 242, 489–500.

Ai, Q., Mai, K., Tan, B., Xu, W., Zhang, W., Ma, H., et al., 2006. Effects of dietary vitamin C on survival, growth, and immunity of large yellow croaker, *Pseudosciaena crocea*. Aquaculture 261, 327–336.

Alam, M.S., Teshima, S., Ishikawa, M., Koshio, S., 2000. Methionine requirement of juvenile Japanese flounder *Paralichthys olivaceus*. J. World Aquacult. Soc. 31, 618–626.

Alam, M.S., Teshima, S.I., Ishikawa, M., Koshio, S., 2002a. Effects of dietary arginine and lysine levels on growth performance and biochemical parameters of juvenile Japanese flounder *Paralichthys olivaceus*. Fish. Sci. 68, 509–516.

Alam, M.S., Teshima, S., Koshio, S., Ishikawa, M., 2002b. Arginine requirement of juvenile Japanese flounder *Paralichthys olivaceus* estimated by growth and biochemical parameters. Aquaculture 205, 127–140.

Alam, Md. S., Teshima, S.-I., Ishikawa, M., Hasegawa, D., Koshio, S., 2004. Dietary arginine requirement of juvenile kuruma shrimp Marsupenaeus japonicus (Bate). Aquacult. Res. 35, 842–849.

Al-Amoudi, M.M., El-Nakkadi, A.M.N., El-Nouman, B.M., 1992. Evaluation of optimum dietary requirement of vitamin C for the growth of *Oreochromis spilurus* fingerlings in water from the Red Sea. Aquaculture 105, 165–173.

Alava, V.R., Lim, C., 1983. The quantitative dietary protein requirements of *Penaeus monodon* juveniles in a controlled environment. Aquaculture 30, 53–61.

Alava, V.R., Kanazawa, A., Teshima, S., Koshio, S., 1993. Effects of dietary vitamins A, E and C on the ovarian development of *Penaeus japonicus*. Nippon Suisan Gakkaishi 59, 1235–1241.

Alcaraz, G., Espinoza, V., Vanegas, C., Carrara, X.C, 1999. Acute effect of ammonia and nitrite on respiration of *Penaeus setiferus* postlarvae under different oxygen levels. J. World Aquacult. Soc. 30, 98–106.

Alexander, C., Sahu, N.P., Pal, A.K., Akhtar, M.S., Saravanan, S., Xavier, B., et al., 2011. Higher water temperature enhances dietary carbohydrate utilization and growth performance in *Labeo rohita* (Hamilton) fingerlings. J. Anim. Physiol. Anim. Nutr. 95, 642–652.

Alexis, M.N., Karanikolas, K.K., Richards, R.H., 1997. Pathological findings owing to the lack of ascorbic acid in cultured gilthead seabream (*Sparus aurata* L.). Aquaculture 151, 209–218.

Ali, M.Z., Jauncey, K., 2005. Approaches to optimizing dietary protein to energy ratio for African catfish *Clarias gariepinus* (Burchell, 1822). Aquacult. Nutr. 11, 95–101.

Ambasankar, K., Ali, A.S., 2006. Effect of dietary phosphorus on growth and phosphorus excretion in Indian white shrimp. J. Aquacult. Trop. 17, 119–126.

Ambasankar, K.S., Ali, A.S., Dayal, J.D., 2006. Effect of dietary phosphorus on growth and its excretion in tiger shrimp, *Penaeus monodon*. Asian Fish. Soc. 19, 21–26.

Amezaga, M.R., Knox, D., 1990. Riboflavin requirements in on-growing rainbow trout. *Oncorhynchus mykiss*. Aquaculture 88, 87–98.

Amirkolaie, A.K., Leenhouwers, J.I., Verreth, J.A.J., Schrama, J.W., 2005. Type of dietary fibre (soluble versus insoluble) influences digestion, faeces characteristics and faecal waste production in Nile tilapia (*Oreochromis niloticus* L.). Aquacult. Res. 36, 1157–1166.

Amouroux, J.M., Cuzón, G., Gremare, A., 1997. Association of pulse chase design, compartmental analysis and analog modeling to assess absorption and assimilation efficiencies in *Penaeus stylirostris* fed an artificial diet. Aquaculture 149, 71–86.

Andersen, F., Maage, A., Julshamn, K., 1996. An estimation of dietary iron requirement of Atlantic salmon, *Salmo salar* L., parr. Aquacult. Nutr. 2, 41–47.

Anderson, A.J., Lipovsek, Z.S., 1998. A comparative study of digestive enzyme activity in aquaculture species in Australia. In: Fish Meal Replacement in Aquaculture: In vitro Studies on Feed Ingredients for Aquaculture Species. Final report to fisheries research and development corporation. NSW Fisheries, Taylors Beach, NSW, Australia, pp. 40–50.

Anderson, J.S., Lall, S.P., Anderson, D.M., McNiven, M.A., 1993. Quantitative dietary lysine requirement of Atlantic salmon (*Salmo salar*) fingerings. Can. J. Fish. Aquat. Sci. 50, 316–322.

Andrews, J.W., Murai, T., 1975. Studies on the vitamin C requirements of channel catfish (*Ictalums punctatus*). J. Nutr. 105, 557–561.

Andrews, J.W., Murai, T., 1978. Dietary niacin requirements of channel catfish. J. Nutr. 108, 1508–1511.

Andrews, J.W., Murai, T., 1979. Pyridoxine requirements of channel catfish. J. Nutr. 109, 533–537.

Andrews, J.W., Sick, L., Baptist, G., 1972. The influence of dietary protein and energy level on growth and survival of penaeid shrimp. Aquaculture 1, 341–347.

Andrews, J.W., Murai, T., Campbell, C., 1973. Effects of dietary calcium and phosphorus on growth, food conversion, bone ash and hematocrit levels of catfish. J. Nutr. 103, 766–771.

Andrews, J.W., Murai, T., Page, J.W., 1980. Effects of dietary cholecalciferol and ergocalciferol on catfish. Aquaculture 19, 49–54.

Aoe, H., Masuda, I., 1967. Water-soluble vitamin requirements of carp. 2. Requirements for p-aminobenzoic acid and inositol. Bull. Jpn. Soc. Sci. Fish. 33, 674–680.

Aoe, H., Masuda, I., Takada, T., 1967a. Water-soluble vitamin requirements of carp. 3. Requirement for niacin. Bull. Jpn. Soc. Sci. Fish. 33, 681–685.

Aoe, H., Masuda, I., Saito, T., Takada, T., 1967b. Water-soluble vitamin requirements of carp. 5. Requirement for folic acid. Bull. Jpn. Soc. Sci. Fish. 33, 1068–1071.

Aoe, H., Masuda, I., Mimura, T., Saito, T., Komo, A., 1968. Requirement of young carp for vitamin A. Bull. Jpn. Soc. Sci. Fish. 34, 959–964.

Aoe, H., Masuda, I., Mimura, T., Saito, T., Komo, A., Kitamura, S., 1969. Water-soluble vitamin requirements for carp. 6. Requirements for thiamin and effects of antithiamins. Bull. Jpn. Soc. Sci. Fish. 35, 459–465.

Appleford, P., Anderson, T.A., 1996. The effect of inclusion level and time on digestibility of starch for common carp (*Cyprinus carpio*, Cyprinidae). Asian Fish. Sci. 9, 121–126.

Aquacop, 1977. Reproduction in captivity and growth of *Penaeus monodon*, Fabricius in Polynesia. Proc. World Maricult. Soc. 8, 927–945.

Aquacop, 1978. Study of nutritional requirements and growth of *Penaeus merguiensis* in tanks by means of purified and artificial diets. In: Proceedings of the Annual Meeting – World Mariculture Society, vol. 9, pp. 225–234.

Aranyakananda, P., Lawrence, A.L., 1993. Dietary protein and energy requirements of the white-legged shrimp, *Penaeus vannamei* and the optimal protein to energy ratio From Discovery to Commercialization. European Aquaculture Soc., Oostende, Belgium, p. 21.

Arnesen, P., Krogdahl, Å., 1993. Crude and pre-extruded products of wheat as nutrient sources in extruded diets for Atlantic salmon (*Salmo salar*, L) grown in sea water. Aquaculture 118, 105–117.

Arnesen, P., Krogdhal, A., Sundby, A., 1995. Nutrient digestibilities, weight gain and plasma and liver levels of carbohydrate in Atlantic salmon (*Salmo salar*, L.) fed diets containing oats and maize. Aquacult. Nutr. 1, 151–158.

Arshad, U., Takami, G.A., Sadeghi, M., Bai, S., Pourali, H.R., Lee, S., 2011. Influence of dietary L-selenomethionine exposure on growth and survival of juvenile *Huso huso*. J. Appl. Ichthyol. 27, 761–765.

Åsgård, T., Shearer, K.D., 1997. Dietary phosphorus requirement of juvenile Atlantic salmon, *Salmo salar* L. Aquacult. Nutr. 3, 17–23.

Austreng, E., Refstie, T., 1979. Effect of varying dietary protein level in different families of rainbow trout. Aquaculture 18, 145–156.

Azarm, H.M., Kenari, A.A., Hedayati, M., 2013. Effect of dietary phospholipid sources and levels on growth performance, enzymes activity, cholecystokinin and lipoprotein fractions of rainbow trout (*Oncorhynchus mykiss*) fry. Aquacult. Res. 44 (4), 634–644.

Bach Knudsen, K.E., 1997. Carbohydrate and lignin contents of plant materials used in animal feeding. Anim. Feed Sci. Technol. 67, 319–338.

Bach Knudsen, K.E., Munck, L., Eggum, B.O., 1988. Effect of cooking, ph and polyphenol level on carbohydrate composition and nutritional quality of a sorghum (*Sorghum bicolor* (L.) Moench) food, ugali. Br. J. Nutr. 59, 31–47.

Bae, J.-Y., Park, G.H., Yoo, K.-Y., Lee, J.-Y., Kim, D.-J., Bai, S.C., 2013. Evaluation of optimum dietary vitamin E requirements using DL-α-tocopheryl acetate in the juvenile eel, *Anguilla japonica*. J. Appl. Ichthyol. 29, 213–217.

Bages, M., Sloane, L., 1981. Effects of dietary protein and starch levels on growth and survival of *Penaeus monodon* (Fabricius) postlarvae. Aquaculture 25, 117–128.

Bai, S.C., 2001. Requirements of L-ascorbic acid in a viviparous marine teleost, Korean rockfish, *Sebastes schlegeli* (Hulgendorf). In: Dabrowski, K. (Ed.), Ascorbic Acid in Aquatic Organisms. CRC Press, Boca Raton. FL, pp. 69–85.

Bai, S.C., Lee, K.J., 1998. Different levels of dietary DL-ot-tocopheryl acetate affect the vitamin E status of juvenile Korean rockfish, *Sebastes schlegeli*. Aquaculture 161, 405–414.

Balazs, G.H., Ross, E., 1976. Effect of protein source and level on growth and performance of the captive freshwater prawn, *Macrobrachium rosenbergii*. Aquaculture 7, 299–313.

Bamett, B.J., Cho, C.Y., Slinger, S.J., 1982. Relative biopotency of ergocalciferol and cholecalciferol and the role of and requirement for vitamin D in rainbow trout (*Salmo gairdneri*). J. Nutr. 112, 2011–2019.

Bautista, M.N., 1986. The response of *P. monodon* juveniles to varying protein/energy ratios in test diets. Aquaculture 53, 229–242.

Bautista, M.N., de la Cruz, M.C., 1988. Linoleic and linolenic acids in the diet of fingerling milkfish (*Chanos chanos* Forsskal). Aquaculture 71, 347–359.

Beamish, F.W.H., Trippel, E.A., 1990. Heat increment: a static or dynamic dimension in bioenergetic models? Trans. Am. Fish. Soc. 119, 649–661.

Bell, J.G., Cowey, C.B., 1989. Digestibility and bioavailability of dietary selenium from fishmeal, selenite, selenomethionine and selenocysteine in Atlantic salmon (*Salmo salar*). Aquaculture 81, 61–68.

Bell, J., Cowey, C., Adron, J., Pirie, B., 1987. Some effects of selenium deficiency on enzyme activities and indices of tissue peroxidation in Atlantic salmon parr (*Salmo salar*). Aquaculture 65, 43–54.

Berge, G.E., Lied, E., Sveier, H., 1997. Nutrition of Atlantic salmon (*Salmo salar*): the requirement and metabolism of arginine. Comp. Biochem. Phys. A 117, 501–509.

Berge, G.E., Sveier, H., Lied, E., 1998. Nutrition of Atlantic salmon (*Salmo salar*): the requirement and metabolic effect of lysine. Comp. Biochem. Physiol. A 120, 477–485.

Berntssen, M.H.G., Lundebye, A.K., Maage, A., 1999. Effects of elevated dietary copper concentrations on growth, feed utilisation and nutritional status of Atlantic salmon (*Salmo Salar* L.) fry. Aquaculture 174, 167–181.

Bhaskar, T.I.C., Ali, S.A., 1984. Studies on the protein requirements of postlarvae of the penaeid prawn *Penaeid indicus* H. Milne Edwards, using purified diets. Indian J. Fish. 31, 74–81.

Bjerregaard, P., Andersen, B.W., Rankin, J.C., 1999. Retention of methyl mercury and inorganic mercury in rainbow trout *Oncorhynchus mykiss* (W): effect of dietary selenium. Aquatic Toxicol. 45, 171.

Blasco, J., Marimon, I., Viaplana, I., Fernández-Borrás, J., 2001. Fate of plasma glucose in tissues of brown trout *in vivo*: effects of fasting and glucose loading. Fish Physiol. Biochem. 24, 247–258.

Blazer, V.S., 1992. Nutrition and disease resistance in fish. Annu. Rev. Fish. Dis. 2, 309–323.

Blom, J.H.K., Dabrowski, Ebeling, J., 2000. Vitamin C requirements of the angelfish *Pterophylum scalare*. J. World Aquacult. Soc. 31, 115–118.

Bodin, N., Mambrini, M., Wauters, J.P., Abboudi, T., Ooghe, W., Le Bou-lenge, E., et al., 2008. Threonine requirements in rainbow trout (*Oncorhynchus inykiss*) and Atlantic salmon (*Salmo salar*) at the fry stage are similar. Aquaculture 274, 353–365.

Bomfim, M.A.D., Lanna, E.A., Donzele, J.L., de Abreu, M.L., Ribeiro, F.B., Quadros, M., 2008. Reduction of crude protein with amino acid supplementation, based on ideal protein concept, in diets for Nile tilapia fingerlings. Rev. Bras. Zootecnia 37, 1713–1720.

Boonyaratpalin, M., 1998. Nutrition of *Penaeus merguiensis* and *Penaeus indicus*. Rev. Fish. Sci. 6, 69–78.

Boonyaratpalin, M., New, M.B., 1982. Evaluation of diets for *Macrobrachium rosenbergii* reared in concrete ponds. In: New, M.B. (Ed.), Giant Prawn Farming. Elsevier Science Publ. Corp., Amsterdam, Oxford, New York, pp. 249–256.

Boonyaratpalin, M., Phromkunthong, W., 2001. Bioavailability of ascorbyl phosphate calcium in hybrid catfish. *Ciarías macrocephalus* (Gunther) × *Ciarías gariepinus* (Burchell) feed. Aquacult. Res. 32 (Suppl. 1), 126–134.

Borgut, I., Bukvic, Z., Steiner, Z., Milakovic, Z., Stevic, I., 1998. Influence of linolenic fatty acid (18:3w3) additive on European sheat fish (*Silurus glanis*) growth bred in cages. Czechoslovakian. J. Anim. Sci. 43, 133–137.

Borlongan, I., Satoh, S., 2001. Dietary phosphorus requirement of juvenile milkfish, *Chanos chanos* (Forsskal). Aquacult. Res. 32, 26–32.

Boscolo, W.R., Signor, A., Signor, A.A., Feiden, A., Reidel, A., 2008. Starch inclusion in the *Nilo tilapia* larvae diet. Rev. Bras. Zootecnia 37, 177–180.

Briggs, M.R.P., Jauncey, K., Brown, J.H., 1988. The cholesterol and lecithin requirements of juvenile prawn (*Macrobrachium rosenbergii*) fed semi-purified diets. Aquaculture 70, 121–129.

Brown, C.R., Cameron, J.N., 1991a. The induction of specific dynamic action in channel catfish by infusion of essential amino acids. Physiol. Zool. 64, 276–297.

Brown, C.R., Cameron, J.N., 1991b. The relationship between specific dynamic action (SDA) and protein synthesis rate in the channel fish. Physiol. Zool. 64, 298–309.

Brown, M.L., Jaramillo Jr., F., Gatlin III, D.M., 1993. Dietary phosphorus requirement of juvenile sunshine bass, *Morone chrysops* × *M. saxatilis*. Aquaculture 113, 355–363.

Buentello, J.A., Goff, J.B., Gatlin III, D.M., 2009. Dietary zinc requirement of hybrid striped bass, *Morone chrysops* × *Morone saxatilis*, and bioavailability of two chemically different zinc compounds. J. World Aquacult. Soc. 40, 687–694.

Bureau, D.P., Kaushik, S.J., Cho, C.Y., 2002. Bioenergetics. In: Halver, J.E., Hardy, R.W. (Eds.), Fish Nutrition. Academic Press, San Diego, CA, pp. 1–56.

Bureau, D.P., Gunther, S.J., Cho, C.Y., 2003. Chemical composition and preliminary theoretical estimates of waste outputs of rainbow trout reared on commercial cage culture operations in Ontario. N. Am. J. Aquacult. 65, 33–38.

Burtle, G.J., Lovell, R.T., 1989. Lack of response of channel catfish (*Ictalunts punctatus*) to dietary myo-inositol. Can. J. Fish. Aquat. Sci. 46, 218–222.

Caffall, K.H., Mohnen, D., 2009. The structure, function, and biosynthesis of plant cell wall pectic polysaccharides. Carbohydr. Res. 344, 1879–1900.

Cahu, C.L., Zambonino Infante, J.L., Barbosa, V., 2003. Effect of dietary phospholipid level and phospholipid:neutral lipid value on the development of sea bass (*Dicentrarchus labrax*) larvae fed a compound diet. Br. J. Nutr. 90, 21–28.

Camara, M.R., 1994. Dietary phosphatidylcholine requirements of *Penaeus japonicas* Bate and *Penaeus vannamei* Boone (Crustacea, Decapoda, Penaeidae) (Doctoral dissertation). University of Ghent, Belgium.

Carrillo, O., González, R., 1998. Control de la digestión en camarones. En: Cruz-Suárez, L.E., Ricque-Marie, D., Tapia-Salazar, M., Olvera-Novoa, M.A. y Civera-Cerecedo, R., (Eds.). Avances en Nutrición Acuícola IV. Memorias del IV Simposium Internacional de Nutrición Acuícola. La Paz, B. C. S., México.

Carriquiriborde, P., Handy, R.D., Davies, S.J., 2004. Physiological modulation of iron metabolism in rainbow trout (*Oncorhynchus mykiss*) fed low and high iron diets. J. Exp. Biol. 207, 75–86.

Castell, J.D., Covey, J.F., 1976. Dietary lipid requirements of adult lobsters, *Homarus americanus* (M.E.). J. Nutr. 106, 1159–1165.

Castell, J.D., Sinnhuber, R.O., Wales, J.H., Lee, D.J., 1972. Essential fatty acids in the diet of rainbow trout: lipid metabolism and fatty acid composition. J. Nutr. 102, 77–86.

Castell, J.D., Mason, E.G., Covey, J.F., 1975. Cholesterol requirements of juvenile American lobster (*Homarus americanus*). J. Fish. Res. Board Can. 32, 1431–1435.

Castell, J.D., Bell, J.G., Tocher, D.R., Sargent, J.R., 1994. Effects of purified diets containing different combinations of arachidonic and docosahexaenoic acid on survival, growth and fatty acid composition of juvenile turbot (*Scophthalmus maximus*). Aquaculture 128, 315–333.

Castille, F., Lawrence, A., Buisman, P., Drost, R., 2004. Effects of sterol supplements (cholesterol FG, cholesterol SF, and sterols M1M) on growth and survival of the shrimp, *Litopenaeus vannamei* Boone. In: Cruz Suárez, L.E., Ricque Marie, D., Nieto López, M.G., Villarreal, D., Scholz, U., González, M. (Eds), Avances en Nutrición Acuícola VII. Memorias del VII Simposium Internacional de Nutrición Acuícola. 16–19 Noviembre, 2004. Hermosillo, Sonora, México, pp. 504–517.

Catacutan, M.R., 1991. Apparent digestibility of diet with varius carbohydrate levels and the growth response of *P. monodon*. Aquaculture 95, 89–96.

Catacutan, M.R., Lavilla-Pitogo, C.R., 1994. L-ascorbyl-2-phosphate Mg as a source of vitamin C for juvenile *Penaeus monodon*. Isr. J. Aquacult. Bamidgeh 46, 40–47.

Chaimongkol, A., Boonyaratpalin, M., 2001. Effects of ash and inorganic phosphorus in diets on growth and mineral composition of seabass *Lates calcarifer* (Bloch). Aquacult. Res. 32, 53–59.

Chakraborty, S.C., Ross, L.G., Ross, B., 1992. Specific dynamic action and feeding metabolism in common carp, *Cyprinus carpio* L. Comp. Biochem. Physiol. 103A, 809–815.

Chamberlain, G.W., 1988. Stepwise investigation of environmental and nutritional requirements for reproduction of penaeid shrimp (PhD dissertation). Department of Wildlife and Fisheries Science, Texas A&M University, TX.

Chance, R.E., Mertz, E.T., Halver, J.E., 1964. Nutrition of salmonoid fishes: XII. Isoleucine, leucine, valine and phenylalanine requirements of chinook salmon and interrelations between isoleucine and leucine for growth. J. Nutr. 83, 177–185.

Chavan, S.L., Dhaker, H.S., Barve, S.K., 2003. Influence of dietary biotin level on growth and survival of fry of *Liza parsia*. Appl. Fish. Aquacult. 3, 42–44.

Chavez de Martinez, M.C., 1990. Vitamin C requirement of the Mexican native cichlid (*Cichlasoma urophthalmus*, Gunther). Aquaculture 86, 409–416.

Chávez-Sánchez, M.C., Olvera-Novoa, M.A., Osuna-Durán, B., Abdo de la Parra, I., Abad-Rosales, S.M., Martínez-Rodríguez, I., 2014. Ascorbic acid requirement and histopathological changes due to its deficiency in juvenile spotted rose snapper *Lutjanus guttatus* (Steindachner, 1869). Aquacult. Int. 22, 1891–1909.

Chen, H.Y., 1993. Requirements of marine shrimp, *Penaeus monodon*, juveniles for phosphatidylcholine and cholesterol. Aquaculture 109, 165–176.

Chen, H.Y., Chang, C.F., 1994. Quantification of vitamin C requirements for juvenile shrimp (*Penaeus monodon*) using polyphosphorylated L-ascorbic acid. J. Nutr. 124, 2033–2038.

Chen, H.Y., Hwang, G., 1992. Estimation of the dietary riboflavin required to maximize tissue riboflavin concentration in juvenile shrimp (*Penaeus monodon*). J. Nutr. 122, 2474–2478.

Chen, H.Y., Jenn, J.S., 1991. Combined effects of dietary phosphatidylcholine and cholesterol on the growth, survival and body lipid composition of marine shrimp, *Penaeus penicillatus*. Aquaculture 96, 167–178.

Chen, S., Li, A., 1994. Investigation on nutrition of vitamin A for shrimp *Penaeus chinensis*: 1. Effects of vitamin A on shrimp's growth and visual organ. Acta Zool. Sin. Dongwu. Xuebao 40, 266–273.

Chen, H.-Y., Tsai, J.-C., 1994. Optimal dietary protein level for the growth of juvenile grouper, *Epinephelus malabaricus*, fed semipurified diets. Aquaculture 119, 265–271.

Chen, H.Y., Wu, F.C., Tang, S.Y., 1991. Thiamin requirement of juvenile shrimp (*Penaeus monodon*). J. Nutr. 121, 1984–1989.

Chen, H.Y., Len, Y.T., Roelants, I., 1992. Quantification of arginine requirements of juvenile marine shrimp *Penaeus monodon* using microencapsulated arginine. Mar. Biol. 114, 229–233.

Chen, C., Sun, B., Li, X., Li, P., Guan, W., Bi, Y., et al., 2013. N-3 essential fatty acids in Nile tilapia, *Oreochromis niloticus*: quantification of optimum requirement of dietary linolenic acid in juvenile fish. Aquaculture 416–417, 99–104.

Cheng, Z.J., Hardy, R.W., Usry, J.L., 2003. Effects of lysine supplementation in plant protein-based diets on the performance of rainbow trout (*Oncorhynchus mykiss*) and apparent digestibility coefficients of nutrients. Aquaculture 215, 255–265.

Cheng, K.-M., Hu, C.-Q., Liu, Y.-N., Zheng, S.-X., Qi, X.-J., 2005. Dietary magnesium requirement and physiological responses of marine shrimp *Litopenaeus vannamei* reared in low salinity water. Aquacult. Nutr. 11, 385–393.

Cheng, K.-M., Hu, C.-Q., Liu, Y.-N., Zheng, S.-X., Qi, X.-J., 2006. Effects of dietary calcium, phosphorus and calcium/phosphorus ratio on the growth and tissue mineralization of *Litopenaeus vannamei* reared in low-salinity water. Aquaculture 251, 472–483.

Chi, S., Tan, B., Dong, X., Yang, Q., Liu, H., 2014. Effects of supplemental coated or crystalline methionine in low-fishmeal diet on the growth performance and body composition of juvenile cobia *Rachycentron canadum* (Linnaeus). Chin. J. Oceanol. Limnol. 32, 1297–1306.

Chiou, J.Y., Ogino, C., 1975. Digestibility of starch in carps. Bull. Jpn. Soc. Sci. Fish. 41, 465–466.

Chiu, Y.N., Austic, R.E., Rumsey, G.L., 1988. Effect of feeding level and dietary electrolytes on the arginine requirement of rainbow trout (*Salmo gairdneri*). Aquaculture 69, 79–91.

Cho, C.Y., Kaushik, S.J., 1985. Effects of protein intake on metabolizable and net energy value of fish diets. In: Cowey, C.B., Mackie, A.M., Bell, J.G. (Eds.), Nutrition and Feeding in fish. Academic Press, London, pp. 95–117.

Cho, C.Y., Woodward, B., 1989. Studies on the protein-to-energy ratio in diets for rainbow trout (*Salmo gairdneri*). In: Proceedings of the Eleventh Symposium on Energy Metabolism, pp. 37–40.

Cho, C.Y., Woodward, B., 1990. Dietary pantothenic acid requirements of young rainbow trout (*Oncorhynchus mykiss*) (abstract). FASEB J. 4, 3747.

Cho, C.Y., Kaushik, S.J., Woodward, B., 1992. Dietary arginine requirement of young rainbow trout (*Oncorhynchus mykiss*). Comp. Biochem. Physiol. A 102, 211–216.

Choct, M., 1997. Feed non-starch polysaccharides: chemical structures and nutritional significance. Feed Milling Int. 6, 13–26.

Choi, S.M., Kim, K.W., Kang, Y.J., Wang, X.J., Kim, J.W., Yoo, G.Y., et al., 2005. Re-evaluation of the phosphorus requirement of juvenile olive flounder *Paralichthys olivaceus* and the bioavailability of various inorganic phosphorus sources. J. World Aquacult. Soc. 36, 217–222.

Choo, P.-S., Smith, T.K., Cho, C.Y., Ferguson, H.W., 1991. Dietary excesses of leucine influence growth and body composition of rainbow trout. J. Nutr. 121, 1932–1939.

Coloso, R.M., Murillo-Gurrea, D.P., Borlongan, I.G., Catacutan, M.R., 1999. Sulphur amino acid requirement of juvenile Asian sea bass Lates calcarifer. J. Appl. Ichthyol. 15, 54–58.

Coloso, R.M., King, K., Fletcher, J.W., Hendrix, M.A., Subramanyam, M., Weis, P., et al., 2003. Phosphorus utilization in rainbow trout (Oncorhynchus mykiss) fed practical diets and its consequences on effluent phosphorus levels. Aquaculture 220, 801–820.

Colvin, P.M., 1976. Nutritional studies on penaeid prawns: protein requirements in compounded diets for juvenile Penaeus indicus (Milne Edwards). Aquaculture 7, 315–326.

Colvin, L.B., Brand, C.W., 1977. The protein requirement of penaeid shrimp at various life cycle stages in controlled environment systems. Proc. World Maricult. Soc. 8, 821–840.

Coneição, L.E.C., Dersjant-Li, Y., Verreth, J.A.J., 1998. Cost of growth in larva and juvenile African catfish (Clarias gariepinus) in relation to growth rate, food intake and oxygen consumption. Aquaculture 161, 95–106.

Cook, H.W., McMaster, C.R., 2002. Fatty acid desaturation and chain elongation in eukaryotes. In: New Comprehensive Biochemistry, pp. 181–204.

Cook, J.T., McNiven, M.A., Sutterlin, A.M., 2000. Metabolic rate of pre-smolt growth-enhanced transgenic Atlantic salmon (Salmo salar). Aquaculture 188, 33–45.

Cornish, I.M.E., Moon, T.W., 1986. The glucose and lactate kinetics of American eels, Anguilla rostrata (LeSueur), under MS 222 anesthesia. J. Fish Biol. 28, 1–8.

Cotter, P.A., Craig, S.R., McLean, E., 2008. Hyperaccumulation of selenium in hybrid striped bass: a functional food for aquaculture? Aquacult. Nutr. 14, 215–222.

Cousin, M., Cuzón, G., Blandrit, E., Ruelle, F., AQUACOP, 1991. Protein requirements following an optimum dietary energy to protein ratio for Penaeus vannamei juveniles. In: Kaushik, Luquet, P. (Eds.), Fourth International Symposium on Fish Nutrition and Feeding, Biarrits, France, June 24–27, 1991, Colloq. INRA, vol. 6, pp. 599–606.

Cousin, M., Cuzon, G., Blanchet, E., Ruelle, F., Aquacop, 1993. Protein requirements following and optimum dietary energy to protein ratio for P. vannamei juveniles. In: Kaushik, S.J., Luquet, P. (Eds.), Fish nutrition in practice (France), June 24–27, 1991, INRA Paris, France, pp. 599–606.

Cousin, M., Cuzon, G., Guillaume, J., Aquacop, 1996. Digestibility of starch in Penaeus vannamei: in vivo and in vitro study on eight samples of various origin. Aquaculture 140, 361–372.

Couto, A., Enes, P., Peres, H., Oliva-Teles, A., 2008. Effect of water temperature and dietary starch on growth and metabolic utilization of diets in gilthead sea bream (Sparus aurata) juveniles. Comp. Biochem. Physiol. A 151, 45–50.

Coutteau, P., VanStappen, G., Sorgeloos, P., 1996a. A standard experimental diet for the study of fatty acid requirements of weaning and first ongrowing stages of the European sea bass Dicentrarchus labrax L: comparison of extruded and extruded/coated diets. Arch. Anim. Nutr. 49, 49–59.

Coutteau, P., Camara, M.R., Sorgeloos, P., 1996b. The effect of different levels and sources of dietary phosphatidylcholine on the growth, survival, stress resistance, and fatty acid composition of postlarval Penaeus vannamei. Aquaculture 147, 261–273.

Coutteau, P., Geurden, I., Camara, M., Bergot, P., Sorgeloos, P., 1997. Review on the dietary effects of phospholipids in fish and crustacean larviculture. Aquaculture 155, 149–164.

Cowey, C.B., 1994. Amino acid requirements of fish: a critical appraisal of present values. Aquaculture 124, 1–11.

Cowey, C.B., Cho, C.Y., 1993. Nutritional requirements of fish. Proc. Nutr. Soc. 52, 417–426.

Cowey, C.B., Sargent, J.R., 1979. Nutrition. In: Hoar, W.S. Randall, J. (Eds.), Fish Physiology, vol. III. Academic Press, New York, NY, pp. 1–69.

Cowey, C.B., Walton, M.J., 1989. Intermediary metabolism. In: Halver, J.E. (Ed.), Fish Nutrition, second ed. Academic Press, New York, NY, pp. 259–329.

Cowey, C.B., Woodward, B., 1993. The dietary requirement of young rainbow trout (*Oncorhynchus mykiss*) for folic acid. J. Nutr. 123, 1594–1600.

Cowey, C.B., Adron, J.W., Youngson, A., 1983. The vitamin E requirement of rainbow trout (*Salmo gairdneri*) given diets containing polyunsaturated fatty acids derived from fish oil. Aquaculture 30, 85.

Cowey, C.B., Cho, C.Y., Sivak, J.G., Weerheim, J.A., Stuart, D.D., 1992. Methionine intake in rainbow trout (*Oncorhynchus mykiss*), relationship to cataract formation and the metabolism of methionine. J. Nutr. 122, 1154–1163.

Craig, S.R., Gatlin III, D.M., 1996. Dietary choline requirement of juvenile red drum (*Sciaenops ocellahis*). J. Nutr. 126, 1696–1700.

Cruz, E.M., 1975. Determination of nutrient digestibility in various classes of natural and purified feed materials for channel catfish (PhD thesis). Auburn University, Auburn, AL.

Cui, Y., Liu, J., 1990. Comparison of energy budget among six teleosts—II. Metabolic rates. Comp. Biochem. Physiol. 97A, 169–174.

D'Abramo, L.R., 1998. Nutritional requirements of the freshwater prawn *Macrobrachium rosenbergii*: Comparisons with species of penaeid shrimp. Rev. Fish. Sci. 6, 153–163.

D'Abramo, L., Daniels, W., 1994. Sterols requirements of the freshwater prawn *Macrobrachium rosenbergii*. Abstracts of World Aquaculture'94. New Orleans. World Aquaculture Society, Baton Rouge, LA, p. 200.

D'Abramo, L.R., Sheen, S.-S., 1993. Polyunsaturated fatty acid nutrition in juvenile freshwater prawn *Macrobrachium rosenbergii*. Aquaculture 115, 63–86.

D'Abramo, L.R., Bordner, C.E., Conklin, D.E., 1981. Essentiality of dietary phosphatidycholine for the survival of juvenile lobsters. J. Nutr. 111, 425–431.

D'Abramo, L.R., Bordner, C.E., Conklin, D.E., 1982. Relationship between dietary phosphatidylcholine and serum cholesterol in the lobster *Homarus* sp. Mar. Biol. 67, 231–235.

D'Abramo, L., Bordner, C., Conklin, D., Baum, N., 1984. Sterol requirement of juvenile lobsters, *Homarus* sp. Aquaculture 42, 13–25.

D'Abramo, L.R., Wright, J.S., Wright, K.H., Bordner, C.E., Conklin, D.E., 1985. Sterol requirements of cultured juvenile crayfish *Pacifasticus leniusculus*. Aquaculture 49, 245–255.

D'Abramo, L.R., Moncreiff, C.A., Holcomb, F.R., Montanez, J.L., Buddington, R.K., 1994. Vitamin C requirement of the juvenile freshwater prawn, *Macrobrachium rosenbergii*. Aquaculture 128, 269–275.

Dabrowska, H., Meyer-Burgdorff, K., Günther, K.-D., 1989. Interaction between dietary protein and magnesium level in tilapia (*Oreochromis niloticus*). Aquaculture 76, 277–291.

Dabrowska, H., Meyer-Burgdorff, K.H., Gunther, K.D., 1991. Magnesium status in freshwater fish, common carp (*Cyprinus carpio*, L.) and the dietary protein–magnesium interaction. Fish Physiol. Biochem. 9, 165–172.

Dabrowski, K.R., 1977. Protein requirements of grass carp fry (*Ctenopharyngodonidella*). Aquaculture 12, 63–73.

Dabrowski, K., Zhang, Y.F., Kwasek, K., Hliwa, P., Ostaszewska, T., 2010. Effects of protein-, peptide- and free amino acid-based diets in fish nutrition. Aquacult. Res. 41, 668–683.

Dall, W., Smith, D.M., 1986. Oxygen consumption and ammonia-N excretion in fed and starved tiger prawns, *Penaeus esculentus* Haswell. Aquaculture 55, 23–33.

Dalla Via, J., Villani, P., Gasteiger, E., Nieders, H., 1998. Oxygen consumption in sea bass fingerling *Dicentrarchus labrax* exposed to acute salinity and temperature changes: metabolic basis for maximum stocking density estimations. Aquaculture 169, 303–313.

Daniels, W.H., Robinson, E.H., 1986. Protein and energy requirements of juvenile red drum (*Sciaenops ocellatus*). Aquaculture 53, 243–252.

Davis, D.A., 1990. Dietary mineral requirements of *Penaeus vannamei*: evaluation of the essentiality for thirteen minerals and the requirements for calcium, phosphorus, copper, iron, zinc, and selenium (PhD Dissertation). Texas A&M University, College Station, TX.

Davis, D.A., Arnold, C.R., 1993. Evaluation of five carbohydrate sources for *P. vannamei*. Aquaculture 114, 285–292.

Davis, A.T., Stickney, R.R., 1978. Growth responses of *Tilapia aurea* to dietary protein quality and quantity. Trans. Am. Fish. Soc. 107, 479–483.

Davis, D.A., Robinson, E.H., 1987. Dietary phosphorus requirement of juvenile red drum *Sciaenops ocellatus*. J. World Aquacult. Soc. 18, 129–136.

Davis, D.A., Lawrence, A.L., Gatlin III, D.M., 1992. Evaluation of the dietary iron requirement of *Penaeus vannamei*. J. World Aquacult. Soc. 23, 15–22.

Davis, D.A., Lawrence, A.L., Gatlin III, D.M., 1993a. Response of *Penaeus vannamei* to dietary calcium, phosphorus, and calcium:phosphorus ratio. J. World Aquacult. Soc. 24, 504–515.

Davis, D.A., Lawrence, A.L., Gatlin III, D.M., 1993b. Dietary copper requirement of *Penaeus vannamei*. Nippon Suisan Gakkaishi 59, 117–122.

Davis, D.A., Lawrence, A.L., Gatlin III, D.M., 1993c. Evaluation dietary zinc requirement of *Penaeus vannamei* and effects of phytic acid on zinc and phosphorus bioavailability. J. World Aquacult. Soc. 24, 40–47.

De la Gándara, F., García-Gómez, A., Jover, M., 2002. Effect of feeding frequency on the daily oxygen consumption rhythms in young Mediterranean yellowtails (*Seriola dumerili*). Aquacult. Eng. 26, 27–39.

DeLong, D.C., Halver, J.E., Mertz, E.T., 1958. Nutrition of salmonid fishes. 6. Protein requirements of chinook salmon at two water temperatures. J. Nutr. 65, 589–599.

Deng, D.F., Wilson, R.P., 2003. Dietary riboflavin requirement of juvenile sunshine bass (*Morone chrysops* ♀ × *Morone saxatilis* ♂). Aquaculture 218, 695–701.

Deng, D.F., Hemre, G.I., Wilson, R.P., 2002. Juvenile sunshine bass (*Morone chrysops* ♀ × *Morone saxatilis* ♂) do not require dietary myo-inositol. Aquaculture 213, 387–393.

Deng, J.X., Zhang, L., Tao, H., Bi, L., Kong, Lei, X., 2011. D-lysine can be effectively utilized for growth by common carp (*Cyprinus carpio*). Aquacult. Nutr. 17, e467–e475.

Deshimaru, O., Kuroki, K., 1974a. Studies on a purified diet for prawn II. Optimum contents of cholesterol and glucosamine in the diet. Nippon Suisan Gakkaishi 40, 421–424.

Deshimaru, O., Kuroki, K., 1974b. Studies on a purified diet for prawn. I. Basal composition of the diet. Bull. Jpn. Soc. Sci. Fish. 40, 413–419.

Deshimaru, O., Kuroki, K., 1976. Studies on a purified diet for prawn. VII. Adequate dietary levels of ascorbic acid and inositol. Bull. Jpn. Soc. Sci. Fish. 42, 571–576.

Deshimaru, O., Yone, Y., 1978a. Requirement of prawn for dietary minerals *1. Bull. Jpn. Soc. Sci. Fish. 44, 907–910.

Deshimaru, O., Yone, Y., 1978b. Optimum level of dietary protein for prawn. Bull. Jpn. Soc. Sci. Fish. 44, 1395–1397.

Deshimaru, O., Kuroki, K., 1979. Requirement of prawn for dietary thiamin, pyridoxine, and choline chloride. Bull. Jpn. Soc. Sci. Fish. 45, 363–367.

Desjardins, L., Hicks, B., Hilton, J., 1987. Iron catalyzed oxidation of trout diets and its effect on the growth and physiological response of rainbow trout. Fish Physiol. Biochem. 3, 173–182.

Dougall, D.S., Curry Woods III, L., Douglas, L.W., Soares, J.H., 1996. Dietary phosphorus requirement of juvenile striped bass *Morone saxatilis*. J. World Aquacult. Soc. 27, 82–91.

Du, L., Niu, C.-J., 2002. Effects of dietary protein level on bioenergetics of the giant freshwater prawn, *Macrobranchium rosenbergii* (De Man, 1879) (Decapoda, Natantia). Crustaceana 75, 875–889.

Du Preez, H.H., Chen, H.-Y., Hsieh, C.-S., 1992. Apparent specific dynamic action of food in the grass shrimp, *Penaeus monodon* Fabricius. Comp. Biochem. Physiol. 103A, 173–178.

Duan, Y., Zhu, X., Han, D., Yang, Y., Xie, S., 2012. Dietary choline requirement in slight methionine-deficient diet for juvenile gibel carp (*Carassius auratus gibelio*). Aquacult. Nutr. 18, 620–627.

Duerr, E.O., Walsh, W.A., 1996. Evaluation of cholesterol additions to a soyabean meal-based diet for juvenile Pacific white shrimp, *Penaeus vannamei* (Boone), in an outdoor growth trial. Aquacult. Nutr. 2, 111–116.

Dumas, A., France, J., France, J., 2008. Mathematical modelling in animal nutrition: a centenary review. J. Agric. Sci. 146, 123–142.

Duncan, P.L., Lovell, R.T., 1991. Effect of folic acid on growth, survival and hematology in channel catfish (*Ictalurus punctatus*). In: Twenty-Second Annual Conference of the World Aquaculture Society, June 16–20, 1991, San Juan, Puerto Rico.

Duncan, P.L., Lovell, R.T., 1994. Influence of vitamin C on the folate requirement of channel catfish, *Ictalurus punctatus*, for growth, hematopoiesis, and resistance to *Edwardsiella ictaluri* infection. Aquaculture 127, 233–244.

Duncan, P.L., Lovell, R.T., Butterworth, C.E., Freeberg, L.F., Tamura, T., 1993. Dietary folate requirement determined for channel catfish. *Ictalunts punctatus*. J. Nutr. 123, 1888–1897.

Dupree, H.K., 1966. Vitamins Essential for Growth of Channel Catfish, *Ictalurus punctatus*. Technical Paper No. 7. U.S. Bureau of Sport Fisheries and Wildlife, Washington, DC.

Dupree, H.K., 1970. Dietary requirement of vitamin A acetate and beta carotene Progress in Sport Fishery Research. 1969. Resource Publication No. 88. U.S. Bureau of Sport Fisheries and Wildlife, Washington, DC, pp. 148–150.

Durruty, C., Maldonado, L., Gaxiola, G., García, T., Pedroza, R., 2000. Requerimientos de proteína dietética en larvas de *Litopenaeus setiferus* y *L. vannamei*. En: Cruz-Suárez, L.E., Ricque-Marie, D., Tapia-Salazar, D., Gaxiola-Cortés y, G., Simoes, N. (Eds.), Avances en Nutrición Acuícola VI. Memorias del VI Simposium Internacional de Nutrición Acuícola, 3–6 November, Cancún, México.

Durve, V.S., Lovell, R.T., 1982. Vitamin C and disease resistance in channel catfish. Can. J. Fish. Aquat. Sci. 39, 948–951.

Ebrahimnezhadarabi, M., Saad, C.R., Harmin, S.A., Abdul Satar, M.K., Kenari, A.A., 2011. Effects of phospholipids in diet on growth of sturgeon fish (*Huso-huso*) Juveniles. J. Fish. Aquat. Sci. 6, 247–255.

Eid, A.E., Ghonim, S.I., 1994. Dietary zinc requirement of fingerling *Oreochromis niloticus*. Aquaculture 119, 259–264.

Einen, O., Roem, A.J., 1997. Dietary protein/energy ratios for Atlantic salmon in relation to fish size: growth, feed utilization and slaughter quality. Aquacult. Nutr. 3, 115–126.

El Naggar, G.O., Lovell, R.T., 1991. L-ascorbyl-2-monophosphate has equal antiscorbutic activity as L-ascorbic acid but L-ascorbyl-2-sulfate is inferior to L-ascorbic acid for channel catfish. J. Nutr. 121, 1622–1626.

Elangovan, A., Shim, K., 1998. Dietary phosphorus requirement of juvenile tiger barb, *Barbus tetrazona* (Bleeker, 1855). Aquarium Sci. Conserv. 2, 9–19.

El-Dahhar, A.A., Lovell, R.T., 1995. Effect of protein to energy ratio in purified diets on growth performance, feed utilization and body composition of *Mozambique tilapia, Oreochromis mossambicus* (Peters). Aquacult. Res. 26, 451–457.

El-Mowafi, A., Maage, A., 1998. Magnesium requirement of Atlantic salmon (*Salmo salar* L.) parr in seawater-treated fresh water. Aquacult. Nutr. 4, 31–38.

El-Sayed, A.F.M., 1987. Protein and energy requirement of *Tilapia zilli* (Doctoral dissertation). Michigan State University, East Lansing, MI.

El-Sayed, A.M., Teshima, S., 1992. Protein and energy requirements of Nile tilapia, *Oreochromis niloticus*, fry. Aquaculture 103, 55–63.

El-Zibdeh, M., Ide, K., Furuichi, M., 1996a. Effects of the deletion of Mg or Fe from semi purified diets on growth and efficiency of feed utilization of yellow croaker nibea albiflora. J. Fac. Agric. Kyushu Univ. 40, 391–397.

El-Zibdeh, M., Yoshimatsu, T., Matsui, S., Furuichi, M., 1996b. Effects of the deletion of K, Mg or Fe from purified diets on growth and efficiency of feed utilization of redlip mullet. J. Fac. Agric., Kyushu Univ. 40, 383–390.

Encamacao, P., de Lange, C., Rodehutscord, M., Hoehler, D., Bureau, W., Bureau, D.P., 2004. Diet digestible energy content affects lysine utilization, but not dietary lysine requirements of rainbow trout (*Oncorhynchus mykiss*) for maximum growth. Aquaculture 235, 569–586.

Enes, P., Panserat, S., Kaushik, S., Oliva-Teles, A., 2009. Nutritional regulation of hepatic glucose metabolism in fish. Fish Physiol. Biochem. 35, 519–539.

Enes, P., Pousão-Ferreira, P., Salmerón, C., Capilla, E., Navarro, I., Gutiérrez, J., et al., 2013. Effect of guar gum on glucose and lipid metabolism in white sea bream *Diplodus sargus*. Fish Physiol. Biochem. 39, 159–169.

Eo, J., Lee, K.J., 2008. Effect of dietary ascorbic acid on growth and non-specific immune responses of tiger puffer, *Takifugu rubripes*. Fish Shellfish Immunol. 25, 611–616.

Erfanullah, Jafri, A.K., 1998. Evaluation of digestibility coefficients of some carbohydrate-rich feedstuffs for Indian major carp fingerlings. Aquacult. Res. 29, 511–519.

Espe, M., Lemme, A., Petri, A., El-Mowafi, A., 2007. Assessment of lysine requirement for maximal protein accretion in Atlantic salmon using plant protein diets. Aquaculture 263, 168–178.

Espe, M., Hevroy, E.M., Liaset, B., Lemme, A., El-Mowafi, A., 2008. Methionine intake affect hepatic sulphur metabolism in Atlantic salmon, *Salmo salar*. Aquaculture 274, 132–141.

Eya, J.C., 1996. "Broken-skull disease" in African catfish *Ciarías gariepinus* is related to a dietary deficiency of ascorbic acid. J. World Aquacult. Soc. 27, 493–498.

Felip, O., Ibarz, A., Fernandez-Borras, J., Beltran, M., Martin-Perez, M., Planas, J.V., et al., 2012. Tracing metabolic routes of dietary carbohydrate and protein in rainbow trout (*Oncorhynchus mykiss*) using stable isotopes ([13C]starch and [15N]protein): effects of gelatinisation of starches and sustained swimming. Br. J. Nutr. 107, 834–844.

Felton, S.P., Landolt, M.L., Grace, R., Palmisano, A., 1996. Effects of selenium dietary enhancement on hatchery-reared coho salmon, *Oncorhynchus kisutch* (Walbaum), when compared with wild coho: hepatic enzymes and seawater adaptation evaluated. Aquacult. Res. 27, 135–142.

Feng, Z., Dong, C., Wang, L., Hu, Y., Zhu, W., 2013. Optimal content and ratio of lysine to arginine in the diet of Pacific white shrimp, *Litopenaeus vannamei*. J. Chin. J. Oceanol. Limnol. 31, 789–795.

Fenucci, J.L., Lawrence, A.L., Zein-Eldin, Z.P., 1981. The effects of fatty acid and shrimp meal composition of prepared diets on growth of juvenile shrimp, *Penaeus stylirostris*. J. World Maricult. Soc. 12, 315–324.

Fineman-Kalio, A.S., Camacho, A.S., 1987. The effects of supplemental feeds containing different protein: energy ratios on the growth and survival of *Oreochromis niloticus* (L.) in brackish water ponds. Aquacult. Fish. Manage. 18, 139–149.

Fontagné, S., Geurden, I., Escaffre, A.-M., Bergot, P., 1998. Histological changes induced by dietary phospholipids in intestine and liver of common carp (*Cyprinus carpio* L.) larvae. Aquaculture 161, 213–223.

Fontagné, S., Silva, N., Bazin, D., Ramos, A., Aguirre, P., Surget, A., et al., 2009. Effects of dietary phosphorus and calcium level on growth and skeletal development in rainbow trout (*Oncorhynchus mykiss*) fry. Aquaculture 297, 141–150.

Forster, I., Ogata, H.Y., 1998. Lysine requirement of juvenile Japanese flounder *Paralichthys olivaceus* and juvenile red sea bream *Pagrus major*. Aquaculture 161, 131–142.

Forster, I.P., Dominy, W.G., Obaldo, L.G., Hartnell, G.F., Sanders, E.F., Hickamn, T.C., et al., 2011. The effect of soybean oil containing stearidonic acid on growth performance, n-3 fatty acid deposition and sensory characteristics of pacific white shrimp (*Litopenaeus vannamei*). Aquacult. Nutr. 17, 200–213.

Foumier, V., Gouillou-Coustans, M.F., Kaushik, S.J., 2000. Hepatic ascorbic acid saturation is the most stringent response criterion for determining the vitamin C requirement of juvenile European sea bass (*Dicentrarchus labrax*). J. Nutr. 130, 617–620.

Fountoulaki, E., Morgane, H., Rigos, G., Antigoni, V., Mente, E., Sweetman, J., et al., 2010. Evaluation of zinc supplementation in European sea bass (*Dicentrarchus labrax*) juvenile diets. Aquacult. Res. 41, e208–e216.

Fournier, V., Gouillou-Coustans, M.F., Metailler, R., Vachot, C., Moriceau, J., Le Delliou, H., et al., 2003. Excess dietary arginine affects urea excretion but does not improve N utilisation in rainbow trout *Oncorhynchus mykiss* and turbot *Psetta maxima*. Aquaculture 217, 559–576.

Fox, J.M., Lawrence, A.L., Li-Chan, E., 1995. Dietary requirement for lysine by juvenile *Penaeus vannamei* using intact and free amino acid sources. Aquaculture 131, 279–290.

Fraga, I., Alvarez, J.S., Galindo, J., 1992. Requerimientos nutricionales y respuesta a varias relaciones proteína/energía en juveniles de camarón blanco *Penaeus schmitti*. Result. Invest. Pesqueras 16, 13–20.

Frischknecht, R., Wahli, T., Meier, W., 1994. Comparison of pathological changes due to deficiency of vitamin C, vitamin E and combinations of vitamins C and E in rainbow trout, *Oncorhynchus mykiss* (Walbaum). Fish Dis. 17, 31–45.

Fu, S.J., Xie, X.J., Cao, Z.D., 2005. Effect of meal size on postprandial metabolic response in southern catfish (*Silurus meridionalis*). Comp. Biochem. Physiol. 140A, 445–451.

Furuichi, M., Yone, Y., 1981. The utilization of carbohydrate by fishes. 3. Change of blood sugar and plasma insulin levels of fishes in glucose tolerance test. Bull. Jpn. Soc. Sci. Fish. 47, 761–764.

Furuita, H., Takeuchi, T., Toyota, M., Watanabe, T., 1996a. EPA and DHA requirements in early juvenile red sea bream using HUFA enriched *Artemia* Nauplii. Fish. Sci. 62, 246–251.

Furuita, H., Takeuchi, T., Watanabe, T., Fujimoto, H., Sekiya, S., Imaizumi, K., 1996b. Requirements of larval yellowtail for eicosapentaenoic, docosahexaenoic acid, and n-3 highly unsaturated fatty acid. Fish Sci. 62, 372–379.

Furuya, W.M.C., Hayashi, V.R.B., Furuya, D., Botaro, L.C., da Silva, Neves, P.R., 2001. Exigencias de metionina + cistina para alevinos revertidos de tilápia do Nilo (*Oreochromis niloticus*), baseadas no conceito de proteína ideal. Maringa 23, 885–889.

Furuya, W.M., Botaro, D., Neves, P.R., Silva, L.C.R., Hayashi, C., 2004. Lysine requirement of Nile tilapia, (*Oreochromis niloticus*), for grow-out phase. Cienc. Rural 34, 1571–1577.

Furuya, W.M., dos Santos, V.G., Silva, L.C.R., Furuya, V.R.B., Sakaguti, E.S., 2006. Digestible lysine requirements of Nile tilapia juveniles. Rev. Bras. Zootecnia 35, 937–942.

Furuya, W.M., Fujii, K.M., dos Santos, L.D., Castro Silva, T.S., Rosa da Silva, L.C., Pinseta Sales, P.J., 2008. Available phosphorus requirements of juvenile Nile tilapia. Braz. J. Anim. Sci. 37, 1517–1522.

Galindo, J., Fraga, I., Arazoza, M., Alvarez, J., Ramos, D., González, R., 2002. Requerimientos nutricionales de juveniles de camarón blanco (*Litopenaeus schmitti*): evaluación de dietas prácticas. CIVA, 84–94. <http://www.civa2002.org>.

Galindo, J., Fraga, I., Artiles, M., Arazoza, M., Alvarez, J., Pelegrin, E., 2003. Efecto de niveles de proteína en la dieta sobre el crecimiento de juveniles de camarón rosado (*Farfantepenaeus notialis*). CIVA, 575–586. <http://www.civa2003.org>.

Gallagher, M.L., Matthews, A.M., 1987. Oxygen consumption and ammonia excretion of the American eel *Anguilla rostrata* fed diets with varying protein energy ratios and proteins levels. J. World Aquacult. Soc. 18, 107–112.

García, T., Galindo, J., 1990. Requerimientos de proteína en las postlarvas de camarón blanco *Penaeus schmitti*. Rev. Inv. Mar. 11, 247–250.

García, T., Gaxiola, G., García, T., Pedroza, R., Soto, L., López, N., et al., 1998. Influencia de las proteínas dietéticas sobre el crecimiento, la sobrevivencia y el rendimiento de las postlarvas del camarón blanco (*Penaeus setiferus*) y del camarón rosado (*P. duorarum*) del Golfo de México. Aquatic 2 <http://aquatic.unizar.es>.

Garling Jr., D.L., Wilson, R.P., 1976. Optimum dietary protein-to-energy ratios for channel catfish fingerling, *Ictalurus punctatus*. J. Nutr. 106, 1368–1375.

Gatesoupe, F.J., Leger, C., Metailler, R., Luquet, P., 1977. Alimentation lipidique de turbot (*Scophthalmus maximus* L.) I. Influence de la longueur de chaine des acides gras de la serie ω-3. Ann. Hydrobiol. 8, 89–97.

Gatlin, D.M., Wilson, R.P., 1983. Dietary zinc requirement of fingerling channel catfish. J. Nutr. 113, 630–635.

Gatlin, D.M., Wilson, R.P., 1984a. Dietary selenium requirement of fingerling channel catfish. J. Nutr. 114, 627.

Gatlin, D.M., Wilson, R.P., 1984b. Studies on the manganese requirement of fingerling channel catfish. Aquaculture 41, 85–92.

Gatlin, D.M., Wilson, R.P., 1984c. Zinc supplementation of practical channel catfish diets. Aquaculture 41, 31–36.

Gatlin, D.M., Wilson, R.P., 1986a. Characterization of iron deficiency and the dietary iron requirement of fingerling channel catfish. Aquaculture 52, 191–198.

Gatlin, D.M., Wilson, R.P., 1986b. Dietary copper requirement of fingerling channel catfish. Aquaculture 54, 277–285.

Gatlin, D.M., Robinson, E.H., Poe, W.E., Wilson, R.P., 1982. Magnesium requirement of fingerling channel catfish and signs of magnesium deficiency. J. Nutr. 112, 1182.

Gatlin, D.M., Phillips, H.F., Torrans, E.L., 1989. Effects of various levels of dietary copper and zinc on channel catfish. Aquaculture 76, 127–134.

Gatlin, D.M., O'Connell, J.P., Scarpa, J., 1991. Dietary zinc requirement of the red drum, *Sciaenops ocellatus*. Aquaculture 92, 259–265.

Gatlin, D.M., Brown, M.L., Keembiyehetty, C.N., Jaramillo, F., Nematipour, G.R., 1994. Nutritional requirements of hybrid striped bass (*Morone chrysops* × *M. saxatilis*). Aquaculture 124, 127.

Gatlin, D.M., Barrows, F.T., Brown, P., Dabrowski, K., Gaylord, T.G., Hardy, R.W., et al., 2007. Expanding the utilization of sustainable plant products in aquafeeds: a review. Aquacult. Res. 38, 551–579.

Gaxiola, G., Cuzon, G., García, T., Taboada, G., Brito, R., Chimal, M.E., et al., 2005. Factorial effects of salinity, dietary carbohydrate and moult cycle on digestive carbohydrases and hexokinases in *Litopenaeus vannamei* (Boone, 1931). Comp. Biochem. Physiol. A Mol. Integr. Physiol. 140, 29–39.

Gaylord, T.G., Rawles, S.D., David, K.B., 2005. Dietary tryptophan requirements of hybrid striped bass (*Morone chrysops* × *M. saxatilis*). Aquacult. Nutr. 11, 367–374.

Gaylord, T.G., Barrows, F.T., Rawles, S.D., Liu, K., Bregitzer, P., Hang, A., et al., 2009. Apparent digestibility of nutrients and energy in extruded diets from cultivars of barley and wheat selected for nutritional quality in rainbow trout *Oncorhynchus mykiss*. Aquacult. Nutr. 15, 306–312.

Geurden, I., Coutteau, P., Sorgeloos, P., 1995a. Dietary phospholipids and body deformities in carp *Cyprinus carpio* L. larvae. Lavens, P. Jaspers, E. Roelants, I. (Eds.), Larvi '95 – Fish and Shellfish Symposium, Gent, Belgium. EAS Spec. Publ., vol. 24. European Aquaculture Society, pp. 162–165.

Geurden, I., Coutteau, P., Sorgeloos, P., 1995b. Dietary phospholipids for European sea bass (*Dicentrarchus labrax* L.) during first ongrowing. In: Lavens, P. Jaspers, E. Roelants, I. (Eds.), Larvi '95 – Fish and Shellfish Symposium, Ghent, Belgium. EAS Spec. Publ., vol. 24. European Aquaculture Society, pp. 175–178.

Geurden, I., Radünz-Neto, J., Bergot, P., 1995c. Essentiality of dietary phospholipids for carp (*Cyprinus carpio* L.) larvae. Aquaculture 131, 303–314.

Geurden, I., Charlon, N., Marion, D., Bergot, P., 1997a. Influence of purified soybean phospholipids on early development of common carp. Aquacult. Int. 5, 137–149.

Geurden, I., Coutteau, P., Sorgeloos, P., 1997b. Effect of a dietary phospholipid supplementation on growth and fatty acid composition of European sea bass (*Dicentrarchus labrax* L.) and turbot (*Scophthalmus maximus* L.) juveniles from weaning onwards. Fish Physiol. Biochem. 16, 259–272.

Ghioni, C., Tocher, D.R., Bell, M.V., Dick, J.R., Sargent, J.R., 1999. Low C18 to C20 fatty acid elongase activity and limited conversion of stearidonic acid, 18:4(n-3), to eicosapentaenoic acid, 20:5(n-3), in a cell line from the turbot, *Scophthalmus maximus*. Biochim. Biophys. Acta, Mol. Cell. Biol. Lipids 1437, 170–181.

Gisbert, E., Villeneuve, L., Zambonino-Infante, J.L., Quazuguel, P., Cahu, C.L., 2005. Dietary phospholipids are more efficient than neutral lipids for long-chain polyunsaturated fatty acid supply in European sea bass *Dicentrarchus labrax* larval development. Lipids 40, 609–618.

Glencross, B.D., Smith, D.M., 1999. The dietary linoleic and linolenic fatty acids requirements of the prawn *Penaeus monodon*. Aquacult. Nutr. 5, 53–63.

Glencross, B.D., Smith, D.M., 2001a. Optimizing the essential fatty acids, eicosapentaenoic and docosahexaenoic acid, in the diet of the prawn, *Penaeus monodon*. Aquacult. Nutr. 7, 101–112.

Glencross, B.D., Smith, D.M., 2001b. A study of the arachidonic acid requirements of the giant tiger prawn, *Penaues monodon*. Aquacult. Nutr. 7, 59–69.

Glencross, B., Smith, D., Thomas, M., Williams, K., 2002. The effect of dietary n-3 and n-6 fatty acid balance on the growth of the prawn *Penaeus monodon*. Aquacult. Nutr. 8, 43–51.

Glencross, B.D., Hien, T.T.T., Phuong, N.T., Cam Tu, T.L., 2011. A factorial approach to defining the energy and protein requirements of Tra catfish, *Pangasianodon hypothalamus*. Aquacult. Nutr. 17, e396–e405.

Goff, J.B., Gatlin III, D.M., 2004. Evaluation of different sulphur amino acid compounds in the diet of red drum, *Sciaenops ocellatus*, and sparing value of cystine for methionine. Aquaculture 241, 465–477.

Gong, H., Lawrnce, A.L., Jiang, D.-H., Castille, F.L., Gatlin III, D.M., 2000. Lipid nutrition of juvenile *Litopenaeus vannamei* I. Dietary cholesterol and de-oiled soy lecithin requirements and their interaction. Aquaculture 190, 305–324.

Gong, H., Lawrence, A.L., Gatlin III, D.M., Jiang, D.H., Zhang, F., 2001. Comparison of different types and levels of commercial soybean lecithin supplemented in semipurified diets for juvenile *Litopenaeus vannamei* Boone. Aquacult. Nutr. 7, 11–17.

Gong, H., Jiang, D., Lightner, D.V., Collins, C., Brock, D., 2004a. A dietary modification approach to improve the osmoregulatory capacity of *Litopenaeus vannamei* cultured in the Arizona desert. Aquacult. Nutr. 10, 227–236.

Gong, H., Jiang, D., Lawrence, A.L., González-Félix, M.L., Perez-Velazquez, M., 2004b. Nuevos avances en el estudio de fosfolípidos nutrimentales para camarón. En Cruz Suárez, L.E., Rique Marie, D., Nieto Lopez, M.G., Villareal, D. Scholz, U. and González, M. Avances en Nutrición Acuícola VII. Simposium Internacional de Nutrición Acuícola. 16–19 Noviembre, Hermosillo, Sonora, México, 329–343.

Gong, W., Lei, W., Zhu, X., Yang, Y., Han, D., Xie, S., 2014. Dietary myo-inositol requirement for juvenile gibel carp (*Carassius auratus gibelio*). Aquacult. Nutr. 20, 514–519.

González Félix, M.L., Gatlin III, D.M., Lawrence, A.L., Pérez Velázquez, M., 2002a. Effect of various dietary lipid levels on quantitative essential fatty acid requirements of juvenile Pacific white shrimp *Litopenaeus vannamei*. J. World Aquacult. Soc. 33, 330–340.

González Félix, M.L., Lawrence, A.L., Gatlin III, D.M., Pérez Velázquez, M., 2002b. Growth survival and fatty acid composition of juvenile *Litopenaeus vannamei* fed different oils in the presence and absence of phospholipids. Aquaculture 205, 325–343.

González Félix, M.L., Gatlin III, D.M., Lawrence, A.L., Pérez Velázquez, M., 2003. Nutritional evaluation of fatty acids for the open thelycum shrimp, *Litopenaeus vannamei*: II. Effect of dietary n-3 and n-6 polyunsaturated and highly unsaturated fatty acids on juvenile shrimp growth, survival and fatty acid. Aquacult. Nutr. 9, 115–122.

Gouillou-Coustans, M.F., Bergot, P., Kaushik, S.J., 1998. Dietary ascorbic acid needs of common carp *Cyprinus carpió* larvae. Aquaculture 161, 453–461.

Grahl-Madsen, E., Lie, Ø., 1997. Effects of different levels of vitamin K in diets for cod (*Gadus morhua*). Aquaculture 151, 269–274.

Grant, B.F., Seib, P.A., Liao, M.L., Corpron, J.A., 1989. Polyphos-phorylated L-ascorbic acid: a stable form of vitamin C for aquaculture feeds. J. World Aquacult. Soc. 20, 143–157.

Green, C.J., 2001. Fiber in enteral nutrition. Clin. Nutr. 20, 23–39.

Griffin, M.E., Brown, P.B., Grant, A.L., 1992. The dietary lysine requirement of juvenile hybrid striped bass. J. Nutr. 122, 1332–1337.

Griffin, M.E., Wilson, K.A., White, M.R., Brown, P.B., 1994. Dietary choline requirement of juvenile hybrid striped bass. J. Nutr. 124, 1685–1689.

Grisdale-Helland, B., Helland, S.J., 1997. Replacement of protein by fat and carbohydrate in diets for Atlantic salmon (*Salmo salar*) at the end of the freshwater stage. Aquaculture 152, 167–180.

Grisdale-Helland, B., Helland, S.J.,1998. Macronutrient utilization by Atlantic halibut (*Hippoglossus hippoglossus*): diet digestibility and growth of 1 kg fish. Aquaculture 166, 57–65.

Grisdale-Helland, B., Gatlin, D.M., Corrent, E., Helland, S.J., 2011. The minimum dietary lysine requirement, maintenance requirement and efficiency of lysine utilization for growth of Atlantic salmon smolts. Aquacult. Res. 42, 1509–1529.

Gu, M., Zhang, W.B., Bai, N., Mai, K.S., Xu, W., 2013. Effects of dietary crystalline methionine or oligo-methionine on growth performance and feed utilization of white shrimp (*Litopenaeus vannamei*) fed plant protein-enriched diets. Aquacult. Nutr. 19, 39–46.

Guary, M, Kanazawa, A., Tanaka, N., Ceccaldi, H.J., 1976. Nutritional requirements of prawn: VI. Requirement for ascorbic acid. Mem. Fac. Fish. Kagoshima Univ. 25, 53–57.

Guillaume, J., Ceccaldi, H.J., 2001. Digestive physiology of shrimps. In: Guillaume, J., Kaushik, S., Bergot, P., Métailler, R. (Eds.), Nutrition and Feeding of Fish and Crustaceans, INRA Editions. Springer-Praxis, Chichester, UK, pp. 239–263.

Guinea, J., Fernandez, F., 1997. Effect of feeding frequency, feeding level and temperature on energy metabolism in *Sparus aurata*. Aquaculture 148, 125–142.

Gundogdu, A., Harmantepe, F.B., Dogan, G., Karsli, Z., Asci, M.Y., 2009. Effects of diet-borne copper on accumulation in the tissues and organs, growth and feed utilization of rainbow trout (*Oncorhynchus mykiss*, Walbaum, 1792) Juvenile. J. Anim. Vet. Adv. 8, 2495–2502.

Gurure, R.M., Moccia, R.D., Atkinson, J.L., 1995. Optimal protein requirements of young Arctic charr (*Salvelinus alpinus*) fed practical diets. Aquacult. Nutr. 1, 227–234.

Hajra, A., Ghosh, A., Mandal, S.K., 1988. Biochemical studies on the determination of optimum dietary protein to energy ratio for Tiger prawn, *Penaeus monodon* (Fab.), juveniles. Aquaculture 71, 71–79.

Halver, J.E., 1972. The vitamins. In: Halver, J.E. (Ed.), Fish Nutrition. Academic Press, New York, NY, pp. 29–103.

Halver, J.E., Ashley, L.M., Smith, R.R., 1969. Ascorbic acid requirements of coho salmon and rainbow trout. Trans. Am. Fish. Soc. 90, 762–771.

Hamada, A., Maeda, W., 1983. Oxygen uptake due to specific dynamic action of the carp, *Cyprinus carpio*. Jpn. J. Limnol. 44, 225–239.

Hamre, K., 2011. Metabolism, interactions, requirements and functions of vitamin E in fish. Aquacult. Nutr. 17, 98–115.

Hamre, K., Lie, Ø., 1995. a-Tocopherol levels in different organs of Atlantic salmon (*Salmo salar* L. Effects of smoltification. Dietary levels of n-3 polyunsaturated fatty acid and vitamin E. Comp. Biochem. Physiol. A IIIA, 547–554.

Hamza, N., Mhetli, M., Khemis, I.B., Cahu, C., Kestemont, P., 2008. Effect of dietary phospholipids levels on performance, enzyme activities and fatty acid composition of pikeperch (*Sander lucioperca*) larvae. Aquaculture 275, 274–282.

Hamza, N., Kestemont, P., Khemis, I.B., Mhetli, M., Cahu, C., 2012. Effect of different sources and levels of dietary phospholipids on performances and fatty acid composition of pikeperch (*Sander lucioperca*) larvae. Aquacult. Nutr. 18, 249–257.

Han, D., Xie, S., Liu, M., Xiao, X., Liu, H., Zhu, X., et al., 2011. The effects of dietary selenium on growth performances, oxidative stress and tissue selenium concentration of gibel carp (*Carassius auratus* gibelio). Aquacult. Nutr. 74, 1–749.

Han, D., Liu, H., Liu, M., Xiao, X., Zhu, X., Yang, Y., et al., 2012. Effect of dietary magnesium supplementation on the growth performance of juvenile gibel carp, *Carassius auratus* gibelio. Aquacult. Nutr. 18, 512–520.

Harán, N.S., Fenucci, J.L., 1996. Efectos del colesterol en la dieta del langostino argentino *Pleoticus muelleri* Bate. Rev. Cubana Invest. Pesqueras 20, 40–43.

Harding, D.E., Allen Jr., O.W., Wilson, R.P., 1977. Sulfur amino acid requirement of channel catfish: L-methionine and L-cystine. J. Nutr. 107, 2031–2035.

Hardy, R., Oram, L., Möller, G., 2010. Effects of dietary selenomethionine on cutthroat trout (*Oncorhynchus clarki* bouvieri) growth and reproductive performance over a life cycle. Arch. Environ. Contam. Toxicol. 58, 237–245.

Hari, B., Kurup, B.M., 2002. Vitamin C (ascorbyl 2 polyphosphate) requirement of freshwater prawn *Macrobrachium rosenbergii* (de Man). Asian Fish. Sci. 15, 145–154.

Harrison, K.E., 1990. The role of nutrition in maturation, reproduction and embryonic development of decapod crustaceans: a review. J. Shellfish Res. 9, 1–28.

He, H., Lawrence, A.L., 1991. Estimation of dietary pyridoxine requirement for the shrimp. *Penaeus vannamei* (Abstr.). In: Paper Presented at the 22nd Annual Conference, World Aquaculture Society, June 16–20, 1991, San Juan. Puerto Rico.

He, H., Lawrence, A.L., 1993a. Vitamin C requirements of the shrimp *Penaeus vannamei*. Aquaculture 114, 305–316.

He, H., Lawrence, A.L., 1993b. Vitamin E requirements of *Penaeus vannamei*. Aquaculture 118, 245–255.

He, H., Lawrence, A.L., Liu, R., 1992. Evaluation of dietary essentiality of fat-soluble vitamins. A. D, E and K. for penaeid shrimp (*Penaeus vannamei*). Aquaculture 103, 177–185.

He, W., Zhou, X.Q., Feng, L., Jiang, J., Liu, Y., 2009. Dietary pyridoxine requirement of juvenile Jian carp (*Cyprinus carpio* var. Jian). Aquacult. Nutr. 15, 402–408.

He, J.-Y., Tian, L.-X., Lemme, A., Gao, W., Yang, H.-J., Niu, J., et al., 2013. Methionine and lysine requirements for maintenance and efficiency of utilization for growth of two sizes of tilapia (*Oreochromis niloticus*). Aquacult. Nutr. 19, 629–640.

He, J., Han, B., Tian, L., Yang, H., Zeng, S., Liu, Y., 2014. The sparing effect of cystine on methionine at a constant TSAA level in practical diets of juvenile Nile tilapia *Oreochromis niloticus*. Aquacult. Res. n/a–n/a http://dx.doi.org/10.1111/are.12657.

Hemre, G., Lie, O., Lied, E., Lambertsen, G., 1989. Starch as an energy source in fed for cod (*Gadus morhua*): digestibility and retention. Aquaculture 80, 261–270.

Hemre, G.-I., Mommsen, T.P., Krogdahl, Å., 2002. Carbohydrates in fish nutrition: effects on growth, glucose metabolism and hepatic enzymes. Aquacult. Nutr. 8, 175–194.

Hemre, G.I., Deng, D.F., Wilson, R.P., Berntssen, M.H.G., 2004. Vitamin A metabolism and early biology responses in juvenile sunshine bass (*Morone chrysops* × *M. saxatilis*) fed graded levels of vitamin A. Aquaculture 235, 645–658.

Henrique Gomes Cornélio, F., da Cunha, D.A., Silveira, J., Alexandre, D., Silva, C.P., Fracalossi, D.M., 2014. Dietary protein requirement of Juvenile Cachara Catfish, *Pseudoplatystoma reticulatum*. J. World Aquacult. Soc. 45, 45–54.

Hernandez, L.H.H., Teshima, S.I., Ishikawa, M., Alam, S., Koshio, S., Tanaka, Y., 2005. Dietary vitamin A requirements of juvenile Japanese flounder *Paralichthys olivaceus*. Aquacult. Nutr. 11, 3–9.

Hew, M., Cuzon, G., 1982. Effect of dietary lysine and arginine levels, and their ratio, on the growth of *Penaeus japonicas* juveniles. J. World Maricult. Soc. 13, 154.

Hewitt, D.R., Irving, M.G., 1990. Oxygen consumption and ammonia excretion of the brown tiger pawn *Penaeus esculentus* fed diets of varying protein content. Comp. Biochem. Physiol. 96A, 373–378.

Hidalgo, M.C., Urea, E., Sanz, A., 1999. Comparative study of digestive enzymes in fish with different nutritional habits. Proteolytic and amylase activities. Aquaculture 170, 267–283.

Hidalgo, L.E., Taboada, G., Rosas, C., Gaxiola, G., Mascaró, M., Jones, D., et al., 2000. Requerimiento proteico relativo de juveniles de camarón rojo del Caribe, *Farfantepenaeus brasiliensis* (La Treille 1817), en un sistema cerrado de recirculación. En: Cruz-Suárez, L.E., Ricque-Marie, D., Tapia-Salazar, D., Olvera y, M., Civera, R., (Eds.), Avances en Nutrición Acuícola V. Memorias del V Simposium Internacional de Nutrición Acuícola, November 19–22, 2000, Mérida, México.

Hilton, J.W., Cho, C.Y., Slinger, S.J., 1978. Effect of graded levels of supplemental ascorbic acid in practical diets fed to rainbow trout (*Salmo gairdneri*). J. Fish. Res. Board Can. 35, 431–436.

Hilton, J.W., Hodson, P.V., Slinger, S.J., 1980. The requirement and toxicity of selenium in rainbow trout (*Salmo gairdneri*). J. Nutr. 110, 2527–2535.

Hilton, J.W., Harrison, K.E., Slinger, S.J., 1984. A semi-purified test diet for *Macrobrachium rosembergii* and the lack of need for supplemental lecithin. Aquaculture 37, 209–215.

Holsapple, D.R., 1990. The effect of dietary sodium chloride on red drum (*Sciaenops oceliatus*) in fresh and brackish water (Master's thesis). Texas A and M University, College Station, TX.

Hossain, M.A., Furuichi, M., 1998. Availability of environmental and dietary calcium in tiger puffer. Aquacult. Int. 6, 121–132.

Hossain, M.A., Furuichi, M., 1999. Necessity of dietary calcium supplement in black sea bream. Fish. Sci. 65, 893–897.

Hossain, M.A., Furuichi, M., 2000a. Essentiality of dietary calcium supplement in fingerling scorpion fish (*Sebastiscus marmoratus*). Aquaculture 189, 155–163.

Hossain, M.A., Furuichi, M., 2000b. Essentiality of dietary calcium supplement in redlip mullet *Liza haematocheila*. Aquacult. Nutr. 6, 33–38.

Hossain, M.A., Furuichi, M., 2000c. Necessity of calcium supplement to the diet of Japanese flounder. Fish. Sci. 66, 660–664.

Hossain, M.A., Yoshimatsu, T., 2014. Dietary calcium requirement in fishes. Aquacult. Nutr. 20, 1–11.

Houlihan, D.F., Waring, C.P., Mathers, E., Gray, C., 1990. Protein synthesis and oxygen consumption of the shore crab *Carcinus maenas* after a meal. Physiol. Zool. 63, 735–756.

Hsu, T.S., Shiau, S.Y., 1997. Comparison of L-ascorbyl-2-polyphosphate with L-ascorbyl-2-sulfate in meeting vitamin C requirements of juvenile grass shrimp. *Penaeus monodon*. Fish. Sci. 63, 958–962.

Hsu, T.S., Shiau, S.Y., 1998. Comparison of vitamin C requirement for maximum growth of grass shrimp. *Penaeus monodon* using L-ascorbyl-2-monophosphate-Na and L-ascorbyl-2-monophosphate-Mg. Aquaculture 163, 203–213.

Hu, C.J., Chen, S.M., Pan, C.H., Huang, C.H., 2006. Effects of dietary vitamin A or ^-carotene concentrations on growth of juvenile hybrid tilapia. *Oreochromis niloticus* × *O. aureus*. Aquaculture 253, 602–607.

Hua, K., 2013a. Estimating maintenance amino acids requirements of fish through a nonlinear mixed modelling approach. Aquacult. Res. 44, 542–553.

Hua, K., 2013b. Investigating the appropriate mode of expressing lysine requirement of fish through non-linear mixed model analysis and multilevel analysis. Br. J. Nutr. 109, 1013–1021.

Huai, M.-Y., Tian, L.-X., Liu, Y.-J., Xu, A.-L., Liang, G.-Y., Yang, H.-J., 2009. Quantitative dietary threonine requirement of juvenile Pacific white shrimp, *Litopenaeus vannamei* (Boone) reared in low-salinity water. Aquacult. Res. 40, 904–914.

Huang, F., Jiang, M., Wen, H., Wu, F., Liu, W., Tian, J., et al., 2015. Dietary zinc requirement of adult Nile tilapia (*Oreochromis niloticus*) fed semi-purified diets, and effects on tissue mineral composition and antioxidant responses. Aquaculture 439, 53–59.

Hughes, S.G., Rumsey, G.L., Nichum, J.G., 1981. Riboflavin requirement of fingerling rainbow trout. Prog. Fish Cult. 43, 167–172.

Hung, S.S.O., 1991. Carbohydrate utilization by white sturgeon as assessed by oral administration tests. J. Nutr. 121, 1600–1605.

Hung, S.S., Lutes, P.B., 1988. A preliminary study on the non-essentiality of lecithin for hatchery-produced juvenile white sturgeon (*Acipenser transmontanus*). Aquaculture 68, 353–360.

Hung, S.S.O., Cho, C.Y., Slinger, S.J., 1980. Measurement of oxidation in fish oil and its effect on vitamin E nutrition of rainbow trout (*Salmo gairdneri*). Can. J. Fish. Aquat. Sci. 37, 1248–1253.

Hunter, B., Magarelli, P.C., Lightner, D.V., Colvin, L.B., 1979. Ascorbic acid-dependent collagen formation in penaeid shrimp. Comp. Biochem. Physiol. 64B, 381–385.

Ibeas, C., Izquierdo, M.S., Lorenzo, A., 1994. Effect of different levels of n-3 highly unsaturated fatty acids on growth and fatty acid composition of juvenile gilthead seabream (*Sparus aurata*). Aquaculture 127, 177–188.

Ibeas, C., Cejas, J.R., Fores, R., Badía, P., Gómez, T., Lorenzo, A., 1997. Influence of eicosapentaenoic acid to docosahexaenoic acid ratio (EPA/DHA) of dietary lipids on growth and fatty acid composition of gilthead seabrem (*Sparus auratta*) juveniles. Aquaculture 150, 91–102.

Ibiyo, L.M.O., Atteh, J.O., Omotosho, J.S., Madu, C.T., 2007. Vitamin C (ascorbic acid) requirements of *Heterobranchus longifilis* fingerlings. Afr. J. Biotechnol. 6, 1559–1567.

Ishac, M., Dollar, A., 1968. Studies on manganese uptake in *Tilapia mossambica* and *Salmo gairdnerii*. Hydrobiologia 31, 572–584.

Ishibashi, Y., Ikeda, S., Murata, O., Nasu, T., Harada, T., 1992. Optimal supplementary ascorbic acid level in the Japanese parrot fish diet. Nippon Suisan Gakkaishi 58, 267–270.

Jaramillo Jr., F., Peng, L.I., Gatlin III, D.M., 2009. Selenium nutrition of hybrid striped bass (*Morone chrysops* × *M. saxatilis*) bioavailability, toxicity and interaction with vitamin E. Aquacult. Nutr. 15, 160–165.

Jauncey, K., 1982. The effects of varying dietary protein level on the growth, food conversion, protein utilization and body composition of juvenile tilapias (*Sarotherodon mossambicus*). Aquaculture 27, 43–54.

Jiang, M., Wang, W., Wen, H., Wu, F., Zhao, Z., Liu, A., et al., 2007. Effects of dietary vitamin K3 on growth, carcass composition and blood coagulation time for grass carp fingerling (*Ctenopharyngodon idellus*) (in Chinese). Freshw. Fish. 37, 61–64.

Jiang, M., Huang, F., Zhao, Z., Wen, H., Wu, F., Liu, W., et al., 2014a. Dietary thiamin requirement of Juvenile Grass Carp, *Ctenopharyngodon idella*. J. World Aquacult. Soc. 45, 461–468.

Jiang, M., Huang, F., Wen, H., Yang, C., Wu, F., Liu, W., et al., 2014b. Dietary niacin requirement of GIFT Tilapia, *Oreochromis niloticus*, reared in freshwater. J. World Aquacult. Soc. 45, 333–341.

Jobling, M., 1982. A study of some factors affecting rates of oxygen uptake of plaice, *Pleuronectes platessa* L. J. Fish Biol. 20, 501–516.

Jobling M., 1994. Fish bioenergetics. Fish Bioenergetics. Fish and Fisheries Series 13. London.

Jobling, M., Davies, P.S., 1980. Effects of feeding on metabolic rate, and the specific dynamic action in plaice, *Pleuronectes platessa* L. J. Fish Biol. 16, 629–638.

Johansson, M.W., Söderhäll, K., 1989. Cellular immunity in crustaceans the pro-PO system. Parasitol. Today 5, 171–176.

Kai, H., Kanazawa, A., 1989. Optimum contents of cholesterol in the purified diet for grass prawn *Penaeus monodon*. Abst. 2nd Asian Fish. Forum, Asian Fish. Soc., Japan. 68.

Kalogeropoulos, N., Alexis, M.N., Henderson, R.J., 1992. Effects of dietary soyabean and cod-liver oil levels on growth and body composition of gilthead bream (*Sparus aurata*). Aquaculture 104, 293–308.

Kanazawa, A., 1983. Penaeid nutrition. In: Pruder, G.D., Conklin, D.E., Langdon, C. (Eds.), Proceedings of the Second International Conference on Aquaculture Nutrition: Biochemical and Physiological Approaches to Shellfish Nutrition. Louisiana State University, Division of Continuing Education, Baton Rouge, LA.

Kanazawa, A., 1985. Essential fatty acid and lipid requirement of fish. In: Cowey, C.B., Mackie, A.M., Bell, J.G. (Eds.), Nutrition and Feeding of Fish. Academic Press, London, pp. 281–298.

Kanazawa, A., 1993. Essential phospholipids of fish and crustaceans. In: Kaushik, S.J., Luquet, P. (Eds.), Fish Nutrition in Practice. IV International Symposium on Fish Nutrition and Feeding, INRA, France National Institute for Agricultural Research, pp. 519–530.

Kanazawa, A., 1997. Effects of docosahexaenoic acid and phospholipids on stress tolerance of fish. Aquaculture 155, 129–134.

Kanazawa, A., 2001. Sterols in marine invertebrates. Fish. Sci. 67, 997–1007.

Kanazawa, A., Teshima, S., 1979. Biosynthesis of fatty acids from acetate in the prawn, *Penaeus japonicas*. Mem. Fac. Fish. Kagoshima Univ. 28, 21–26.

Kanazawa, A., Tanaka, N., Teshima, S., Kashiwada, K., 1971a. Nutritional requirements of prawn-II: Requirements for sterols. Bull. Jpn. Soc. Sci. Fish. 37, 211–215.

Kanazawa, A., Tanaka, N., Teshima, S., Kashiwada, K., 1971b. Nutritional requirements of prawn-III: Utilization of dietary sterols. Nippon Suisan Gakkaishi 37, 1015–1019.

Kanazawa, A., Teshima, S., Tanada, N., 1976. Nutritional requirements for choline and inositol. Mem. Fac. Fish. Kagoshima Univ. 28, 27–33.

Kanazawa, A., Tokiwa, S., Kayama, M., Hirata, M., 1977. Essential fatty acids in the diet of prawn. I. Effects of linoleic and linolenic acids on growth. Bull. Jpn. Soc. Sci. Fish. 43, 1111–1114.

Kanazawa, A., Teshima, S., Endo, M., Kayama, M., 1978. Effects of eicosapentaenoic acid on growth and fatty acid composition of the prawn. Mem. Fac. Fish. Kagoshima Univ. 27, 35–40.

Kanazawa, A., Teshima, S., Tokiwa, M., Kayama, Hirata, M., 1979a. Essential fatty acids in the diet of prawn-II. Effect of docohexaenoic acid on growth. Bull. Jpn. Soc. Sci. Fish. 45, 1151–1153.

Kanazawa, A.S., Teshima, S., Tokiwa, M., Endo, M., Abdel Razek, F.A., 1979b. Effects of short-necked clam phospholipids on the growth of the prawn. Bull. Jpn. Soc. Sci. Fish. 45, 961–965.

Kanazawa, A., Teshima, S., Sakamoto, M., Awal, M.A., 1980. Requirements of *Tilapia zillii* for essential fatty acids. Bull. Jpn. Soc. Sci. Fish. 46, 1353–1356.

Kanazawa, A.S., Teshima, S., Inamori, S., Iwashita, T., Nagao, A., 1981. Effects of phospholipids on growth, survival rate, and incidence of malformation in the larval ayu. Mem. Fac. Fish. Kagoshima Univ. 30, 301–309.

Kanazawa, A., Teshima, S., Sakamoto, M., 1982. Requirements of essential fatty acids for the larval ayu. Bull. Jpn. Soc. Sci. Fish. 48, 586–590.

Kanazawa, A., Teshima, S., Kobayashi, T., Takae, M., Iwashita, T., Uehara, R., 1983a. Necessity of phospholipids for growth of the larval ayu. Mem. Fac. Fish. Kagoshima Univ. 32, 115–120.

Kanazawa, A., Teshima, S., Inamori, S., Matsubara, H., 1983b. Effects of dietary phospholipids on growth of the larval red sea bream and knife jaw. Mem. Fac. Fish. Kagoshima Univ. 32, 109–114.

Kanazawa, A., Teshima, S.-I., Sasaki, M., 1984. Requirements of juvenile prawn for calcium, phosphorus, magnesium. potassium, copper, manganese, and iron. Mem. Fac Fish. Kagoshima Univ. 33, 63–71.

Kanazawa, A., Teshima, S., Sakamoto, M., 1985. Effects of dietary lipids, fatty acids, and phospholipids on growth and survival of prawn (*Penaeus japonicus*) larvae. Aquaculture 50, 39–49.

Kanazawa, A., Teshima, S.I., Koshio, S., Higashi, M., Itoh, S., 1992. Effect of L-ascorbyl-2-phosphate-Mg on the yellowtail. *Seriola quia-queradiata* as a vitamin C source. Nippon Suisan Gakkaishi 58, 337–341.

Kashiwada, K., Teshima, S., Kanazawa, A., 1970. Studies on the production of B vitamins by intestinal bacteria of fish. 5. Evidence of the production of vitamin B_p by microorganisms in the intestinal canal of carp. *Cyprinus carpio*. Bull. Jpn. Soc. Sci. Fish. 36, 421–424.

Kasper, C.S., Brown, P.B., 2003. Growth improved in juvenile Nile tilapia fed phosphatidyl-choline. N. Am. J. Aquacult. 65, 39–43.

Kaushik, S.J., 2001. Carbohydrate nutrition: importance and limits of carbohydrate supplies. In: Guillaume, J., Kaushik, S., Bergot, P., Métailler, R. (Eds.), Nutrition and Feeding of Fish and Crustaceans, INRA Editions. Springer-Praxis, Chichester, UK, pp. 131–144.

Kaushik, S.J., Dabrowski, K., 1983. Postprandial metabolic change inlarval and juvenile carp (*Cyprinus carpio*). Reprod. Nutr. Dev. 23, 223–234.

Kaushik, S.J., Fauconneau, B., 1984. Effects of lysine administration on plasma arginine and on some nitrogenous catabolites in rainbow trout. Comp. Biochem. Physiol. A 79, 459–462.

Kean, J.C., Castell, J.D., Boghen, A.G., D'Abramo, L.R., Conklin, D.E., 1985. A re-evaluation of the lecithin and cholesterol requirements of juvenile lobster (*Homarus americanus*) using crab protein-based diets. Aquaculture 47, 143–149.

Keembiyehetty, C.N., Gatlin, D.M., 1992. Dietary lysine requirement of juvenile hybrid striped bass (*Morone chrysops* × *M. saxatilis*). Aquaculture 104, 271–277.

Keembiyehetty, C.N., Gatlin, D.M., 1993. Total sulfur amino acid requirement of juvenile hybrid striped bass (*Morone chrysops* × *M. saxatilis*). Aquaculture 110, 331–339.

Keembiyehetty, C.N., Gatlin III, D.M., 1997. Dietary threonine requirement of juvenile hybrid striped bass (*Morone chrysops* × *M. saxatilis*). Aquacult. Nutr. 3, 217–221.

Keshavanath, P., Gangadhara, B., Khadri, S., 2003. Growth enhancement of carp and prawn through dietary sodium chloride supplementation. Aquacult. Asia 8, 4–8.

Ketola, H., 1975. Requirement of Atlantic salmon for dietary phosphorus. Trans. Am. Fish. Soc. 104, 548–551.

Ketola, H.G., 1976. Choline metabolism and nutritional requirement of lake trout (*Salvelinus namaycush*). J. Anim. Sci. 43, 474–477.

Ketola, H.G., Richmond, M.E., 1994. Requirement of rainbow trout for dietary phosphorus and its relationship to the amount discharged in hatchery effluents. Trans. Am. Fish. Soc. 123, 587–594.

Khajarern, J., Khajarem, S., 1997. Stability and bioavailability of vitamin C-glucose in *Clarias* hybrid catfish (*Clarias gariepinus* × *Clarias macrocephalus*). Aquaculture 151, 219–224.

Khan, M.A., Abidi, S.F., 2007a. Dietary isoleucine requirement of fingerling Indian major carp, *Labeo whita* (Hamilton). Aquacult. Nutr. 13, 424–430.

Khan, M.A., Abidi, S.F., 2007b. Total aromatic amino acid requirement of Indian major carp *Labeo rohita* (Hamilton) fry. Aquaculture 267, 111–118.

Khosravi, S., Lim, S.-J., Rahimnejad, S., Kim, S.-S., Lee, B.-J., Kim, K.-W., et al., 2015. Dietary myo-inositol requirement of parrot fish, *Oplegnathus fasciatus*. Aquaculture 436, 1–7.

Kim, K.-I., 1993. Requirement for phenylalanine and replacement value of tyrosine for pheny-lalanine in rainbow trout (*Oncorhynchus mykiss*). Aquaculture 113, 243–250.

Kim, S.G., Kang, J.C., 2004. Effect of dietary copper exposure on accumulation, growth and hematological parameters of the juvenile rockfish, *Sebastes schlegeli*. Mar. Environ. Res. 58, 65–82.

Kim II, K., Kayes, T.B., Amundson, C.H., 1991. Purified diet development and re-evaluation of the dietary protein requirement of fingerling rainbow trout (*Oncorhynchus mykiss*). Aquaculture 96, 57–67.

Kim, K.-I., Kayes, T.B., Amundson, C.H., 1992a. Requirements for sulfur amino acids and utili-zation of D-methionine by rainbow trout (*Oncorhynchus mykiss*). Aquaculture 101, 95–103.

Kim, K.-I., Kayes, T.B., Amundson, C.H., 1992b. Requirements for lysine and arginine by rainbow trout (*Oncorhynchus mykiss*). Aquaculture 106, 333–344.

Kim, K.I., Grimshaw, T.W., Kayes, T.B., Amudson, C.H., 1992c. Effect of fasting or feeing diets containing different levels of protein or amino acids on the activities of the liver amino acid and degrading enzyme and amino acid oxidation in rainbow trout (*Oncorhynchus mykiss*). Aquaculture 107, 89–105.

Kitamura, S., Suwa, T., Ohara, S., Nakagawa, K., 1967a. Studies on vitamin requirements of rainbow trout. 3. Requirement for vitamin A and deficiency symptoms. Bull. Jpn. Soc. Sci. Fish. 33, 1126–1131.

Kjær, M.A., Vegusdal, A., Gjøen, T., Rustan, A.C., Todorčević, M., Ruyter, B., 2008. Effect of rapeseed oil and dietary n-3 fatty acids on triacylglycerol synthesis and secretion in Atlantic salmon hepatocytes. Biochim. Biophys. Acta, Mol. Cell. Biol. Lipids 1781, 112–122.

Kjørsvik, E., Olsen, C., Wold, P.-A., Hoehne-Reitan, K., Cahu, C.L., Rainuzzo, J., et al., 2009. Comparison of dietary phospholipids and neutral lipids on skeletal development and fatty acid composition in Atlantic cod (Gadus morhua). Aquaculture 294, 246–255.

Klein, R.G., Halver, J.E., 1970. Nutrition of salmonoid fishes: arginine and histidine requirements of chinook and coho salmon. J. Nutr. 100, 1105–1109.

Knox, D., Cowey, C., Adron, J., 1981. The effect of low dietary manganese intake on rainbow trout (*Salmo gairdneri*). Br. J. Nutr. 46, 495–501.

Kocabas, A.M., Gatlin III, D.M., 1999. Dietary vitamin E requirement of hybrid striped bass (*Morone chrysops* female × *M. saxatilis* male). Aquacult. Nutr. 5, 3–7.

Köprücü, K., 2012. Effects of dietary protein and lipid levels on growth, feed utilization and body composition of juvenile grass carp (*Ctenopharyngodon idella*). J. Fish. Sci. Com 6, 243–251.

Koshio, S., Castell, J.D., O'Dor, R.K., 1992. The effect of different dietary energy levels in crab-protein-based diets on digestibility, oxygen consumption, and ammonia excretion of bilaterally eyestalkablated and intact juvenile lobsters, *Homarus americanus*. Aquaculture 108, 285–297.

Kousoulaki, K., Fjelldal, P.G., Aksnes, A., Albrektsen, S., 2010. Growth and tissue mineralisation of Atlantic cod (*Gadus morhua*) fed soluble P and Ca salts in the diet. Aquaculture 309, 181–192.

Krossøy, C., Waggbø, R., Fjelldal, P.G., Wargelius, A., Lock, E.J., Graff, I.E., et al., 2009. Dietary menadione nicotinamide bisulphite (vitamin K_3) does not affect growth or bone health in first-feeding fry of Adantic salmon (*Salmo salar* L.). Aquacult. Nutr. 15, 638–649.

Krossøy, C.R., Waggbø, P.G., Ørnsrud, R., 2011. Vitamin K in fish nutrition. Aquacult. Nutr. 17, 585–594.

Küçükbay, F.Z., Yazlak, H., Karaca, I., Sahin, N., Tuzcu, M., Cakmak, M.N., et al., 2009. The effects of dietary organic or inorganic selenium in rainbow trout (*Oncorhynchus mykiss*) under crowding conditions. Aquacult. Nutr. 15, 569–576.

Kureshy, N., Davis, D.A., 2002. Protein requirement for maintenance and maximum weight gain for the Pacific white shrimp *Litopenaeus vannamei*. Aquaculture 204, 125–143.

Lall, S.P., 2002. The minerals. In: Halver, J.E., Hardy, R.W. (Eds.), Fish Nutrition, third ed. Academic Press, San Diego, CA, pp. 259–308.

Lall, S.P., Bishop, F.J., 1977. Studies on mineral and protein utilization by Atlantic salmon (*Salmo salar*) grown in seawater. Fisheries and Marine Service. Environment Canada. Technical report No. 688: 21 pp.

Lall, S.P., Olivier, G., 1993. Role of micronutrients in immune response and disease resistance in fish. In: INRA ŽEd., Fish Nutrition in Practice (Les Colloques, No. 61), pp. 101–118.

Lall, S.P., Weerakoon, D.E.M., 1990. Vitamin B_6 requirement of Atlantic salmon (*Salmo salar*) (abstract). FASEB J. 4, 3749.

Lall, S.P., Olivier, G., Hines, J.A., Ferguson, H.W., 1988. The role of vitamin E in nutrition and immune response of Atlantic salmon (*Salmo salar*). Bull. Aquacult. Assoc. Can. 88-2, 76.

Lall, S.P., Olivier, G., Weerakoon, D.E.M., Hines, J.A., 1991. The effect of vitamin C deficiency and excess on immune response of Atlantic salmon (Salmo salar L.). In: Takeda, M., Watanabe, T. (Eds.), The Current Status of Fish Nutrition in Aquaculture. Proceedings of the Third International Symposium on Feeding and Nutrition in Fish, 28 August–1 September 1989, Toba, Japan, pp. 421–441.

Lall, S.P., Kaushik, S.J., Le Bail, P.Y., Keith, R., Anderson, J.S., Plisetskaya, E., 1994. Quantitative arginine requirement of Atlantic salmon (*Salmo salar*) reared in sea water. Aquaculture 124, 13–25.

Lanno, R.P., Slinger, S.J., Hilton, J.W., 1985. Maximum tolerable and toxicity levels of dietary copper in rainbow trout *Salmo gairdneri* Richardson. Aquaculture 49, 257–268.

Larumbe-Moran, E., Hernandez-Vergara, M.P., Olvera-Novoa, M.A., Rostro, C.I.P., 2010. Protein requirements of Nile tilapia (*Oreochromis niloticus*) fry cultured at different salinities. Aquacult. Res. 41, 1150–1157.

Lavens, P., Merchie, G., Ramos, X., Kujan, A.L.-H., Van Hauwaert, A., Pedrazzoli, A., et al., 1999. Supplementation of ascorbic acid 2-monophosphate during the early postlarval stages of the shrimp *Penaeus vannamei*. Aquacult. Nutr. 5, 205–209.

Lee, D.L., 1971. Studies on the protein utilization related to growth of *P. monodon* Fab. Aquaculture 1, 1–13.

Lee, P.G., Lawrence, A.L., 1985. Effect of diet and size on growth, feed digestibility and digestive enzyme activities of the marine shrimp *Penaeus setiferus*. Proc. World Maricult. Soc. 16, 275–287.

Lee, P.G., Lawrence, A.L., 1997. DigestibilityD'Abramo, L.R. Concklin, D.E. Akiyama, D.M. (Eds.), Crustacean Nutrition. Advances in World Aquaculture, vol. 6. World Aquaculture Society, Baton Rouge, Louisiana, USA, pp. 194–260.

Lee, M.-H., Shiau, S.-Y., 2002. Dietary copper requirement of juvenile grass shrimp, *Penaeus monodon*, and effects on non-specific immune responses. Fish Shellfish Immunol. 13, 259–270.

Lee, M.H., Shiau, S.Y., 2004. Vitamin E requirements of juvenile grass shrimp, *Penaeus monodon*, and effects on immune responses. Fish Shellfish Immunol. 16, 475–485.

Lee, S.-M., Lee, J.-Y., Hur, S.B., 1993a. Essentiality of dietary eicosapentaenoic acid and docosahexaenoic acid in Korean rockfish, *Sebastes schlegeli*. Bull. Korean Fish. Soc. 27, 712–726.

Lee, S.-M., Lee, J.-Y., Kang, Y.J., Yoon, H.-D., Hur, S.B., 1993b. n-3 Highly unsaturated fatty acid requirement of the Korean rockfish *Sebastes schlegeli*. Bull. Korean Fish. Soc. 26, 477–492.

Lee, K.J., Kim, K.W., Bai, S.C., 1998. Effects of different dietary levels of L-ascorbic acid on growth and tissue vitamin C concentration in juvenile Korean rockfish. *Sebastes schlegeli* (Hilgendorf). Aquacult. Res. 29, 237–244.

Lee, S.M., Lee, J.H., Kim, K.D., 2003. Effect of dietary essential fatty acids on growth, body composition and blood chemistry of juvenile starry flounder (*Platichthys stellatus*). Aquaculture 225, 269–281.

Lee, B.J., Lee, K.J., Lim, S.J., Lee, S.M., 2009. Dietary myo-inositol requirement for olive flounder. *Paralichthys oliváceas* (Temminch et Schlegel). Aquacult. Res. 40, 83–90.

Leenhouwers, J.I., Adjei-Boateng, D., Verreth, J.A.J., Schrama, J.W., 2006. Digesta viscosity, nutrient digestibility and organ weights in African catfish (*Clarias gariepinus*) fed diets supplemented with different levels of a soluble non-starch polysaccharide. Aquacult. Nutr. 12, 111–116.

Leenhouwers, J.I., Ortega, R.C., Verreth, J.A.J., Schrama, J.W., 2007. Digesta characteristics in relation to nutrient digestibility and mineral absorption in Nile tilapia (*Oreochromis niloticus* L.) fed cereal grains of increasing viscosity. Aquaculture 273, 556–565.

Legeay, A., Massabuau, J.-C., 1999. Blood oxygen requirements in resting crab (*Carcinus maenas*) 24 h after feeding. Can. J. Zool. 77, 784–794.

Lester, R., Carey, M., Little, J., Cooperstein, L., Dowd, S., 1975. Crustacean intestinal detergent promotes sterol solubilization. Science 189, 1098–1100.

Leu, M.-Y., Yang, S.-D., Wu, C.-H., Liou, C.-H., 1994. Effect of dietary n-3 highly unsaturated fatty acids on growth, feed efficiency and fatty acid composition of juvenile silver bream *Rhabdosargus sarba* (Sparidae). Asian Fish. Sci. 7, 233–240.

Li, M., Lovell, R.T., 1992. Comparison of satiate feeding and restricted feeding of channel catfish with various concentrations of dietary protein in production ponds. Aquaculture 103, 165–175.

Li, M.H., Robinson, E.H., 1996. Comparison of chelated zinc and zinc sulfate as zinc sources for growth and bone mineralization of channel catfish (*Ictalurus punctatus*) fed practical diets. Aquaculture 146, 237–243.

Li, M.H., Wise, D.J., Robinson, E.H., 1998. Effect of dietary vitamin C on weight gain, tissue ascorbate concentration, stress response, disease resistance of channel catfish *Ictalunts punctalus*. J. World Aquacult. Soc. 29, 1–8.

Li, J., Li, W., Yan, Y., Luo, X., Liu, Z., Wang, J., et al., 2008. Requirement of calcium and phosphorus in the diet for juvenile *Pelteobagrus fulvidraco*. J. HebeiFish 5, 13–16.

Li, W., Zhou, X.Q., Feng, L., Liu, Y., Jiang, J., 2010a. Effect of dietary riboflavin on growth, feed utilization, body composition and intestinal enzyme activities of juvenile Jian carp (*Cyprinus carpio* var. Jian). Aquacult. Nutr. 16, 137–143.

Li, J., Zhang, L., Mai, K.S., Ai, Q.H., Wan, J., Zhang, C., Zhang, J., et al., 2010b. Estimation of dietary biotin requirement of Japanese seabass *Lateolabrax japonicus* C. Aquacult. Nutr. 16, 231–236.

Li, E., Yu, N., Chen, L., Zeng, C., Liu, L., Qin, J.G., 2010c. Dietary vitamin B6 requirement of the pacific white shrimp, *Litopenaeus vannamei*, at low salinity. J. World Aquacult. Soc. 41, 756–763.

Liang, J.J., Liu, Y.J., Yang, Z.N., Tian, L.X., Yang, H.J., Liang, G.Y., 2012a. Dietary calcium requirement and effects on growth and tissue calcium content of juvenile grass carp (*Ctenopharyngodon idella*). Aquacult. Nutr. 18, 544–550.

Liang, J.J., Liu, Y.J., Tian, L.X., Yang, H.J., Liang, G.Y., 2012b. Dietary available phosphorus requirement of juvenile grass carp (*Ctenopharyngodon idella*). Aquacult. Nutr. 18, 181–188.

Liang, J.J., Tian, L.X., Liu, Y.J., Yang, H.J., Liang, G.Y., 2012c. Dietary magnesium requirement and effects on growth and tissue magnesium content of juvenile grass carp (*Ctenopharyngodon idella*). Aquacult. Nutr. 18, 56–64.

Liang, J.J., Yang, H.J., Liu, Y.J., Tian, L.X., Liang, G.Y., 2012d. Dietary zinc requirement of juvenile grass carp (*Ctenopharyngodon idella*) based on growth and mineralization. Aquacult. Nutr. 18, 380–387.

Liang, J.J., Yang, H.J., Liu, Y.J., Tian, L.X., 2014. Dietary potassium requirement of juvenile grass carp (*Ctenopharyngodon idella* Val.) based on growth and tissue potassium content. Aquacult. Res. 45, 701–708.

Lightner, D.V., Redman, R., 1977. Histochemical demonstration of melanin in cellular inflammatory processes of penaeid shrimp. J. Invertebr. Pathol. 30, 298–302.

Lightner, D.V., Colvin, L.B., Brand, C., Donald, D.A., 1977. Black death, a disease syndrome of penaeid shrimp related to a dietary deficiency of ascorbic acid. Proc. World Maricult. Soc. 8, 611–623.

Lightner, D.V., Hunter, B., Magarelli Jr., P.C., Colvin, L.B., 1979. Ascorbic acid: nutritional requirement and role in wound repair in penaeid shrimp. Proc. World Maricult. Soc. 10, 513–519.

Lim, C., Lovell, R.T., 1978. Pathology of the vitamin C deficiency syndrome in channel catfish *(Ictalurus punctatus)*. J. Nutr. 108, 1137–1146.

Lim, C., Sukhawongs, S., Pascual, F.P., 1979. A preliminary study on the protein requirements of *Chanos chanos* (Forskal) fry in a controlled environment. Aquaculture 17, 195–201.

Lim, C., LeaMaster, B.R., Brock, J.A., 1993. Riboflavin requirement of fingerlings red hybrid tilapia grown in seawater. J. World Aquacult. Soc. 24, 451–458.

Lim, C., LeaMaster, B.R., Brock, J.A., 1995. Pyridoxine requirement of fingerling red hybrid tilapia growth in seawater. J. Appl. Aquacult. 5, 49–60.

Lim, C., Klesius, P.H., Li, M.H., Robinson, E.H., 2000. Interaction between dietary levels of iron and vitamin C on growth, hematology, immune response and resistance of channel catfish *(Ictalurus punctatus)* to *Edwardsiella ictaluri* challenge. Aquaculture 185, 313–327.

Lim, C., Shoemaker, C.A., Klesius, P., 2001. The effect of ascorbic acido in the immune response in fish. In: Dabrowski, K. (Ed.), Ascorbic Acid in Aquatic Organisms. CRC Press, Boca Raton, FL, pp. 149–166.

Lim, C., Yildirim-Aksoy, M., Barros, M.M., Klesius, P., 2011. Thiamin requirement of Nile Tilapia, *Oreochromis niloticus*. J. World Aquacult. Soc. 42, 824–833.

Limsuwan, T., Lovell, R.T., 1981. Intestinal synthesis and absorption of vitamin B_{12} in channel catfish. J. Nutr. 111, 2125–2132.

Lin, D., 1991. Grass carp, *Ctenopharyngodon idella*. In: Wilson, R.P. (Ed.), Handbook of Nutrient Requirement of Finfish. CRC Press, Boca Raton, FL, USA, pp. 89–96.

Lin, M.F., Shiau, S.Y., 2004. Requirements of vitamin C (L-ascorbyl-2-monophosphate-Mg and L-ascorbyl-2-monophosphate-Na) and its effects on immune responses of grouper. *Epinephelus malabaricus*. Aquacult. Nutr. 10, 327–333.

Lin, Y.H., Shiau, S.Y., 2005. Dietary selenium requirements of juvenile grouper, *Epinephelus malabaricus*. Aquaculture 250, 356–363.

Lin, S.C., Liou, C.H., Shiau, S.Y., 2000. Renal threshold for urinary glucose excretion by tilapia in response to orally administered carbohydrates and injected glucose. Fish Physiol. Biochem. 23, 127–132.

Lin, Y.-H., Shie, Y.-Y., Shiau, S.-Y., 2008a. Dietary copper requirements of juvenile grouper, *Epinephelus malabaricus*. Aquaculture 274, 161–165.

Lin, Y.H., Lin, S.M., Shiau, S.Y., 2008b. Dietary manganese requirements of juvenile tilapia, *Oreochromis niloticus* × *O. aureus*. Aquaculture 284, 207–210.

Lin, Y.-H., Shih, C.-C., Kent, M., Shiau, S.-Y., 2010. Dietary copper requirement reevaluation for juvenile grouper, *Epinephelus malabaricus*, with an organic copper source. Aquaculture 310, 173–177.

Lin, Y.-H., Ku, C.-Y., Shiau, S.-Y., 2013. Estimation of dietary magnesium requirements of juvenile tilapia, *Oreochromis niloticus* × *Oreochromis aureus*, reared in freshwater and seawater. Aquaculture 380-383, 47–51.

Lin, M.E., Shiau, S.Y., 2005a. Dietary L-ascorbic acid affects growth, non-specific immune responses and disease resistance in juvenile grouper, *Epinephelus malabaricus*. Aquaculture 244, 215–221.

Lin, M.F., Shiau, S.Y., 2005b. Requirements of vitamin C (L-ascorby1-2-sulfate and L-ascorbyl-2-polyphosphate) and its effects on immune responses of grouper, *Epinephelus malabaricus*. Aquacult. Nutr. 11, 183–189.

Lin, Y.H., Shiau, S.Y., 2005c. Dietary vitamin E requirements of grouper, *Epinephelus malabaricus*, under two lipid levels and their effects on immune responses. Aquaculture 248, 235–244.

Lin, Y.-H., Lin, H.-Y., Shiau, S.-Y., 2011. Dietary folic acid requirement of grouper, *Epinephelus malabaricus*, and its effects on non-specific immune responses. Aquaculture 317, 133–137.

Lin, Y.-H., Lin, H.-Y., Shiau, S.-Y., 2012. Estimation of dietary pantothenic acid requirement of grouper, *Epinephelus malabaricus* according to physiological and biochemical parameters. Aquaculture 324–325, 92–96.

Ling, J., Feng, L., Liu, Y., Jiang, J., Jiang, W.D., Hu, K., et al., 2010. Effect of dietary iron levels on growth, body composition and intestinal enzyme activities of juvenile Jian carp (*Cyprinus carpio* var. Jian). Aquacult. Nutr. 16, 616–624.

Liu, F., Liang, D., Sun, F., Li, H., Lan, X., 1990. Effects of dietary copper on the prawn *Penaeus orientalis*. Oceanologia Et Limnologia Sinica 21, 404–410.

Liu, T.B., Li, A.J., Zhang, J.M., 1993. Studies on vitamin nutrition for the shrimp Penaeus chinensis: 10. Studies on the choline chloride and inositol requirements in the shrimp *Penaeus chinensis*. J. Ocean Univ. Qingdao 23, 67–74.

Liu, T., Jiameng, Z., Li, A.J., 1995. Studies on the optimal requirements of pantothenic acid, biotin. folic acid and vitamin B12 in the shrimp *Penaeus chinensis*. J. Fish. Sci. China 2, 48–55.

Liu, Y., Wang, W.N., Wang, A.L., Wang, J.M., Sun, R.Y., 2007. Effects of dietary vitamin E supplementation on antioxidant enzyme activities in *Litopenaeus vannamei* (Boone, 1931) exposed to acute salinity changes. Aquaculture 265, 351–358.

Liu, K., Wang, X.J., Ai, Q., Mai, K., Zhang, W., 2010. Dietary selenium requirement for juvenile cobia, *Rachycentron canadum* L. Aquacult. Res. 41, e594–e601.

Liu, K., Ai, Q.H., Mai, K.S., Zhang, W.B., Zhang, L., Zheng, S.X., 2013. Dietary manganese requirement for juvenile cobia, *Rachycentron canadum* L. Aquacult. Nutr. 19, 461–467.

Liu, F.-J., Liu, Y.-J., Tian, L.-X., Li, X.-F., Zhang, Z.-H., Yang, H.-J., et al., 2014a. Quantitative dietary isoleucine requirement of juvenile Pacific white shrimp, *Litopenaeus vannamei* (Boone) reared in low-salinity water. Aquacult. Int. 22, 1481–1497.

Liu, F.-J., Liu, Y.-J., Tian, L.-X., Chen, W.-D., Yang, H.-J., Du, Z.-Y., 2014b. Quantitative dietary leucine requirement of juvenile Pacific white shrimp, *Litopenaeus vannamei* (Boone) reared in low-salinity water. Aquacult. Nutr. 20, 332–340.

Liu, X., Mai, K., Liufu, Z., Ai, Q., 2014c. Effects of dietary protein and lipid levels on growth, nutrient utilization, and the whole-body composition of Turbot, *Scophthalmus maximus*, Linnaeus 1758, at different growth stages. J. World Aquacult. Soc. 45, 355–366.

Lochmann, R.T., Gatlin III, D.M., 1993. Essential fatty acid requirement of juvenile red drum (*Sciaenops ocellatus*). Fish Physiol. Biochem. 12, 221–235.

Lochmann, R.T., Phillips, H., 1994. Dietary protein requirement of juvenile golden shiners (*Notemigonus crysoleucas*) and goldfish (*Carassius auratus*) in aquaria. Aquaculture 128, 277–285.

Lock, E.J., Waagbø, R., Wendelaar Bonga, S., Flik, G., 2010. The significance of vitamin D for fish: a review. Aquacult. Nutr. 16, 100–116.

Lorentzen, M., Maage, A., Julshamn, K., 1994. Effects of dietary selenite or selenomethionine on tissue selenium levels of Atlantic salmon (*Salmo salar*). Aquaculture 121, 359–367.

Lorentzen, M., Maage, A., Julshamn, K., 1996. Manganese supplementation of a practical, fish meal based diet for Atlantic salmon parr. Aquacult. Nutr. 2, 121–125.

Lorentzen, M., Maage, A., Julshamn, K., 1998. Supplementing copper to a fish meal based diet fed to Atlantic salmon parr affects liver copper and selenium concentrations. Aquacult. Nutr. 4, 67.

Lovell, R.T., 1978. Dietary phosphorus requirement of channel catfish (*Ictalurus punctatus*). Trans. Am. Fish. Soc. 107, 617–621.

Lovell, R.T., Lim, C., 1978. Vitamin C in pond diets for channel catfish. Trans. Am. Fish. Soc. 107, 321–325.

Lovell, R.T., Limsuwan, T., 1982. Intestinal synthesis and dietary nonessentiality of vitamin B_p for *Tilapia nilotica*. Trans. Am. Fish. Soc. 111, 485–490.

Lu, L., Wang, R.L., Zhang, Z.J., Steward, F.A., Luo, X., Liu, B., 2010. Effect of dietary supplementation with copper sulfate or tribasic copper chloride on the growth performance, liver copper concentrations of broilers fed in floor pens, and stabilities of vitamin E and phytase in feeds. Biol. Trace Elem. Res., 181–189.

Luna-González, A., Almaraz-Salas, J.C., Fierro-Coronado, J.A., Flores-Miranda, M.C., González-Ocampo, H.A., Peraza-Gómez, V., 2012. The prebiotic inulin increases the phenoloxidase activity and reduces the prevalence of WSSV in whiteleg shrimp (*Litopenaeus vannamei*) cultured under laboratory conditions. Aquaculture 362–363, 28–32.

Luo, Y., Xie, X., 2008. Effects of temperature on the specific dynamic action of the southern catfish, *Silurus meridionalis*. Comp. Biochem. Physiol. A Mol. Integr. Physiol. 149, 150–156.

Luo, Z., Tan, X.Y., Liu, X., Wang, W.M., 2010. Dietary total phosphorus requirement of juvenile yellow catfish *Pelteobagrus fulvidraco*. Aquacult. Int. 18, 897–908.

Luo, Z., Tan, X.-Y., Zheng, J.-L., Chen, Q.-L., Liu, C.-X., 2011. Quantitative dietary zinc requirement of juvenile yellow catfish *Pelteobagrus fulvidraco*, and effects on hepatic intermediary metabolism and antioxidant responses. Aquaculture 319, 150–155.

Luzzana, U., Hardy, R.W., Halver, J.E., 1998. Dietary arginine requirement of fingerling coho salmon (*Oncorhynchus kisutch*). Aquaculture 163, 137–150.

Maage, A., Julshamn, K., 1993. Assessment of zinc status in juvenile Atlantic salmon (Salmo salar) by measurement of whole body and tissue levels of zinc. Aquaculture 117, 179–191.

Maage, A., Lygren, B., El Mowafic, A.F.A., 2000. Manganese requirement of Atlantic salmon (*Salmo salar*) fry. Fish. Sci. 66, 1–8.

Maenz, D.D., Engele-Schaan, C.M., Newkirk, R.W., Classen, H.L., 1999. The effect of minerals and mineral chelators on the formation of phytase-resistant and phytase-susceptible forms of phytic acid in solution and in a slurry of canola meal. Anim. Feed Sci. Technol. 81, 177–192.

Magarelli Jr., P.C., Hunter, B., Lightner, D.V., Colvin, L.B., 1979. Black death: an ascorbic acid deficiency disease in penaeid shrimp. Comp. Biochem. Physiol. A 63A, 103–108.

Mahajan, C.L., Agrawal, N.K., 1980. Nutritional requirements of ascorbic acid by Indian major carp *Cirrhina mrigala* during early growth. Aquaculture 19, 37–48.

Mai, K.S., Zhang, C.X., Ai, Q.H., Duan, Q.Y., Xu, W., Zhang, L., et al., 2006. Dietary phosphorus requirement of large yellow croaker, *Pseudosciaena crocea* R. Aquaculture 251, 346–353.

Mai, K., Xiao, L., Ai, Q., Wang, X., Xu, W., Zhang, W., 2009. Dietary choline requirement for juvenile cobia, *Rachycentron canadum*. Aquaculture 289, 124–128.

Mambrini, M., Guillaume, J., 2001. Protein nutrition. In: Guillaume, J., Kaushik, S., Bergot, P., Métailler, R. (Eds.), Nutrition and Feeding of Fish and Crustaceans, INRA Editions. Springer-Praxis, Chichester, UK, pp. 81–109.

Mamun, S.M., Focken, U., Becker, K., 2007. Comparison of metabolic rates and feed nutrient digestibility in conventional, genetically improved (GIFT) and genetically male (GMNT) Nile tilapia, *Oreochromis niloticus* (L.). Comp. Biochem. Physiol. 148A, 214–222.

Mangalik, A., 1986. Dietary energy requirements of channel catfish (PhD dissertation). Auburn University, Auburn, AL.

Martínez Romero, P., Casal de Fenucci, A., Fenucci, J., 1991. Dietary cholesterol influence on the growth and survival of the Argentine prawn *Artemesia longinaris* Bate. J. Aquacult. Trop. 6, 111–117.

Martins, M.L., 1995. Effect of ascorbic acid deficiency on the growth, gill filament lesions and behavior of pacu fry (*Piarctus mesopotamicus* Holmberg. 1987). Braz. J. Med. Biol. Res. 28, 563–568.

Mazid, M.A., Tanaka, Y., Katayama, T., Rahman, M.A., Simpson, K.L., Chichester, C.O., 1979. Growth response of *Tilapia zillii* fingerlings fed isocaloric diets with variable protein levels. Aquaculture 18, 115–122.

Mazur, C.N., Higgs, D.A., Plisetskaya, E.M., March, B.E., 1992. Utilization of dietary starch and glucose tolerance in juvenile chinook salmon (*Oncorhynchus tshawytscha*) of different strains in seawater. Fish Physiol. Biochem. 10, 303–313.

McClain, W., Gatlin, D.M., 1988. Dietary zinc requirement of *Oreochromis aureus* and effects of dietary calcium and phytate on zinc bioavailability. J. World Aquacult. Soc. 19, 103–108.

McGaw, I.J., 2006. Feeding and digestion in low salinity in an osmoconforming crab, *Cancer gracilis* I. Cardiovascular and respiratory responses. J. Exp. Biol. 209, 3766–3776.

McLaren, B.A., Keller, E., ODonnell, D.J., Elvehjem, C.A., 1947. The nutrition of rainbow trout. 1. Studies of vitamin requirements. Arch. Biochem. Biophys. 15, 169–178.

Médale, F., Poli, J.M., Vallée, F., Blanc, D., 1999. Utilization of a carbohydrate-rich diet by common carp reared at 18 and 25°C. Cybium 23, 139–152.

Meena, D.K., Das, P., Kumar, S., Mandal, S.C., Prusty, A.K., Singh, S.K., et al., 2013. Beta-glucan: an ideal immunostimulant in aquaculture (a review). Fish Physiol. Biochem. 39, 431–457.

Merchie, G., Lavens, P., Storch, V., Ubel, U., Nelis, H., Deleenheer, A., et al., 1996. Influence of dietary vitamin C storage on turbot (*Scophthalmus maximus*) and European seabass (*Dicentrarchus labrax*) nursery stages. Comp. Biochem. Physiol. A 114A, 117–121.

Merican, Z.O., Shim, K.F., 1997. Quantitative requirements of linolenic and docosahexaenoic acid for juvenile *Penaeus monodon*. Aquaculture 157, 277–295.

Metón, I., Caseras, A., Mediavilla, D., Fernández, F., Baanante, I.V., 1999. Molecular cloning of a cDNA encoding 6-phosphofructo-2-kinase/fructose-2,6-bisphosphatase from liver of *Sparus aurata*: nutritional regulation of enzyme expression1. Biochim. Biophys. Acta, Gene Struct. Expression 1444, 153–165.

Miles, R.D., O'keefe, S.F., Henry, P.R., Ammerman, C.B., Luo, X.G., 1998. The effect of dietary supplementation with copper sulfate or tribasic copper chloride on broiler performance, relative copper bioavailability, and dietary prooxidant activity. Poult. Sci. 77, 416–425.

Millamena, O.M., Bautista, M.N., Kanazawa, A., 1996a. Methionine requirement of juvenile tiger shrimp *Penaeus monodon* Fabricius. Aquaculture 143, 403–410.

Millamena, O.M., Bautista-Teruel, M.N., Kanazawa, A., 1996b. Valine requirement of postlarval tiger shrimp, *Penaeus monodon* Fabricius. Aquacult. Nutr. 2, 129–132.

Millamena, O.M., Bautista, M., Reyes, O., Kanazawa, A., 1997. Threonine requirement of juvenile marine shrimp *Penaeus monodon*. Aquaculture 151, 9–14.

Millamena, O.M., Bautista-Teruel, M.N., Reyes, O.S., Kanazawa, A., 1998. Requirements of juvenile marine shrimp, *Penaeus monodon* (Fabricius) for lysine and arginine. Aquaculture 164, 95–104.

Millamena, O.M., Teruel, M.B., Kanawaza, A., Teshima, S., 1999. Quantitative dietary requirements of postlarval tiger shrimp. *Penaeus monodon*, for histidine, isoleucine, leucine, phenylalanine and tryptophan. Aquaculture 179, 169–179.

Miller, E.R. 1980. Bioavailability of minerals. In: Proc. of the Minnesota Nutr. Conf. Univ. Minnesota, St. Paul, MN, pp. 144–154.

Millikin, M.R., Fortner, A.R., Fair, P.H., Sick, L.V., 1980. Influence of dietary protein concentration on growth, feed conversion and general metabolism of juvenile prawn (*Macrobrachium rosenbergii*). Proc. World Maricult. Soc. 11, 382–391.

Minoso, M.G.G., Borlongan, I.G., Satoh, S., 1999. Essentiality of phosphorus, magnesium, iron, zinc, and manganese in milkfish diet. Fish. Sci. 65, 721–725.

Mishra, S., Mukhopadhyay, P.K., 1996. Ascorbic acid requirement of catfish fry *Clasrias batrachus* (Linn). Indian J. Fish. 43, 157–162.

Misra, C.K., Das, B.K., Mukherjee, S.C., Pradhan, J., 2007. Effects of dietary vitamin C on immunity, growth and survival of Indian major carp *Labeo rohita*, fingerlings. Aquacult. Nutr. 13, 35–44.

Miura, T., Suzuki, N., Nagoshi, M., Yamamura, K., 1976. The rate of production and food consumption of the biwamasu, *Oncorhynchus rhodurus*, population in Lake Biwa. Res. Popul. Ecol. 17, 135–154.

Moe, Y.Y., Koshio, S., Teshima, S.I., Ishikawa, M., Matsunaga, Y., Panganiban, A., 2004. Effect of vitamin C derivatives on the performance of larval kuruma shrimp, *Marsupenaeus japónicas*. Aquaculture 242, 501–512.

Moe, Y.Y., Koshio, S., Ishikawa, M., Teshima, S.I., Panganiban, A., Thu, M., et al., 2005. Vitamin C requirement of kuruma shrimp postlarvae, *Marsupenaeus japonicus* (Bate), using L-ascorbyl-2-monophosphate–Na–Ca. Aquacult. Res. 36, 739–745.

Molina-Poveda, C., Gómez, G., 2002. Digestibility of different carbohydrates in the diet of the juvenile *Litopenaeus vannamei*. World Aquaculture Society, (book of abstracts). pp. 518. Beijing, China.

Mommsen, T.P., Plisetskaya, E.M., 1991. Insulin in fishes and agnathans: history, structure, and metabolic regulation. Rev. Aquat. Sci. 4, 225–259.

Montoya, N., Molina, C., 1995. Optimum supplemental level of L-ascorbyl-2-phosphate-Mg to diet for white shrimp *Penaeus vannamei*. Fish. Sci. 61, 1045–1046.

Moon, T.W., Busby, E.R., Cooper, G.A., Mommsen, T.P., 1999. Fish hepatocyte glycogen phosphorylase – a sensitive indicator for hormonal modulation. Fish Physiol. Biochem. 21, 15–24.

Moore, B.J., Hung, S.S.O., Medrano, J.F., 1988. Protein requirement of hatchery-produced juvenile white sturgeon (*Acipenser transmontanus*). Aquaculture 71, 235–245.

Moreau, R., Dabrowski, K., 2001. Gluconolactone oxidase presence in fishes: activity and significance. In: Dabrowski, K. (Ed.), Ascorbic Acid in Aquatic Organisms. CRC Press, Boca Raton, FL, pp. 13–34.

Moren, M., Opstad, I., Berntssen, M.H.G., Infante, J.-L.Z., Hamre, K., 2004. An optimum level of vitamin A supplements for Atlantic halibut (*Hippoglossus hippoglossus* L.) juveniles. Aquaculture 235, 587–599.

Morito, C.L.H., Conrad, D.H., Hilton, J.W., 1986. The thiamin deficiency signs and requirement of rainbow trout (*Salmo gairdneri*. Richardson). Fish Physiol. Biochem. 1, 93–104.

Morris, P.C., Baker, R.T.M., Davies, S.J., 1998. Nicotinic acid supplementation of diets for the African catfish. *Ciarías gariepinus* (Burchell). Aquacult. Res. 29, 791–799.

Morris, T., Samocha, T.M., Davis, D., 2011. Cholesterol supplements for Litopenaeus vannamei reared on plant based diets in the presence of natural productivity. Aquaculture 314 (1–4), 140–144.

Mourente, G., Tocher, D.R., 1993. Incorporation and metabolism of c-14-labeled polyunsaturated fatty-acids in juvenile gilthead sea bream *Sparus aurata* L. *in vivo*. Fish Physiol. Biochem. 10, 443–453.

Murai, T., Andrews, J.W., 1974. Interactions of dietary ot-tocopherol, oxidized menhaden oil and ethoxyquin on channel catfish (*Ictalurus punctatus*). J. Nutr. 104, 1416–1431.

Murai, T., Andrews, J.W., 1977. Vitamin K and anticoagulant relationships in catfish diets. Bull. Jpn. Soc. Sci. Fish. 43, 785–794.

Murai, T., Andrews, J.W., 1978a. Riboflavin requirement of channel catfish fingerlings. J. Nutr. 108, 1512–1517.

Murai, T., Andrews, J.W., 1978b. Thiamin requirement of channel catfish fingerlings. J. Nutr. 108, 176–180.

Murai, T., Andrews, J.W., 1979. Pantothenic acid requirement of channel catfish fingerlings. J. Nutr. 109, 1140–1142.

Murai, T.J.W., Andrews, Bauerfeind, J.C., 1978. Use of L-ascorbic acid, ethocel coated ascorbic acid and ascorbic 2-sulfate in diets of channel catfish (*Ictalurus punctatus*). J. Nutr. 108, 1761–1766.

Murillo-Gurrea, D.P., Coloso, R.M., Borlongan, I.G., Serrano Jr., A.E., 2001. Lysine and arginine requirements of juvenile Asian sea bass (*Lates calcarifer*). J. Appl. Ichthyol. 17, 49–53.

Mustin, W.G., Lovell, R.T., 1992. Na-L-ascorbyl-2-monophosphate as a source of vitamin C for channel catfish. Aquaculture 105, 95–100.

Nakamura, Y., 1982. Effects of dietary phosphorus and calcium contents on the absorption of phosphorus in the digestive tract of carp. Bull. Jpn. J. Sci. Fish. 51, 605–608.

Nakamura, Y., Yamada, J., 1980. Effects of dietary calcium levels, Ca/P ratios and calcium components on the calcium absorption rate in carp. Bull. Fac. Fish. Hokkaido Univ. 31, 277–282.

Naser, N., 2000. Role of Iron in Atlantic salmon (*Salmo salar*) Nutrition: Requirement, Bioavailability, Disease Resistance and Immune Response. Dalhousie University, Halifax, Nova Scotia, pp. 282.

Navarro, I., Gutiérrez, J., 1995. Fasting and starvation. In: Hochachka, P.W., Mommsen, T.P. (Eds.), Biochemistry and Molecular Biology of Fishes. Elsevier, New York, NY, pp. 394–434.

Nelson, S.G., Knight, A.W., Li, H.W., 1977. The metabolic cost of food utilization and ammonia production by juvenile *Macrobrachium rosenbergii* (Crustacea: Palaemonidae). Comp. Biochem. Physiol. A 57, 67–72.

Nelson, S.G., Simmons, M.A., Knight, A.W., 1985. Calorigenic effect of diet on the grass shrimp *Crangon franciscorum* (Crustacea: Crangonidae). Comp. Biochem. Physiol. 82A, 373–376.

Nematipour, G.R., Brown, M.L., Gatlin III, D.M., 1992. Effects of dietary energy:protein ratio on growth characteristic and body composition of hybrid striped bass, *Morone chrysops* × *M. saxitalis*. Aquaculture 107, 359–368.

Neu, D.H., Furuya, W.M., Boscolo, W.R., Potrich, F.R., Lui, T.A., Feiden, A., 2013. Glycerol inclusion in the diet of Nile tilapia (*Oreochromis niloticus*) juveniles. Aquacult. Nutr. 19 (2), 211–217.

Ng, W., Serrini, K.G., Zhang, Z., Wilson, R.P., 1997. Niacin requirement and inability of tryptophan to act as a precursor of NAD* in channel catfish. *Ictalurus punctatus*. Aquaculture 152, 273–285.

Nguyen, T.N., Davis, D.A., 2009. Methionine requirement in practical diets of juvenile Nile tilapia, *Oreochromis niloticus*. J. World Aquacult. Soc. 40, 410–416.

Niu, J., Tian, L.X., Liu, Y.J., Mai, K.S., Yang, H.J., Ye, C.X., et al., 2009. Nutrient values of dietary ascorbic acid (L-ascorbyl-2-poly-phosphate) on growth, survival and stress tolerance of larval shrimp. *Lilopenaeus vannamei*. Aquacult. Nutr. 15, 194–201.

Niu, J., Lin, H.-Z., Jiang, S.-G., Chen, X., Wu, K.-C., Tian, L.-X., et al., 2012a. Effect of seven carbohydrate sources on juvenile *Penaeus monodon* growth performance, nutrient utilization efficiency and hepatopancreas enzyme activities of 6-phosphogluconate dehydrogenase, hexokinase and amylase. Anim. Feed Sci. Technol. 174 (1–2), 86–95.

Niu, J., Chen, P.-F., Tian, L.-X., Liu, Y.-J., Lin, H.-Z., Yang, H.-J., et al., 2012b. Excess dietary cholesterol may have an adverse effect on growth performance of early post-larval *Litopenaeus vannamei*. J. Anim. Sci. Biotechnol. 3, 19.

Norrgren, L., Borjeson, H., Forlin, L., Akerblom, N., 2001.. In: Dabrowski, K. (Ed.), Ascorbic Acid in Aquatic Organisms. CRC Press, Boca Raton, FL, pp. 133–147.

Nose, T., 1979. Summary report on the requirements of essential amino acids for carp. In: Halver, J.E., Tiews, K. (Eds.), Finfish Nutrition and Fishfeed Technology. Heenemann GmbH, Berlin, Germany, pp. 145–156.

Nose, T., Arai, S., 1979. Recent advances in studies on mineral nutrition in Japan. In: Pillay, T.V.R., Dill, A. (Eds.), Advances in Aquaculture Fishing. News Books, Faraham, UK, pp. 584–590.

NRC, 2011. Nutrient requirements of fish and shrimp. Animal Nutrition Series National Research Council of the National Academies. The National Academies Press, Washington, DC, 376 pp.

NRC (National Research Council), 1983. Nutrient Requirements of Warmwater Fishes and Shellfishes, Revised Edition. National Academy Press, Washington, DC.

Nunes, A.J.P., Sá, M.V.C., Browdy, C.L., Vazquez-Anon, M., 2014. Practical supplementation of shrimp and fish feeds with crystalline amino acids. Aquaculture 431, 20–27.

Nwanna, L.C., Adebayo, I.A., Omitoyin, B.O., 2009. Phosphorus requirements of African catfish, *Clarias gariepinus*, based on broken-line regression analysis methods. ScienceAsia 35, 227–233.

Nwanna, L.C., Kuehlwein, H., Schwarz, F.J., 2010. Phosphorus requirement of common carp *Cyprinus carpio* L) based on growth and mineralization. Aquacult. Res. 41, 401–410.

Ogino, C., 1965. B vitamin requirements of carp. *Cyprinus carpió*. 1. Deficiency symptoms and requirements of vitamin B5. Bull. Jpn. Soc. Sci. Fish. 31, 546–551.

Ogino, C., 1967. B vitamin requirements of carp. 2. Requirements for riboflavin and pantothenic acid. Bull. Jpn. Soc. Sci. Fish. 33, 351–354.

Ogino, C., Saito, K., 1970. Protein nutrition in fish. 1. The utilization of dietary protein by carp. Bull. Jpn. Soc. Sci. Fish. 36, 250–254.

Ogino, C., Chiou, J.Y., 1976. Mineral requirements in fish-II: magnesium requirement of carp. Bull. Jpn. Soc. Sci. Fish. 42, 71–75.

Ogino, C., Takeda, H., 1976. Mineral requirements in fish. III. Calcium and phosphorus requirements in carp. Bull. Jpn. Soc. Sci. Fish. 42, 793–799.

Ogino, C., Takeda, H., 1978. Requirement of rainbow trout for dietary calcium and phosphorous. Bull. Jpn. J. Sci. Fish. 44, 1019–1022.

Ogino, C., Takashima, F., Chiou, J.Y., 1978. Requirement of Rainbow trout for dietary Magnesium. Bull. Jpn. Soc. Sci. Fish. 44, 1105–1108.

Ogino, O., Yang, G.Y., 1978. Requirement of rainbow trout for dietary zinc. Bull. Jpn. Soc. Sci. Fish. 44, 1015–1018.

Ogino, C., Yang, C., 1979. Requirement of carp for dietary zinc. Bull. Jpn. Soc. Sci. Fish. 45, 967–969.

Ogino, C., Yang, G., 1980. Requirements of carp and rainbow trout for dietary manganese and copper. Bull. Jpn. Soc. Sci. Fish. 46, 455–458.

Ogino, C., Uki, N., Watanabe, T., Iida, Z., Ando, K., 1970a. B vitamin requirements of carp. 4. Requirement of choline. Bull. Jpn. Soc. Sci. Fish. 36, 1140–1146.

Ogino, C., Watanabe, T., Kakino, J., Iwanaga, N., Mizuno, M., 1970b. B vitamin requirements of carp. 3. Requirement for biotin. Bull. Jpn. Soc. Sci. Fish. 36, 734–740.

Ohta, M., Watanabe, T., 1996. Dietary energy budgets in carp. Fish. Sci. 62, 745–753.

Oliva Teles, A., Pimentel Rodrigues, A., 2004. Phosphorus requirement of European sea bass (*Dicentrarchus labrax* L.) juveniles. Aquacult. Res. 35, 636–642.

Olsen, R.E., Ringø, E., 1997. Lipid digestibility in fish: a review. Recent Res. Dev. Lipids Res. 1, 199–265.

Olsen, R.E., Myklebust, R., Kaino, T., Ringø, E., 1999. Lipid digestibility and ultrastructural changes in the enterocytes of Arctic char (*Salvelinus alpinus* L.) fed linseed oil and soybean lecithin. Fish Physiol. Biochem. 21, 35–44.

Ortiz, L.T., Rebolé, A., Velasco, S., Rodríguez, M.L., Treviño, J., Tejedor, J.L., et al., 2012. Effects of inulin and fructooligosaccharides on growth performance, body chemical composition and intestinal microbiota of farmed rainbow trout (*Oncorhynchus mykiss*). Aquacult. Nutr. 19, 475–482.

Ostrowski, A.C., Kim, B.G., 1993. Responses of larval and juvenile mahimahi (*Coryphaena hippurus*) to various dietary lipid sources and n-3 HUFA contents From Discovery to Commercialization. European Aquaculture Society, Oostende, Belgium, Special Publication no. 19, p. 424.

Owen, S.F., 2001. Meeting energy budgets by modulation of behavior and physiology in the eel (*Anguilla anguilla* L.). Comp. Biochem. Physiol. 128A, 631–644.

Page, J.W., Andrews, J.W., 1973. Interactions of dietary levels of protein and energy on channel catfish (*Ictalurus punctatus*). J. Nutr. 103, 1339–1346.

Paibulkichakui, C., Piyatiratitivorakul, S., Kittakpp, P., Vorsnop, V., Fast, A.W., Menasveta, P., 1998. Optimal dietary levels of lecithin and cholesterol for black tiger prawn *Penaeus monodon* larvae and postlarvae. Aquaculture 167, 273–281.

Pan, Q., Chen, X.-Y., Li, F., Bi, Y.-Z., Zheng, S.-X., 2005. Response of juvenile *Litopenaeus vannamei* to varying levels of calcium phosphate monobasic supplemented to a practical diet. Aquaculture 248, 97–102.

Pan, L., Zhu, X., Xie, S., Lei, W., Han, D., Yang, Y., 2008. Effects of dietary manganese on growth and tissue manganese concentrations of juvenile gibel carp, *Carassius auratus* gibelio. Aquacult. Nutr. 14, 459–463.

Pan, L., Xie, S., Zhu, X., Lei, W., Han, D., Yang, Y., 2009. The effect of different dietary iron levels on growth and hepatic iron concentration in juvenile gibel carp (*Carassius auratus* gibelio). J. Appl. Ichthyol. 25, 428–431.

Paripatananont, T., Lovell, R.T., 1995. Chelated zinc reduces the dietary zinc requirement of channel catfish, *Ictalurus punctatus*. Aquaculture 133, 73–82.

Paripatananont, T., Lovell, R.T., 1997. Comparative net absorption of chelated and inorganic trace minerals in channel catfish *Ictalurus punctatus* diets. J. World Aquacult. Soc. 28, 62–67.

Paul, B.N., Sarkar, S., Mohanty, S.N., 2004a. Dietary vitamin E requirement of mrigal. *Cirrhinus mrigala* fry. Aquaculture 242, 529–536.

Paul, B.N., Sarkar, S., Giri, S.S., Rangacharyulu, P.V., Mohanty, S.N., 2004b. Phosphorus requirements and optimum calcium/phosphorus ratio in the diet of mrigal *Cirrhinus mrigala* (Ham.) fingerlings. J. Appl. Ichthyol. 20, 306–309.

Paul, B.N., Sarkar, S., Girl, S.S., Mohanty, S.N., Mukhopadhyay, P.K., 2006. Dietary calcium and phosphorus requirements of rohu *Labeo rohita* fry. Anim. Nutr. Feed Technol. 6, 257–263.

Pedrazzoli, A., Molina, C., Montoya, N., Townsend, S., Leon-Hing, A., Paredes, Y., et al., 1998. Recent advances on nutrition research of *Penaeus vannamei* in Ecuador. Rev. Fish. Sci. 6, 143–151.

Penaflorida, V.D., 1999. Interaction between dietary levels of calcium and phosphorus on growth of juvenile shrimp, *Penaeus monodon*. Aquaculture 172, 281–289.

Peng, L.I., Gadin III, D.M., 2009. Dietary vitamin E requirement of the red drum *Sciaenops ocellatus*. Aquacult. Nutr. 15, 329–337.

Peres, H., Oliva-Teles, A., 2001. Effect of dietary protein and lipid level on metabolic utilization of diets by European sea bass (*Dicentrarchus labrax*) juveniles. Fish. Physiol. Biochem. 25, 269–275.

Peres, H., Oliva-Teles, A., 2008. Lysine requirement and efficiency of lysine utilization in turbot (*Scophthalmus maximus*) juveniles. Aquaculture 275, 283–290.

Peres, H., Lim, C., Klesius, P.H., 2004. Growth, chemical composition and resistance to *Streptococcus iniae* challenge of juvenile Nile tilapia (*Oreochromis niloticus*) fed graded levels of dietary inositol. Aquaculture 235, 423–432.

Perez-Jimenez, A., Guedes, M.J., Morales, A.E., Oliva-Teles, A., 2007. Metabolic responses to short starvation and refeeding in *Dicentrarchus labrax*. Effect of dietary composition. Aquaculture 265, 325–335.

Persson, H., Türk, M., Nyman, M., Sandberg, A.-S., 1998. Binding of $Cu2+$, $Zn2+$, and $Cd2+$ to inositol tri-, tetra-, penta-, and hexaphosphates. J. Agric. Food Chem. 46, 3194–3200.

Petriella, A.M., Muller, M.I., Penucci, J.L., Saez, M.B., 1984. Influence of dietary fatty acids and cholesterol on the growth and survival of the Argentine prawn, *Artemesia longinaris* Bate. Aquaculture 37, 11–20.

Phillips Jr., A.M., 1959. The known and possible role of minerals in trout nutrition and physiology. Trans. Am. Fish. Soc. 88, 133–135.

Phromkunthong, W., Udom, U., 2008. Available phosphorus requirement of sex-reversed red tilapia fed all-plant diets. Songklanakarin J. Sci. Tech. 30, 7–16.

Phromkunthong, W., Boonyaratpalin, M., Storch, V., 1997. Different concentrations of ascoryl-2-monophosphate-magnesium as dietary sources of vitamin C for seabass. *Lates calcarifer*. Aquaculture 151, 225–243.

Pimentel Rodrigues, A., Oliva Teles, A., 2001. Phosphorus requirements of gilthead sea bream (*Sparus aurata* L.) juveniles. Aquacult. Res. 32, 157–161.

Piper, R.G., McElwain, I.B., Orme, L.E., McCraren, J.P., Fowler, L.G., Leonard, J.R., 1982. Fish Hatchery Management. Department of the Interior U.S. Fish and Wildlife Service, Washington, DC, 517 pp.

Polakof, S., Mommsen, T.P., Soengas, J.L., 2011. Glucosensing and glucose homeostasis: from fish to mammals. Comp. Biochem. Physiol. A Mol. Integr. Physiol. 160, 123–149.

Ponat, A., Adelung, D., 1983. Studies to establish an optimal diet for *Carcinus maenas* 3. Vitamin and quantitative lipid requirements. Mar. Biol. 74, 275–279.

Posten, H.A., Rumsey, G.L., 1983. Factors affecting dietary requirement and deficiency signs of L-tryptophan in rainbow trout. J. Nutr. 113, 2568–2577.

Poston, H.A., 1976a. Relative effect of two dietary water-soluble analogues of menaquinone on coagulation and packed cell volume of blood of lake trout. *Salvelinus namaycush*. J. Fish. Res. Board Can. 33, 1791–1793.

Poston, H.A., 1976b. Optimum level of dietary biotin for growth, feed utilization, and swimming stamina of fingerling lake trout (*Salvelinus namaycush*). J. Fish. Res. Board Can. 33, 1803–1806.

Poston, H.A., 1990a. Effect of body size on growth, survival and chemical composition of Atlantic salmon fed soy lecithin and choline. Prog. Fish Cult. 52, 226–230.

Poston, H.A., 1990b. Performance of rainbow trout fed supplemental soybean lecithin and choline. Prog. Fish Cult. 52, 218–225.

Poston, H.A., 1991. Response of Atlantic salmon fry to feed-grade lecithin and choline. Prog. Fish Cult. 53, 224–228.

Poston, H.A., Combs Jr., G.F., Leibovitz, L., 1976. Vitamin E and selenium interrelations in the diet of Atlantic salmon (Salmo salar): gross, histological and biochemical deficiency signs. J. Nutr. 106, 892–904.

Poston, H.A., Wolfe, M.J., 1985. Niacin requirement for optimum growth, feed conversion and protection of rainbow trout, *Salmo gairdneri* Richardson, from ultraviolet-B-irradiation. J. Fish. Dis. 8, 451–460.

Power, D.M., Llewellyn, L., Faustino, M., Nowell, M.A., Björnsson, B.T., Einarsdottir, I.E., et al., 2001. Thyroid hormones in growth and development of fish. Comp. Biochem. Physiol. C Toxicol. Pharmacol. 130, 447–459.

Prabhu, P.A.J., Schrama, J.W., Kaushik, S.J., 2013. Quantifying dietary phosphorus requirement of fish – a meta-analytic approach. Aquacult. Nutr. 19, 233–249.

Qin, Z.H., Li, J., Wang, F., Liu, Q., Wang, Q., 2007. Effect of L-ascorbyl-2-polyphosphate on growth and immunity of shrimp (*Penaeus chinensis*) larvae. Fish. Sci. 26, 21.

Radunzneto, J., Corraze, G., Bergot, P., Kaushik, S.J., 1996. Estimation of essential fatty acid requirements of common carp larvae using semi-purified diets. Arch. Anim. Nutr. 49, 41–48.

Ramos, R., Andreatta, E., 2011. Gross protein and energy requirements for pink shrimp *Farfantepenaeus paulensis* (Pérez-Farfante, 1967) juvenile under different salinities. Latin Am. J. Aquat. Res. 39, 427–438.

Reigh, R.C., Robinson, E.H., Brown, P.B., 1991. Effects of dietary magnesium on growth and tissue magnesium content of blue tilapia *Oreochromis aureus*. J. World Aquacult. Soc. 22, 192–200.

Reitan, K.I., Rainuzzo, J.R., Olsen, Y., 1994a. Influence of lipid composition of live feed on growth, survival and pigmentation of turbot larvae. Aquacult. Int. 2, 33–48.

Reitan, K.I., Rainuzzo, J.R., Olsen, Y., 1994b. Influence of lipid class and fatty acid deficiency on survival, growth and pigmentation of turbot larvae. Aqualt. Int. 2, 33–48.

Ren, T., Koshio, S., Teshima, S., Ishikawa, M., Panganiban Jr., A., Uyan, O., et al., 2008. Effectiveness of L-ascorbyl-2-monophosphate Na/Ca as a vitamin C source for yellowtail *Serióla quinqueradiata* juveniles. Aquacult. Nutr. 14, 416–422.

Ren, M., Ai, Q., Mai, K., 2014. Dietary arginine requirement of juvenile cobia (*Rachycentron canadum*). Aquacult. Res. 45, 225–233.

Richard, L., Blanc, P.-P., Rigolet, V., Kaushik, S.J., Geurden, I., 2010. Maintenance and growth requirements for nitrogen, lysine and methionine and their utilisation efficiencies in juvenile black tiger shrimp, *Penaeus monodon*, using a factorial approach. Br. J. Nutr. 103, 984–995.

Rider, S.A., Davies, S.J., Jha, A.N., Fisher, A.A., Knight, J., Sweetman, J.W., 2009. Supranutritional dietary intake of selenite and selenium yeast in normal and stressed rainbow trout (*Oncorhynchus mykiss*): implications on selenium status and health responses. Aquaculture 295, 282–291.

Rigos, G., Samartzis, A., Henry, M., Fountoulaki, E., Cotou, E., Sweetman, J., et al., 2010. Effects of additive iron on growth, tissue distribution, haematology and immunology of gilthead sea bream, *Sparus aurata*. Aquacult. Int. 18, 1093–1104.

Rinchard, J., Czesny, S., Dabrowski, K., 2007. Influence of lipid class and fatty acid deficiency on survival, growth, and fatty acid composition in rainbow trout juveniles. Aquaculture 264, 363–371.

Rivas-Vega, M.E., Goytortúa-Bores, E., Ezquerra-Brauer, J.M., Salazar-García, M.G., Cruz-Suárez, L.E., Nolasco, H., et al., 2006. Nutritional value of cowpea (*Vigna unguiculata* L. Walp) meals as ingredients in diets for Pacific white shrimp (*Litopenaeus vannamei* Boone). Food Chem. 97, 41–49.

Robertson, R.F., Meagor, J., Taylor, E.W., 2002. Specific dynamic action in the shore crab, *Carcinus maenas* (L.), in relation to acclimation temperature and to the onset of the emersion response. Physiol. Biochem. Zool. 75, 350–359.

Robinson, E., 1990. Réévaluation of the ascorbic acid (vitamin C) requirement of channel catfish (*Ictalurus punctatus*) (abstract). FASEB J. 4, 3745.

Robinson, E.H., Li, M.H., 1997. Low protein diets for channel catfish *Ictalurus punctatus* raised in earthen ponds at high density. J. World Aquacult. Soc. 28, 224–229.

Robinson, E.H., Lovell, R.T., 1978. Essentiality of biotin for channel catfish *Ictalurus punctatus* fed lipid and lipid-free diets. J. Nutr. 108, 1600–1605.

Robinson, E.H., Wilson, R.P., Poe, W.E., 1980a. Re-evaluation of the lysine requirement and lysine utilization by fingerling channel catfish. J. Nutr. 110, 2313–2316.

Robinson, E.H., Wilson, R.P., Poe, W.E., 1980b. Total aromatic amino acid requirement, phenylalanine requirement and tyrosine replacement value for fingerling channel catfish. J. Nutr. 110, 1805–1812.

Robinson, E.H., Wilson, R.P., Poe, W.E., 1981. Arginine requirement and apparent absence of a lysine–arginine antagonist in fingerling channel catfish. J. Nutr. 111, 46–52.

Robinson, E.H., Rawles, S.D., Yette, H.E., Greene, L.W., 1984. An estimate of the dietary calcium requirement of fingerlings *Tilapia aurea* reared in calcium-free water. Aquaculture 41, 389–393.

Robinson, E.H., Rawles, S.D., Brown, P.B., Yette, H.E., Greene, L.W., 1986. Dietary calcium requirement of channel catfish *Ictalurus punctatus*, reared in calcium-free water. Aquaculture 53, 263–270.

Robinson, E.H., LaBomascus, D., Brown, P.B., Linton, T.L., 1987. Dietary calcium and phosphorus requirements of *Oreochromis aureus* reared in calcium-free water. Aquaculture 64, 267–276.

Rocha, C.B., Portelinha, M.K., Fernandes, J.M., Britto, A.C.P., de Piedras, S.R.N., Pouey, J.L.O.F., 2014. Dietary phosphorus requirement of pejerrey fingerlings (*Odontesthes bonariensis*). Rev. Bras. Zootecnia 43, 55–59.

Rodehutscord, M., 1996. Response of rainbow trout (*Oncorhynchus mykiss*) growing from 50 to 200 g to supplements of dibasic sodium phosphate in a semipurified diet. J. Nutr. 126, 324–331.

Rodehutscord, M., Jacobs, S., Pack, M., Pfeffer, E., 1995. Response of rainbow trout (*Oncorhynchus mykiss*) growing from 50 to 150 g to supplements of DL-methionine in a semipurified diet containing low or high levels of cystine. J. Nutr. 125, 964–969.

Rodehutscord, M., Becker, A., Pack, M., Pfeffer, E., 1997. Response of rainbow trout (*Oncorhynchus mykiss*) to supplements of individual essential amino acids in a semipurified diet, including an estimate of the maintenance requirement for essential amino acids. J. Nutr. 127, 1166–1175.

Rodehutscord, M., Gregus, Z., Pfeffer, E., 2000. Effect of phosphorus intake on faecal and non-faecal phosphorus excretion in rainbow trout (*Oncorhynchus mykiss*) and the consequences for comparative phosphorus availability studies. Aquaculture 188, 383–398.

Rodriguez, C., Perez, J.A., Izquierdo, M.S., Mora, J., Lorenzo, A., Fernandez-Palacios, H., 1994a. Essential fatty acid requirements of larval gilthead sea bream, *Sparus aurata* (L.). Aquacult. Res. 25, 295–304.

Rodriguez, C., Perez, J.A., Izquierdo, M.S., Lorenzo, A., Fernandez-Palacios, H., 1994b. The effect of n-3 HUFA proportions in diets for gilthead seabream (*Sparus aurata*) larval culture. Aquaculture 124, 284.

Rodriguez, C., Perez, J.A., Izquierdo, M.S., Mora, J., Lorenzo, A., Fernandez-Palacios, H., 1994c. Essential fatty acid requirements of larval gilthead sea bream *Sparus aurata* (L). Aquacult. Fish. Manage. 25, 295–304.

Rodriguez, C., Perez, J.A., Badia, P., Izquierdo, M.S., Fernandez-Palacios, H., Hernadez, A.L., 1998a. The n-3 highly unsaturated fatty acids requirements of gilthead seabream (*Sparus aurata* L.) larvae when using an appropriate DHA/EPA ratio in the diet. Aquaculture 169, 9–23.

Roem, A.J., Kohler, C.C., Stickney, R.R., 1990. Vitamin E requirements of the blue tilapia. *Oreochromis aureus* (Steindachner), in relation to dietary lipid levels. Aquaculture 87, 155–164.

Rollin, X., Mambrini, M., Abboudi, T., Larondelle, Y., Kaushik, S.J., 2003. The optimum dietary indispensable amino acid pattern for growing Atlantic salmon (*Salmo salar* L.) fry. Br. J. Nutr. 90, 865–876.

Rosas, C., Sanchez, A., Diaz, E., Soto, L.A., Gaxiola, G., Brito, R., 1996. Effect of dietary protein level on apparent heat increment and postprandial nitrogen excretion of *Penaeus setiferus, P. schmitti, P. duorarum,* and *P. notialis* Postlarvae. J. World Aquacult. Soc. 27, 92–102.

Rosas, C., Martinez, E., Gaxiola, G., Brito, R., Diaz-Iglesia, E., Soto, L.A., 1998. Effect of dissolved oxygen on the energy balance and survival of *Penaeus setiferus* juveniles. Mar. Ecol. Process Series 174, 67–75.

Rosas, C., Cuzon, G., Gaxiola, G., Le Priol, Y., Pascual, C., Rossignyol, J., et al., 2001. Metabolism and growth of juveniles of *Litopenaeus vannamei*: effect of salinity and dietary carbohydrate levels. J. Exp. Mar. Biol. Ecol. 259, 1–22.

Rosenlund, G., Jorgensen, L., Waagbo, R., Sandnes, K., 1990. Effect of different dietary levels of ascorbic acid in plaice *Pleuronectes platessa* L. Comp. Biochem. Physiol. A 96A, 395–398.

Ross, L.G., McKinney, R.W., Cardwell, S.K., Fullarton, J.G., Roberts, S.E.J., Ross, B., 1992. The effects of dietary protein content, lipid content and ration level on oxygen consumption and specific dynamic action in *Oreochromis niloticus* L. Comp. Biochem. Physiol. 103A, 573–578.

Roy, P.K., Lall, S.P., 2003. Dietary phosphorus requirement of juvenile haddock *Melanogrammus aeglefinus* L.). Aquaculture 221, 451–468.

Rumsey, G.L., 1991. Choline–betaine requirements of rainbow trout (*Oncorhynchus mykiss*). Aquaculture 95, 107–116.

Ruxton, C.H.S., Calder, P.C., Reed, S.C., Simpson, M.J.A., 2005. The impact of long-chain n-3 polyunsaturated fatty acids on human health. Nutr. Res. Rev. 18, 113–129.

Ruyter, B., Rosjo, C., Einen, O., Thomassen, M.S., 2000a. Essential fatty acids in Atlantic salmon: time course of changes in fatty acid composition of liver, blood and carcass induced by a diet deficient in n-3 and n-6 fatty acids. Aquacult. Nutr. 6, 109–118.

Ruyter, B., Rosjo, C., Einen, O., Thomassen, M.S., 2000b. Essential fatty acids in Atlantic salmon: effects of increasing dietary doses of n-3 and n-6 fatty acids on growth, survival and fatty acid composition of liver, blood and carcass. Aquacult. Nutr. 6, 119–127.

Sa, M., Pezzato, L.E., Lima, M., Padilha, P.D., 2004. Optimum zinc supplementation level in Nile tilapia *Oreochromis niloticus* juveniles diets. Aquaculture 238, 385–401.

Saad, C.R.B., 1989. Carbohydrate metabolism in channel catfish (PhD dissertation). Auburn University, Auburn, AL, 69 pp.

Sabaut, J.J., Luquet, P., 1973. Nutritional requirements of the gilthead bream (*Chrysophyrys aurata*), quantitative protein requirements. Mar. Biol. 18, 50–54.

Sage, 1973. The evolution of thyroidal function in fishes. Integr. Comp. Biol. 13, 899–905.

Sakamoto, S., Yone, Y., 1973. Effect of dietary calcium/phosphorus ratio upon growth, feed efficiency and blood serum Ca and P level in red sea bream. Bull. Jpn. J. Sci. Fish. 39, 343–348.

Sakamoto, S., Yone, Y., 1976a. Requirement of red sea bream for dietary Ca. Rep. Fish. Res. Lab. Kyushu Univ. (Jpn) 3, 59–64.

Sakamoto, S., Yone, Y., 1976b. Requirement of red sea bream for dietary Fe. Rep. Fish. Res. Lab. Kyushu Univ. (Jpn) 3, 53–58.

Sakamoto, S., Yone, Y., 1978. Effect of dietary phosphorous level on chemical composition red sea bream. Bull. Jpn. Soc. Sci. Fish. 44, 227–229.

Saleh, R., Betancor, M.B., Roo, J., Hernandez-Cruz, C.M., Moyano, F.-J., Izquierdo, M., 2013. Optimum soybean lecithin contents in microdiets for gilthead seabream (Sparus aurata) larvae. Aquacult. Nutr. 19, 585–597.

Salhi, M., Hernandez-Cruz, C.M., Bessonart, M., Izquierdo, M.S., Fernandez-Palacios, H., 1999. Effect of different dietary polar lipid levels and different n-3 HUFA content in polar lipids on gut and liver histological structure of gilthead seabream (Sparus aurata) larvae. Aquaculture 179, 253–263.

Sandnes, K., Torrissen, O., Waagø, R., 1992. The minimum dietary requirement of vitamin C in Atlantic salmon Salmo salar fry using Ca ascorbat-2-monophosphate as dietary source. Fish Physiol. Biochem. 10, 315–319.

Santiago, C.B., Lovell, R.T., 1988. Amino acid requirements for growth of Nile tilapia. J. Nutr. 118, 1540–1546.

Santos, T., Connolly, C., Murphy, R., 2015. Trace element inhibition of phytase activity. Biol. Trace Elem. Res. 163, 255–265.

Sapkale, P.H., Singh, R.K., 2011. Dietary zinc and cobalt requirements of fry of seabass (Lates calcarifer) and catfish (Clarias batrachus). Isr. J. Aquacult. Bamidgeh 63, 613–619.

Sargent, J.R., Tocher, D.R., Bell, J.G., 2002. The lipids. In: Halver, J.E., Hardy, R.W. (Eds.), Fish Nutrition, third ed. Academic Press, San Diego, CA, pp. 181–257.

Sarker, P.K., Satoh, S., Fukada, H., Masumoto, T., 2009. Effects of dietary phosphorus level on non-faecal phosphorus excretion from yellowtail (Seriola quinqueradiata Temminck & Schlegel) fed purified and practical diets. Aquacult. Res. 40, 225–232.

Saroglia, M., Scarano, G., 1992. Experimental induction of ascorbic acid deficiency in seabass in intensive aquaculture. Bull. Eur. Ass. Fish Pathol. 12, 96.

Satia, B.P., 1974. Quantitative protein requirements of rainbow trout. Prog. Fish Cult. 36, 80–85.

Sato, M., Kondo, T., Yashinaka, R., Ikeda, S., 1982. Effect of dietary ascorbic acid levels on collagen formation in rainbow trout. Bull. Jpn. Soc. Sci. Fish. 48, 553–556.

Satoh, S., Takeuchi, T., Watanabe, T., 1987. Requirement of tilapia for α-tocopherol. Nippon Suisan Gakkaishi 53, 119–124.

Satoh, S., Izume, K., Takeuchi, T., Watanabe, T., 1987a. Availability to rainbow trout of zinc contained in various types of fish meals. Nippon Suisan Gakkaishi 53, 1861–1866.

Satoh, S., Takeuchi, T., Watanabe, T., 1987b. Availability to carp of manganese in white fish meal and of various manganese compounds. Nippon Suisan Gakkaishi 53, 825–832.

Satoh, S., Tabata, K., Izume, K., Takeuchi, T., Watanabe, T., 1987c. Effect of dietary tricalcium phosphate on availability of zinc to rainbow trout. Nippon Suisan Gakkaishi 53, 1199–1205.

Satoh, S., Takeuchi, T., Watanabe, T., 1987d. Availability to rainbow trout of zinc in white fish meal and of various zinc compounds. Nippon Suisan Gakkaishi 53, 595–599.

Satoh, S., Izume, K., Takeuchi, T., Watanabe, T., 1989a. Availability to carp of manganese contained in various types of fish meals. Nippon Suisan Gakkaishi 55, 313–319.

Satoh, S., Poe, W.E., Wilson, R.P., 1989b. Studies on the essential fatty acid requirement of channel catfish, Ictalurus punctatus. Aquaculture 79, 121–128.

Satoh, S., Takeuchi, T., Watanabe, T., 1991. Availability of manganese and magnesium contained in white fish meal to rainbow trout *Oncorhynchus mykiss*. Nippon Suisan Gakkaishi 57, 99–104.

Satoh, S., Izume, K., Takeuchi, T., Watanabe, T., 1992. Effect of supplemental tricalcium phosphate on zinc and manganese availability to common carp. Nippon Suisan Gakkaishi 58, 539–545.

Sau, S.K., Paul, B.N., Mohanta, K.N., Mohanty, S.N., 2004. Dietary vitamin E requirement, fish performance and carcass composition of rohu (*Labeo rohita*) fry. Aquaculture 240, 359–368.

Scarpa, J., Gatlin, D.M., 1992. Dietary zinc requirements of channel catfish, *Ictalurus punctatus*, swim-up fry in soft and hard water. Aquaculture 106, 311–322.

Schaefer, A., Koppe, W., Meyer-Burgdorff, K.H., Guenther, K., 1995. Effects of P-supply on growth and mineralization in mirror carp (*Cyprinus carpio* L.). J. Appl. Ichthyol. 11, 397–400.

Schamber, C.R., Boscolo, W.R., Natali, M.R.M., Michelato, M., Furuya, V.R.B., Furuya, W.M., 2014. Growth performance and bone mineralization of large Nile tilapia (*Oreochromis niloticus*) fed graded levels of available phosphorus. Aquacult. Int. 22, 1711–1721.

Sealey, W.M., Gatlin III, D.M., 1999. Dietary vitamin C requirement of hybrid striped bass (*Morone chrysops* × *M. saxatilis cf*). J. World Aquacult. Soc. 30, 297–301.

Sealey, W.M., Barrows, F.T., Hang, A., Johansen, K.A., Overturf, K., LaPatra, S.E., et al., 2008. Evaluation of the ability of barley genotypes containing different amounts of b-glucan to alter growth and disease resistance of rainbow trout *Oncorhynchus mykiss*. Anim. Feed Sci. Technol. 141, 115–128.

Secor, S.M., Diamond, J., 1997. Determinants of post-feeding metabolic response in Burmese pythons, *Python molurus*. Physiol. Zool. 70, 202–212.

Secor, S., 2009. Specific dynamic action: a review of the postprandial metabolic response. J. Comp. Physiol. B 179, 1–56.

Secor, S.M., 2011. Cost of digestion and assimilation. In: Farrell, A.P. (Ed.), Energetics, Interactions with the Environment, Lifestyles, and Applications. Encyclopedia of Fish Physiology: From Genome to Environment, vol. 3. Elsevier, San Diego, CA, USA, pp. 1608–1616.

Sedgwick, R.W., 1979. Influence of dietary protein and energy on growth, food consumption and food conversion efficiency in *Penaeus merguiensis* de man. Aquaculture 16, 7–30.

Seiliez, I., Bruant, J.S., Zambonino-Infante, J., Kaushik, S., Bergot, P., 2006. Effect of dietary phospholipid level on the development of gilthead sea bream (*Sparus aurata*) larvae fed a compound diet. Aquacult. Nutr. 12, 372–378.

Serra, R., Isani, G., Cattani, O., Carpené, E., 1996. Effects of different levels of dietary zinc on the gilthead, *Sparus aurata* during the growing season. Biol. Trace Elem. Res. 51, 107–116.

Serrano, J.A., Nematipour, G.R., Gatlin, D.M., 1992. Dietary protein requirement of the red drum (*Sciaenops ocellatus*) and relative use of dietary carbohydrate and lipid. Aquaculture 101, 283–291.

Serrini, G., Zhang, Z., Wilson, R.P., 1996. Dietary riboflavin requirement of fingerling channel catfish (*Ictalurus punctatus*). Aquaculture 139, 285–290.

Shaik Mohamed, J., 2001a. Dietary pyridoxine requirement for Indian catfish, *Heteropneustes fossilis* (Bloch). Aquaculture 194, 327–335.

Shaik Mohamed, J., 2001b. Dietary biotin requirement determined for Indian catfish. *Heteropneustes fossilis* (Bloch). fingerlings. Aquacult. Res. 32, 709–716.

Shaik Mohamed, J., Ibrahim, A., 2001. Quantifying the dietary niacin requirement of the Indian catfish, *Heteropneustes fossilis* (Bloch). fingerlings. Aquacult. Res. 32, 157–162.

Shaik Mohamed, J., Ravisankar, B., Ibrahim, A., 2000. Quantifying the dietary' biotin requirement of the catfish. *Clarias batrachus*. Aquacult. Int. 8, 9–18.

Shaik Mohamed, J., Sivaram, V., Christopher Roy, T.S., Peter Marian, M., Murugadass, S.,
Saffiq Hussain, M., 2003. Dietary vitamin A requirement of juvenile greasy grouper
(*Epinephelus tauvina*). Aquaculture 219, 693–701.

Shao, Q., Ma, J., Xu, Z., Hu, W., Xu, J., Xie, S., 2008. Dietary phosphorus requirement of
juvenile black seabream, *Sparus macrocephalus*. Aquaculture 277, 92–100.

Shao, X., Liu, W., Xu, W., Lu, K., Xia, W., Jiang, Y., 2010. Effects of dietary copper sources
and levels on performance, copper status, plasma antioxidant activities and relative copper
bioavailability in *Carassius auratus* gibelio. Aquaculture 308, 60–65.

Shao, X.-P., Liu, W.-B., Lu, K.-L., Xu, W.-N., Zhang, W.-W., Wang, Y., et al., 2012. Effects of
tribasic copper chloride on growth, copper status, antioxidant activities, immune responses
and intestinal microflora of blunt snout bream (*Megalobrama amblycephala*) fed practical
diets. Aquaculture 338–341, 154–159.

Shearer, K.D., 1988. Dietary potassium requirement of juvenile chinook salmon. Aquaculture
73, 119–129.

Shearer, K.D., 1989. Whole body magnesium concentration as an indicator of magnesium
status in rainbow trout (*Salmo gairdneri*). Aquaculture 77, 201–210.

Shearer, K.D., Åsgård, T., 1992. The effect of water-borne magnesium on the dietary magnesium requirement of the rainbow trout (*Oncorhynchus mykiss*). Fish Physiol. Biochem. 9,
387–392.

Shen, X., Liu, Y., 1992. On the digestibility of protein, oil and starch in *Penaeus orientalis*. Can.
Transl. Fish. Aquatic Sci. 5576, 17.

Shewbart, K.L., Mies, W.L., 1973. Studies on the nutritional requirements of brown shrimp:
the effects of linolenic acid on growth of *Penaeus aztecus*. Proc. World Maricult. Soc. 4,
277–287.

Shiau, S.Y., Chen, Y., 2000. Estimation of the dietary vitamin A requirement of juvenile grass
shrimp. *Penaeus monodon*. J. Nutr. 130, 90–94.

Shiau, S.Y., Chin, Y.H., 1998. Dietary biotin requirement for maximum growth of juvenile
grass shrimp. *Penaeus monodon*. J. Nutr. 128, 2494–2497.

Shiau, S.Y., Chin, Y.H., 1999. Estimation of the dietary biotin requirement of juvenile hybrid
tilapia. *Oreochromis niloticus* × *O. aureus*. Aquaculture 170, 71–78.

Shiau, S., Chou, B., 1991. Effects of dietary protein and energy on growth performance of tiger
shrimp *Penaeus monodon* reared in seawater. Nippon Suisan Gakkaishi 57, 2271–2276.

Shiau, S.Y., Hsieh, H.L., 1997. Vitamin B_6 requirements of tilapia *Oreochmmis niloticus* ×
O. aureus fed two dietary protein concentrations. Fish. Sci. 63, 1002–1007.

Shiau, S.-Y., Hsieh, J.-F., 2001a. Quantifying the dietary potassium requirement of juvenile
hybrid tilapia (*Oreochromis niloticus* × *O. aureus*). Br. J. Nutr. 85, 213–218.

Shiau, S.-Y., Hsieh, J.-F., 2001b. Dietary potassium requirement of juvenile grass shrimp
Penaeus monodon. Fish. Sci. 67, 592–595.

Shiau, S.Y., Hsu, T.S., 1994. Vitamin C requirement of grass shrimp, *Penaeus mondon*. as
determined with L-ascorbyl-2-monophosphate. Aquaculture 122, 347–357.

Shiau, S.Y., Hsu, T.S., 1995. L-Ascorbyl-2-sulfate has equal antiscorbutic activity as L-ascorbyl-
2-monophosphate for tilapia. *Oreochromis niloticus* × *O. aureus*. Aquaculture 133,
147–157.

Shiau, S.Y., Hsu, T.S., 1999a. Quantification of vitamin C requirement for juvenile hybrid
tilapia. *Oreochromis niloticus* × *O. aureus*, with L-ascorbyl-2-monophosphate-Na and
L-ascorbyl-2-monophosphate-Mg. Aquaculture 175, 317–326.

Shiau, S.Y., Hsu, C.W., 1999b. Dietary pantothenic acid requirement of juvenile grass shrimp,
Penaeus monodon. J. Nutr. 129, 718–721.

Shiau, S.Y., Huang, S.L., 1990. Influence of varying energy levels with two protein concentrations in diets for hybrid tilapia (*Oreochromis niloticus* × *Oreochromis aureaus*) reared in seawater. Aquaculture 91, 143–152.

Shiau, S.Y., Huang, S.Y., 2001a. Dietary folic acid requirement for maximal growth of juvenile tilapia, *Oreochromis niloticus* × *O. aureus*. Fish. Sci. 67, 655–659.

Shiau, S.Y., Huang, S.Y., 2001b. Dietary folic acid requirement determined for grass shrimp, *Penaeus monodon*. Aquaculture 200, 339–347.

Shiau, S.Y., Hwang, J.Y., 1993. Vitamin D requirement of juvenile hybrid tilapia *Oreochromis niloticus* × *O. aureus*. Nippon Suisan Gakkaishi 59, 553–558.

Shiau, S.Y., Hwang, J.Y., 1994. The dietary requirement of juvenile grass shrimp. *Penaeus monodon*. for vitamin D. J. Nutr. 124, 2445–2450.

Shiau, S.Y., Jan, F.L., 1992a. Dietary ascorbic acid requirement of juvenile tilapia *Oreochromis niloticus* × *O. aureus*. Nippon Suisan Gakkaishi 58, 671–675.

Shiau, S.Y., Jan, F.L., 1992b. Ascorbic acid requirement of grass shrimp *Penaeus monodon*. Nippon Suisan Gakkaishi 58, 363.

Shiau, S.-Y., Jiang, L.-C., 2006. Dietary zinc requirements of grass shrimp, *Penaeus monodon*, and effects on immune responses. Aquaculture 254, 476–482.

Shiau, S.Y., Lan, C.W., 1996. Optimum dietary protein level and protein to energy ratio for growth of grouper (*Epinephelus malabaricus*). Aquaculture 145, 259–266.

Shiau, S.Y., Liang, H.S., 1995. Carbohydrate utilization and digestibility by tilapia, *Oreochromis niloticus* × *O. aureaus*, are affected by chromic oxide inclusion in the diet. J. Nutr. 125, 976–982.

Shiau, S.Y., Liu, J.S., 1994a. Quantifying the vitamin K requirement of juvenile marine shrimp. *Penaeus monodon*. with menadione. J. Nutr. 124, 277–282.

Shiau, S.Y., Liu, J.S., 1994b. Estimation of the dietary vitamin K requirement of juvenile *Penaeus chinensis* using menadione. Aquaculture 126, 129–135.

Shiau, S.Y., Lo, P.S., 2000. Dietary choline requirements of juvenile hybrid tilapia. *Oreochromis niloticus* × *O. aureus*. J. Nutr. 130, 100–103.

Shiau, S.Y., Lo, P.S., 2001. Dietary choline requirement of juvenile grass shrimp. *Penaeus monodon*. Anim. Sci. 72, 477–482.

Shiau, S.Y., Lung, C.Q., 1993a. No dietary vitamin B, required for juvenile tilapia *Oreochromis niloticus* × *O. aureus*. Comp. Biochem. Physiol. A 105A, 147–150.

Shiau, S.Y., Lung, C.Q., 1993b. Estimation of the vitamin B, requirements of grass shrimp. *Penaeus monodon*. Aquaculture 117, 157–163.

Shiau, S.Y., Peng, C.-Y., 1992. Utilization of different carbohydrates at different dietary levels in grass prawn *Penaeus monodon*, reared in seawater. Aquaculture 101, 241–250.

Shiau, S.-Y., Peng, C.-Y., 1993. Protein-sparing effect by carbohydrates in diets for tilapia, *Oreochromis niloticus* × *O. aureus*. Aquaculture 117, 327–334.

Shiau, S.Y., Shiau, L.F., 2001. Re-evaluation of the vitamin E requirements of juvenile tilapia (*Oreochromis niloticus* × *O. aureus*). Anim. Sci. 72, 529–534.

Shiau, S.Y., Su, L.W., 2003. Ferric citrate is half as effective as ferrous sulfate in meeting the iron requirement of juvenile tilapia, *Oreochromis niloticus* × *O. aureus*. J. Nutr. 133, 483–488.

Shiau, S.Y., Su, S.L., 2004. Dietary inositol requirement determine for juvenile grass shrimp. *Penaeus monodon*. Aquaculture 241, 1–8.

Shiau, S.Y., Su, S.L., 2005. Juvenile tilapia (*Oreochmmis niloticus* × *O. aureus*) requires dietary myo-inositol for maximal growth. Aquaculture 243, 273–277.

Shiau, S.Y., Suen, G.S., 1992. Estimation of the niacin requirements for tilapia fed diets containing glucose or dextrin. J. Nutr. 122, 2030–2036.

Shiau, S.Y., Suen, G.S., 1994. The dietary requirement of juvenile grass shrimp (*Penaeus monodon*) for niacin. Aquaculture 125, 139–145.

Shiau, S.Y., Tseng, H.C., 2007. Dietary calcium requirements of juvenile tilapia, *Oreochromis niloticus* × *O. aureus*, reared in fresh water. Aquacult. Nutr. 13, 298–303.

Shiau, S.Y., Wu, M.H., 2003. Dietary vitamin B_6 requirement determined for grass shrimp. *Penaeus monodon*. Aquaculture 225, 397–404.

Shibuya, N., Nakane, R., 1984. Pectic Polysaccharides of rice endosperm cell walls. Phytochemistry 23, 1425–1429.

Shieh, H.S., 1969. The biosynthesis of phospholipids in the lobsters, *Homarus americanus*. Comp. Biochem. Physiol. 30, 679–684.

Shigueno, K., Itoh, S., 1988. Use of Mg-L-ascorbyl-2-phosphate as a vitamin C source in shrimp diets. World Aquacult. Soc. 19, 168–174.

Shim, K.F., Ho, C.S., 1989. Calcium and phosphorus requirements of guppy *Poecilia reticulata*. Nippon Suisan Gakkaishi 55, 1947–1953.

Shim, K.F., Ng, S.H., 1988. Magnesium requirement of the guppy (*Poecilia reticulata* Peters). Aquaculture 73, 131–141.

Shimeno, S., 1991. Yellowtail. *Seriola quinqueradiata*. In: Wilson, R.P. (Ed.), Handbook of Nutrient Requirements of Finfish. CRC Press, Boca Raton, FL, pp. 181–191.

Shimeno, S., Hosokawa, H., Hirata, H., Takeda, M., 1977. Comparative studies of on carbohydrate metabolism of yellowtail and carp. Bull. Jpn. Soc. Sci. Fish. 43, 213–217.

Shudo, K., Nakamura, K., Ishikawa, S., Kitabayashi, K., 1971. Studies on formula feed for Kuruma prawn-IV. On the growth-promoting effects on both squid liver oil and cholesterol. Bull. Tokai Reg. Fish. Res. Lab. 65, 129–137.

Shurson, G.C., Salzerb, T.M., Koehlerc, D.D., Whitney, M.H., 2011. Effect of metal specific amino acid complexes and inorganic trace minerals on vitamin stability in premixes. Anim. Feed Sci. Technol. 163, 200–206.

Siddiqui, A.Q., Howlader, M.S., Adam, A.A., 1988. Effects of dietary protein levels on growth, feed conversion and protein utilization in fry and young Nile tilapia, *Oreochromis niloticus*. Aquaculture 70, 63–73.

Simkiss, K., Wilbur, K.M., 1989. Biomineralization: Cell Biology and Mineral Deposition. Academic Press, San Diego.

Singh, S., Khan, M.A., 2007. Dietary arginine requirement of fingerling hybrid Ciarías (*Ciarías gariepinus* × *Ciarías macrocephalus*). Aquacult. Res. 38, 17–25.

Singh, R.P., Nose, T., 1967. Digestibility of carbohydrates in young rainbow trout. Bull. Freshw. Fish. Res. Lab. (Japan) 17, 21–25.

Sinha, A.K., Kumar, V., Makkar, H.P.S., De Boeck, G., Becker, K., 2011. Non-starch polysaccharides and their role in fish nutrition – a review. Food Chem. 127, 1409–1426.

Sink, T.D., Lochmann, R.T., 2014. The effects of soybean lecithin supplementation to a practical diet formulation on juvenile channel catfish, *Ictalurus punctatus*: growth, survival, hematology, innate immune activity, and lipid biochemistry. J. World Aquacult. Soc. 45, 163–172.

Skonberg, D.I., Yogev, L., Hardy, R.W., Dong, F.M., 1997. Metabolic response to dietary phosphorus intake in rainbow trout (*Oncorhynchus mykiss*). Aquaculture 157, 11–24.

Smith, R.R., Rumsey, G.L., Scott, M.L., 1978. Heat increment associated with dietary protein, fat, carbohydrate and complete diets in salmonids: comparative energetic efficiency. J. Nutr. 108, 1025–1032.

Smith, L.L., Lee, P.G., Lawrence, A.L., Strawn, K., 1985. Growth and digestibility by three sizes of *Penaeus vannamei* Boone: effects of dietary protein level and protein source. Aquaculture 46, 85–96.

Söderhäll, K., 1982. The prophenoloxidase activating system and melanization – a recognition mechanism of arthropods? A review. Dev. Comp. Immunol. 6, 601–611.

Soengas, J.L., Strong, E.F., Fuentes, J., Veira, J.A.R., Andrés, M.D., 1996. Food deprivation and refeeding in Atlantic salmon. *Salmo salar*. Effects on brain and liver carbohydrate and ketone bodies metabolism. Fish Physiol. Biochem. 15, 491–511.

Soler, M.P., 1996. Nutrientes esenciales. In: Soler, M.P., Rodríguez, H., Daza, P.D. (Eds.), Fundamentos de Nutrición y Alimentación en Acuicultura. Serie III Cal Publicidad, Santa Fe de Bogotá, Colombia, pp 55–116.

Soliman, A.K., Wilson, R.R., 1992a. Water-soluble vitamin requirements of tilapia. II. Riboflavin requirement of blue tilapia, *Oreoehromis aureus*. Aquaculture 104, 309–314.

Soliman, A.K., Wilson, R.P., 1992b. Water-soluble vitamin requirements of tilapia. I. Pantothenic acid requirement of blue tilapia. *Oreoehromis aureus*. Aquaculture 104, 121–126.

Soliman, A.K., Jauncey, K., Roberts, R.J., 1994. Water-soluble vitamin requirements of tilapia: ascorbic acid (vitamin C) requirement of Nile tilapia. *Oreoehromis niloticus* (L.). Aquacult. Fish. Manage. 25, 269–278.

Somboon, S., Semachai, V., 2010. Requirement of protein and digestible energy in diets on growth performance of basa catfish (*Pangasius bocouti*, Sauvage, 1880). In: Proceedings of the 48th Kasetsart University Annual Conference, Kasetsart, March 3–5, 2010.

Sritunyalucksana, K., Intaraprasong, A., Sa-nguanrut, P., Filer, K., 2011. Organic selenium supplementation promotes shrimp growth and disease resistance to Taura syndrome virus. ScienceAsia 37, 24–30.

Stanley, R.W., Moore, L.B., 1983. Effect on growth and apparent digestibility of diets varying in grain source and protein level in *Macrobrachium rosenbergii*. J. World Maricult. Soc. 14, 174–184.

Stickney, R.R., McGeachin, R.B., Lewis, D.H., Marks, J., Riggs, A., Sis, R.F., et al., 1984. Response of *Tilapia aurea* to dietary vitamin C. J. World Maricult. Soc. 15, 179–185.

Stone, D.A.J., 2003. Dietary carbohydrate utilization by fish. Rev. Fish. Sci. 11, 337–369.

Stone, D.A.J., Allan, G.L., Anderson, A.J., 2003. Carbohydrate utilization by juvenile silver perch, *Bidyanus bidyanus* (Mitchell). II. Digestibility and utilization of starch and its breakdown products. Aquacult. Res. 34, 109–121.

Storebakken, T., 1985. Binders in fish feeds: I. Effect of alginate and guar gum on growth, digestibility, feed intake and passage through the gastrointestinal tract of rainbow trout. Aquaculture 47, 11–26.

Storebakken, T., 2002. In: Webster, C.D., Lim, C. (Eds.), Atlantic Salmon, *Salmo salar* Nutrient Requirements and Feeding of Finfish for Aquaculture. CAB International, Wallingford, UK, pp. 79–102.

Su, S.L., Shiau, S.Y., 2004. Requirements of dietary myo-insitol of juvenile grouper, Epinephelus malabaricus. J. Fish. Soc. Taiwan 31, 313–317.

Su, C., Luo, L., Wen, H., Sheng, X., Li, S., 2007. Effects of dietary iron on growth performance, nutritional composition and some blood indices of grass carp (*Ctenopharyngodon idellus*). Freshw. Fish. 37, 48–52.

Sugiura, S.H., Dong, F.M., Hardy, R.W., 2000a. A new approach to estimating the minimum dietary requirement of phosphorus for large rainbow trout based on nonfecal excretions of phosphorus and nitrogen. J. Nutr. 130, 865–872.

Sugiura, S.H., Dong, F.M., Hardy, R.W., 2000b. Primary responses of rainbow trout to dietary phosphorus concentrations. Aquacult. Nutr. 6, 235–245.

Sugiura, S.H., Kelsey, K., Ferraris, R.P., 2007. Molecular and conventional responses of large rainbow trout to dietary phosphorus restriction. J. Comp. Physiol. B 177, 461–472.

Sukumaran, K., Pal, A.K., Sahu, N.P., Debnath, D., Patro, B., 2009. Phosphorus requirement of Catla (*Catla catla* Hamilton) fingerlings based on growth, whole-body phosphorus concentration and non-faecal phosphorus excretion. Aquacult. Res. 40, 139–147.

Suprayudi, M.A., Takeuchi, T., Hamasaki, K., 2012. Cholesterol effect on survival and development of larval mud crab *Scylla serrata*. HAYATI. J. Biosci. 19, 1–5.

Taboada, G., Gaxiola, G., García, T., Pedroza, R., Sánchez, A., Soto, L., et al., 1998. Oxygen consumption and ammonia-N excretion related to protein requirements for growth of white shrimp, *Penaeus setiferus* (L), juveniles. Aquacult. Res. 29, 1–11.

Tackaert, W., Camara, M., Sorgeloos, P., 1991. The effect of dietary phosphatidylcholine in postlarval penaeid shrimp. II. Preliminary culture results. In: Larvi '91-Fish & Crustacean Larviculture Symposium. Lavens, P., Sorgeloos, P., Jaspers, E., Ollevier, F., (Eds.), European Aquaculture Society, Special Publications N°15, Ghent, Belgium, pp. 80–82.

Taketani, Y., Segawa, H., Chikamori, M., et al., 1998. Regulation of type II renal Na+-dependent inorganic phosphate transporters by 1,25-dihydroxyvitamin D3. J. Biol. Chem. 273, 14575–14581.

Takeuchi, T., 1997. Essential fatty acids requirements of aquatic animals with emphasis on fish larvae and fingerlings. Rev. Fish. Sci. 5, 1–25.

Takeuchi, T., Watanabe, T., 1976. Nutritive value of ω-3 highly unsaturated fatty acids in pollock liver oil for rainbow trout. Bull. Jpn. Soc. Sci. Fish. 42, 907–919.

Takeuchi, T., Watanabe, T., 1977. Requirement of carp for essential fatty acids. Bull. Jpn. Soc. Sci. Fish. 43, 541–551.

Takeuchi, T., Yokoyama, M., et al., 1978. Optimum ratio of dietary energy to protein for rainbow trout. Nippon Suisan Gakkaishi 44, 729–732.

Takeuchi, T., Watanabe, T., Nose, T., 1979a. Requirement for essential fatty acids of chum salmon (*Oncorhyncus keta*) in freshwater environment. Bull. Jpn. Soc. Sci. Fish. 45, 1319–1323.

Takeuchi, T., Watanabe, T., Ogino, C., 1979b. Optimum ratio of dietary energy to protein for carp. Bull. Jpn. Soc. Sci. Fish. 45, 983–987.

Takeuchi, T., Arais, S., Watanabe, T., Shimma, Y., 1980a. Requirements of the eel *Anguilla japonica* for essential fatty acids. Bull. Jpn. Soc. Sci. Fish. 46, 345–353.

Takeuchi, L., Takeuchi, T., Ogino, C., 1980b. Riboflavin requirements in carp and rainbow trout. Bull. Jpn. Soc. Sci. Fish. 46, 733–737.

Takeuchi, T., Satoh, S., Watanabe, W., 1983. Requirement of *Tilapia nilotica* for essential fatty acids. Bull. Jpn. Soc. Sci. Fish. 49, 1127–1134.

Takeuchi, T., Toyota, M., Satoh, S., Watanabe, T., 1990. Requirement of juvenile red sea bream (*Pagrus major*) for eicosapentaenoic and docosahexanoic acids. Bull. Jpn. Soc. Sci. Fish. 56, 1263–1269.

Takeuchi, T., Watanabe, K., Yong, W.-Y., Watanabe, T., 1991. Essential fatty acids of grass carp (*Ctenopharyngodon idella*). Nippon Suisan Gakkaishi 57, 467–473.

Takeuchi, T., Arakawa, T., Satoh, S., Watanabe, T., 1992a. Supplemental effect of phospholipids and requirement of eicosapentaenoic acid and docosahexaenoic acid of juvenile striped jack. Nippon Suisan Gakkaishi 58, 707–713.

Takeuchi, T., Watanabe, K., Satoh, S., Watanabe, T., 1992b. Requirement of grass carp fingerlings for α-tocopherol. Nippon Suisan Gakkaishi 58, 1743–1749.

Takeuchi, T., Feng, Z., Yoseda, K., Hirokawa, J., Watanabe, T., 1994. Nutritive value of DHA-enriched rotifer for larval cod. Nippon Suisan Gakkaishi 60, 641–652.

Takeuchi, T., Masuda, R., Ishizaki, Y., Watanabe, T., Kanematsu, M., Imaizumi, K., et al., 1996. Determination of the requirement of larval striped jack for eicosapentaenoic acid and docosahexaenoic acid using enriched *Artemia nauplii*. Fish. Sci. 62, 760–765.

Takeuchi, T., Satoh, S., Kiron, V., 2002. Common carp, *Cyprinus carpio*. In: Webster, C.D., Lim, C. (Eds.), Nutrient Requirements and Feeding of Finfish for Aquaculture. CABI Publishing, New York, NY, pp. 245–261.

Tan, B., Mai, K., 2001. Effects of dietary vitamin K on survival, growth, and tissue concentrations of phylloquinone (PK) and menaquinone-4 (MK-4) for juvenile abalone, *Haliotis discus* hannai Ino. J. Exp. Mar. Biol. Ecol. 256, 229–239.

Tan, L.N., Feng, L., Liu, Y., Jiang, J., Jiang, W.D., Hu, K., et al., 2011a. Growth, body composition and intestinal enzyme activities of juvenile Jian carp (*Cyprinus carpio* var. Jian) fed graded levels of dietary zinc. Aquacult. Nutr. 17, 338–345.

Tan, X.Y., Luo, Z., Liu, X., Xie, C.X., 2011b. Dietary copper requirement of juvenile yellow catfish *Pelteobagrus fulvidraco*. Aquacult. Nutr. 17, 170–176.

Tan, X.Y., Xie, P., Luo, Z., Lin, H.Z., Zhao, Y.H., Xi, W.Q., 2012. Dietary manganese requirement of juvenile yellow catfish *Pelteobagrus fulvidraco*, and effects on whole body mineral composition and hepatic intermediary metabolism. Aquaculture 326–329, 68–73.

Teshima, S., 1971. *In vivo* transformation of ergosterol to cholesterol in crab, *Portunus trituberculatus*. Bull. Jpn. Soc. Sci. Fish. 37, 671–674.

Teshima, S., 1982. Sterol metabolism. In: En: Pruder, Langdon, Conklin (Eds.), Proceedings of the Second International Conference on Aquaculture Nutrition. Louisiana State University, Baton Rouge, LA, pp. 205–216.

Teshima, S., 1997. Phospholipids and sterols. In: D'Abramo, L.R., Conklin, D.E., Akiyama, D.M. (Eds.), Advances in World Aquaculture, Volume 6. Crustacean Nutrition World Aquaculture Society, Baton Rouge, LA, pp. 85–107.

Teshima, S., 1998. Nutrition of Penaeus japonicus. Rev. Fish. Sci. 6, 97–111.

Teshima, S., Kanazawa, A., 1971a. Biosynthesis of sterols in the lobster *Panulirus japonica*, the prawn *Penaeus japonicus* and the crab *Portunus trituberculatus*. Comp. Biochem. Physiol. 38B, 597–602.

Teshima, S., Kanazawa, A., 1971b. Bioconversion of the dietary ergosterol to cholesterol in *Artemia salina*. Comp. Biochem. Physiol. 38B, 603–607.

Teshima, S., Kanazawa, A., 1972. *In vivo* bioversion of β-sitosterol to cholesterol in the crab, *Portunus trituberculatus*. Mem. Fac. Fish. Kagoshima Univ. 21, 91–95.

Teshima, S., Kanazawa, A., 1973a. Bioconversion of brassicasterol to cholesterol in *Artemia salina*. Bull. Jpn. Soc. Sci. Fish. 38, 1305–1310.

Teshima, S., Kanazawa, A., 1973b. Metabolism of desmosterol in the prawn, *Penaeus japonicas*. Mem. Fac. Fish. Kagoshima Univ. 22, 15–19.

Teshima, S., Kanazawa, A., 1984. Effects of protein, lipid and carbohydrate levels in purified diets on growth and survival rates of the prawn larvae. Bull. Jpn. Soc. Sci. Fish. 50, 1709–1715.

Teshima, S., Kanazawa, A., Okamoto, H., 1974. Absorption of sterols and cholesterylesters in a prawn, *Penaeus japonicus*. Bull. Jpn. Soc. Sci. Fish. 40, 1015–1019.

Teshima, S., Ceccaldi, H.J., Patrois, J., Kanazawa, A., 1975. Bioconversion of desmoterol to cholesterol at various stages of molting cycle in *Palaemon serratus* Pennat, Crustacea Decapoda. Comp. Biochem. Physiol. 50B, 485–489.

Teshima, S., Kanazawa, A., Okamoto, H., 1976. Sterol biosynthesis from acetate and the fate of dietary cholesterol and demosterol in crabs. Bull. Jpn. Soc. Sci. Fish. 42, 1273–1280.

Teshima, S., Kanazawa, A., Sasada, H., Kawasaki, M., 1982a. Nutritional value of dietary cholesterol and other sterols to larval prawn, *Penaeus japonicas* Bate. Aquaculture 31, 159–167.

Teshima, S., Kanazawa, A., Sasada, H., Kawasaki, M., 1982b. Requirements of the larval prawn, *Penaeus japonicus*, for cholesterol and soybean phospholipids. Mem. Fac. Fish. 31, 193–199.

Teshima, S., Kanazawa, A., Kakuta, Y., 1986a. Effects of dietary phospholipids on growth and body composition of the juvenile prawn. Bull. Jpn. Soc. Sci. Fish. 52, 155–158.

Teshima, S., Kanazawa, A., Kakuta, Y., 1986b. Growth, survival and body composition of the prawn larvae receiving several dietary phospholipids. Mem. Fac. Fish. Kagoshima Univ. 35, 17–27.

Teshima, S.I., Kanazawa, A., Koshio, S., Itoh, S., 1991. L-ascorbyl-2-phosphate-Mg as vitamin C source for the Japanese flounder (Paralichthys olivaceus). In: Kaushik, J., Lequet, P. (Eds.), Fish Nutrition in Practice, Coll. les Colloq., No. 6IS. INRA, Paris, France, pp. 157–166.

Teshima, S., Ishikawa, M., Koshio, S., Kanazawa, A., 1997. Necessity of dietary cholesterol for the freshwater prawn. Fish. Sci. 63, 596–599.

Teshima, S., Koshio, S., Ishikawa, M., Kanazawa, A., 2001. Protein requirement of the prawn Marsupenaeus japonicus estimated by a factorial method. Hydrobiologia 449, 293–300.

Teshima, S., Alam, M.S., Koshio, S., Ishikawa, M., Kanazawa, A., 2002. Assessment of requirement values for essential amino acids in the prawn, Marsupenaeus japonicus (Bate). Aquacult. Res. 33, 395–402.

Thanuthong, T., Francis, D.S., Manickam, E., Senadheera, S.D., Cameron-Smith, D., Turchini, G.M., 2011. Fish oil replacement in rainbow trout diets and total dietary PUFA content: II) Effects on fatty acid metabolism and in vivo fatty acid bioconversion. Aquaculture 322-323, 99–108.

Thongrod, S., Boonyaratpalin, M., 1998. Cholesterol and lecithin requirement of juvenile banana shrimp, Penaeus merguiensis. Aquaculture 161, 315–321.

Thongrod, S., Takeuchi, T., Satoh, S., Watanabe, T., 1989. Requirement of fingerling white fish Coregonus lavaretus maraena for dietary n-3 fatty acids. Bull. Jpn. Soc. Sci. Fish. 55, 1983–1987.

Thongrod, S., Takeuchi, T., Satoh, S., Watanabe, T., 1990. Requirement of Yamane (Oncorhynchus masou) for essential fatty acids. Nippon Suisan Gakkaishi 56, 1255–1262.

Thorarensen, H., Farrell, A.P., 2006. Postprandial intestinal blood flow, metabolic rates, and exercise in chinook salmon (Oncorhynchus tshawytscha). Physiol. Biochem. Zool. 79, 688–694.

Tibaldi, E., Lanari, D., 1991. Optimal dietary lysine levels for growth and protein utilisation of fingerling sea bass (Dicentrarchus labrax L.) fed semipurified diets. Aquaculture 95, 297–304.

Tibaldi, E., Tulli, F., 1999. Dietary threonine requirement of juvenile European sea bass (Dicentrarchus labrax). Aquaculture 175, 155–166.

Tibaldi, E., Tulli, F., Lanari, D., 1994. Arginine requirement and effect of different dietary arginine and lysine levels for fingerling sea bass (Dicentrarchus labrus). Aquaculture 127, 207–218.

Tocher, D.R., 1995. Glycerophospholipid metabolism. In: Hochachka, P.W. Mommsen, T.P. (Eds.), Biochemistry and Molecular Biology of Fishes. Metabolic and Adaptational Biochemistry, vol. 4. Elsevier Press, Amsterdam, pp. 119–157.

Tocher, D.R., 2003. Metabolism and functions of lipids and fatty acids in teleost fish. Rev. Fish. Sci. 11, 107–184.

Tocher, D.R., 2010. Fatty acid requirements in ontogeny of marine and freshwater fish. Aquacult. Res. 41, 717–732.

Tocher, D.R., Carr, J., Sargent, J.R., 1989. Polyunsaturated fatty acid metabolism in fish cells: differential metabolism of (n-3) and (n-6) series acids by cultured cells originating from a freshwater teleost fish and from a marine teleost fish. Comp. Biochem. Physiol. B Comp. Biochem. 94, 367–374.

Tocher, D.R., Bendiksen, E.A., Campbell, P.J., Bell, J.G., 2008. The role of phospholipids in nutrition and metabolism of teleost fish. Aquaculture 230, 21–34.

Torres, J.J., Brightman, R.I., Donnelly, J., Harvey, J., 1996. Energetics of larval red drum, *Sciaenops occelatus*. Part 1: oxygen consumption, specific dynamic action, and nitrogen excretion. Fish Bull. 94, 756–765.

Torstensen, B.E., Frøyland, L., Lie, Ø., 2004. Replacing dietary fish oil with increasing levels of rapeseed oil and olive oil – effects on Atlantic salmon (*Salmo salar* L.) tissue and lipoprotein lipid composition and lipogenic enzyme activities. Aquacult. Nutr. 10, 175–192.

Tosaka, O., Enei, H., Hirose, Y., 1983. The production of L-lysine by fermentation. Trends Biotechnol. 1, 70–74.

Tran-Duy, A., Smit, B., van Dam, A.A., Schrama, J.W., 2008. Effects of dietary starch and energy levels on maximum feed intake, growth and metabolism of Nile tilapia, *Oreochromis niloticus*. Aquaculture 277, 213–219.

Tuan, L.A., Williams, K.C., 2007. Optimum dietary protein and lipid specifications for juvenile malabar grouper (*Epinephelus malabaricus*). Aquaculture 267, 129–138.

Tulli, E., Messina, M., Calligaris, M., Tibaldi, E., 2010. Response of European sea bass (*Dicentrarchus labrax*) to graded levels of methionine (total sulfur amino acids) in soya protein-based semi-purified diets. Br. J. Nutr. 104, 664–673.

Twibell, R.G., Brown, P.B., 1997. Dietary arginine requirement of juvenile yellow perch. J. Nutr. 127, 1838–1841.

Twibell, R.G., Brown, P.B., 1998. Optimal dietary protein concentration for hybrid tilapia (*Oreochromis niloticus* × *Oreochromis aureus*) fed all-plant diets. J. World Aquacult. Soc. 29, 9–16.

Twibell, R.G., Brown, P.B., 2000. Dietary choline requirement of juvenile yellow perch (*Perca flavescens*). J. Nutr. 130, 95–99.

Twibell, R.G., Wilson, K.A., Brown, P.B., 2000. Dietary sulfur amino acid requirement of juvenile yellow perch fed the maximum cystine replacement value for methionine. J. Nutr. 130, 612–616.

Ufodike, E.B.C., Matty, A.J., 1989. Effect of potato and corn meal on protein and carbohydrate digestibility by rainbow trout. Prog. Fish Cult. 51, 113–114.

Ukawa, M., Takii, K., Nakamura, M., Akutsu, M., Kumai, H., Takeda, M., 1994. Effect of iron supplements on a soy protein concentrate diet on hematology of yellowtail. Fish. Sci. 60, 165–169.

Unestam, T., Beskow, S., 1976. Phenol oxidase in crayfish blood: activation by and attachment on cells of other organisms. J. Invertebr. Pathol. 27, 297–305.

Uyan, O., Koshio, S., Ishikawa, M., Uyan, S., Ren, T., Yokoyama, S., et al., 2007. Effects of dietary phosphorus and phospholipid level on growth, and phosphorus deficiency signs in juvenile Japanese flounder, *Paralichthys olivaceus*. Aquaculture 267, 44–54.

Van Wormhoudt, A., Le Moullac, G., Klein, B., Sellos, D., 1996. Caracterización de las tripsinas y amilasas de *Penaeus vannamei* (Crustacea, Decapoda): Adaptación a la composición del régimen alimenticio. En: Avances en Nutrición Acuícola III. Memorias del III Simposium Internacional de Nutrición Acuícola. UANL. Monterrey, NL, México. 673 pp.

Vázquez-Añón, M., González-Esquerra, R., Saleh, E., Hampton, T., Ritcher, S., Firman, J., et al., 2006. Evidence for 2-hydroxy-4(methylthio) butanoic acid and DL-methionine having different dose responses in growing broilers. Poult. Sci. 85, 1409–1420.

Vedenov, D., Pesti, G.M., 2010. An economic analysis of a methionine source comparison response model. Poult. Sci. 89, 2514–2520.

Vega-villasante, F., Nolasco, H., Rivera, C., 1993. The digestive enzymes of the Pacific brown shrimp *Penaeus californiensis*. I – Properties of amylase activity in the digestive tract. J. Comp. Biochem. Physiol. 106B, 547–550.

Velasco, M., Lawrence, A.L., Castille, F.L., Obaldo, L.G., 2000. Dietary protein require-ment for *Litopenaeus vannamei*. Cruz-Suarez, L.E., Ricque-Marie, D., Tapia-Salazar, M., Olvera-Novoa, M.A., Civera-DCerecedo, R. (Eds.), Advances en Nutricion Aauicola V. Memorias del V. Symposium Internacional de Nutricion Acuicola. Merida, Mexico, pp. 181–192.

Velurtas, S.M., Diaz, A.C., Fernandez-Gimenez, A.V., Fenucci, J.L., 2011. Influence of dietary starch and cellulose levels on the metabolic profile and apparent digestibility in penaeid shrimp. Latin Am. J. Aquat. Res. 39, 214–224.

Venkataramiah, A., Lakshmi, G., Gunther, G., 1975. Effect of protein level and vegetable mat-ter on growth and food conversion efficiency of brown shrimp. Aquaculture 6, 115–125.

Viegas, I., Rito, J., Jarak, I., Leston, S., Carvalho, R.A., Pardal, M.Â., et al., 2012. Hepatic glycogen synthesis in farmed European seabass (*Dicentrarchus labrax* L.) is dominated by indirect pathway fluxes. Comp. Biochem. Physiol. A 163, 22–29.

Vielma, J., Lall, S.P., 1998. Control of phosphorus homeostasis of Atlantic salmon (*Salmo salar*) in fresh water. Fish Physiol. Biochem. 19, 83–93.

Vielma, J., Koskela, J., Ruohonen, K., 2002. Growth, bone mineralization, and heat and low oxygen tolerance in European whitefish (*Coregonus lavaretus* L.) fed with graded levels of phosphorus. Aquaculture 212, 321–333.

Villeneuve, L., Gisbert, E., Le Delliou, H., Cahu, C.L., Zambonino-Infante, J.L., 2005a. Dietary levels of all-trans retinol affect retinoid nuclear receptor expression and skeletal development in European sea bass larvae. Br. J. Nutr. 93, 791–801.

Villeneuve, L., Gisbert, E., Zambonino-Infante, J.L., Quazuguel, P., Cahu, C.L., 2005b. Effect of nature of dietary lipids on European sea bass morphogenesis: implication of retinoid receptors. Br. J. Nutr. 94, 877–884.

Voragen, A.G.J., Coenen, G.-J., Verhoef, R.P., Schols, H.A., 2009. Pectin, a versatile polysac-charide present in plant cell walls. Struct. Chem. 20, 263–275.

WaagbØ, R., 1994. The impact of nutritional factors on the immune system in Atlantic salmon, *Salmo salar* L.: a review. Aquacult. Res. 25, 175–197.

Walton, M.J., Coloso, R.M., Cowey, C.B., Adron, J.W., Knox, D., 1984a. The effects of dietary tryptophan levels on growth and metabolism of rainbow trout (*Salmo gairdneri*). Br. J. Nutr. 51, 279–287.

Walton, M.J., Cowey, C.B., Adron, J.W., 1984b. The effect of dietary lysine levels on growth and metabolism of rainbow trout (*Salmo gairdneri*). Br. J. Nutr. 52, 115–122.

Walton, M., Coloso, R.M., Cowey, C.B., Adron, J.W., 1986. Dietary requirements of rainbow trout for tryptophan, lysine and arginine determined by growth and biochemical measure-ments. Fish Physiol. Biochem. 2, 161–169.

Wang, C., Lovell, R.T., 1997. Organic selenium sources, selenomethionine and selenoyeast, have higher bioavailability than an inorganic selenium source, sodium selenite, in diets for channel catfish (*Ictalurus punctatus*). Aquaculture 152, 223–234.

Wang, K., Takeuchi, T., Watanabe, T., 1985. Effect of dietary protein levels on growth of *Tilapia nilotica*. Bull. Jpn. Soc. Sci. Fish. 51, 133–140.

Wang, D., Zhao, L., Tan, Y., 1995. Requirement of the fingerling grass carp (*Ctenopharyngodon idella*) for choline [in Chinese]. J. Fish. China 19, 132–139.

Wang, C., Lovell, R.T., Klesius, P.H., 1997. Response to *Edwardsiella ictaluri* challenge by channel catfish fed organic and inorganic sources of selenium. J. Aquatic Anim. Health 9, 172–179.

Wang, X., Kim, K.W., Bai, S.C., 2002. Effects of different dietary levels of L-ascorbyl-2-polyphosphate on growth and tissue vitamin C concentrations in juvenile olive flounder. *Paralicluhys olivaceus* (Temminck et Schlegel). Aquacult. Res. 33, 261–267.

Wang, X., Kim, K.W., Bai, S.C., Huh, M.D., Cho, B.Y., 2003a. Effects of the different levels of dietary vitamin C on growth and tissue ascorbic acid changes in parrot fish (*Oplegnathus fasciatus*). Aquaculture 215, 203–211.

Wang, X., Kim, K.W., Bai, S.C., 2003b. Comparison of L-ascorbyl-2-monophosphate-Ca with L-ascorbyl-2-monophosphate-Na/Ca on growth and tissue ascorbic acid concentrations in Korean rockfish (*Sebastes schlegeli*). Aquaculture 225, 387–395.

Wang, X., Kim, K.W., Park, G.J., Choi, S.M., Jun, H.K., Bai, S.C., 2003c. Evaluation of L-ascorbyl-2-glucose as the source of vitamin C for juvenile Korean rockfish *Sebastes schlegeli* (Hilgendorf). Aquacult. Res. 34, 1337–1341.

Wang, X.J., Choi, S., Park, S., Yoo, G., Kim, K., Kang, J.C., et al., 2005. Optimum dietary phosphorus level of juvenile Japanese flounder *Paralichthys olivaceus* reared in the recirculating system. Fish. Sci. 71, 168–173.

Wang, Y., Han, J., Li, W., Xu, Z., 2007. Effect of different selenium source on growth performances, glutathione peroxidase activities, muscle composition and selenium concentration of allogynogenetic crucian carp (*Carassius auratus* gibelio). Anim. Feed Sci. Technol. 134, 243–251.

Wang, S., Encarnacao, P.M., Payne, R.L., Bureau, D.P., 2010a. Estimating dietary lysine requirements for live weight gain and protein deposition in juvenile rainbow trout (*Oncorhynchus mykiss*). In: International Symposium on Fish Nutrition and Feeding, 31 May–4 June 2010, Qingdao, China.

Wang, J.-L., Zhu, X.-M., Lei, W., Han, D., Yang, Y.-X., Xie, S.-Q., 2011. Dietary vitamin B6 requirement of juvenile gibel carp, *carassius auratus gibelio*. Acta Hydrobiol. Sin. 35, 98–104.

Wang, F.B., Luo, L., Lin, S.M., Li, Y., Chen, S., Wang, Y.G., et al., 2011. Dietary magnesium requirements of juvenile grass carp, *Ctenopharyngodon idella*. Aquacult. Nutr. 17, 691–700.

Watanabe, T., Takashima, F., Ogino, C., Hibiya, T., 1970. Requirements of young carp for a-tocopherol. Bull. Jpn. Soc. Sci. Fish. 36, 972–976.

Watanabe, T., Murakami, A., Takeuchi, L., Nose, T., Ogino, C., 1980. Requirement of chum salmon held in freshwater for dietary phosphorus. Bull. Jpn. J. Sci. Fish. 46, 361–367.

Watanabe, T., Takeuchi, T., Wada, M., Vehara, R., 1981. The relationship between dietary lipid levels and a-tocopherol requirement of rainbow trout. Bull. Jpn. Soc. Sci. Fish. 47, 1463–1471.

Watanabe, T., Thongrod, S., Takeuchi, T., Satoh, S., Kubota, S.S., Fujimaki, Y., et al., 1989. Effect of dietary n-6 and n-3 fatty acids on growth, fatty acid composition and histological changes of white fish *Coregonus lavaretus* maraena. Bull. Jpn. Soc. Sci. Fish. 55, 1977–1982.

Waterman, T.H. (Ed.), 1960. The physiology of crustacea, vol. 1. Metabolism and Growth. Academic Press, New York, NY, 670 pp.

Webster, C.D., Lovell, R.T., 1990. Response of striped bass larvae fed brine shrimp from different sources containing different fatty acid compositions. Aquaculture 90, 49–61.

Webster, C.D., Tiu, L.G., Tidwell, J.H., Van Wyk, P., Howerton, R.D., 1995. Effects of dietary protein and lipid levels on growth and body composition of sunshine bass (*Morone chrysops × M. saxatilis*) reared in cages. Aquaculture 131, 291–301.

Wee, K.L., Tuan, N.A., 1988. Effects of dietary protein level on growth and reproduction of Nile tilapia (*Oreochromis niloticus*). In: Pullin, R.S.V., Bhukaswan, T., Tonguthai, K., Maclean, J.L. (Eds.), Proceedings of the Second International Symposium on Tilapia in Aquaculture, ICLARM Conference Proceedings No. 15. Department of Fisheries, Bangkok, Thailand and ICLARM, Manila, Philippines, pp. 401–410.

Wen, H., Zhao, Z., Jiang, M., Liu, A., Wu, F., Liu, W., 2007. Dietary myoinositol requirement for grass carp. *Ctenopkaryngodon idella* fingerling [in Chinese]. J. Fish. China 14, 794–800.

Wen, Z.P., Zhou, X.Q., Feng, L., Jiang, J., Liu, Y., 2009. Effect of dietary pantothenic acid supplement on growth, body composition and intestinal enzyme activities of juvenile Jian carp (*Cyprinus carpio* var. Jian). Aquacult. Nutr. 15, 470–476.

West, T.G., Arthur, P.G., Suarez, R.K., Doll, C.J., Hochachka, P.W., 1993. *In vivo* utilization of glucose by heart and locomotory muscles of exercising rainbow trout (*Oncorhynchus mykiss*). J. Exp. Biol. 177, 63–79.

West, T.G., Schulte, P.M., Hochachka, P.W., 1994. Implications of hyperglycemia for postexercise resynthesis of glycogen in trout skeletal muscle. J. Exp. Biol. 189, 69–84.

Whalen, K.S., Brown, J.A., Parrish, C.C., Lall, S.P., Goddard, J.S., 1999. Effect of dietary n-3 HUFA on growth and body composition of juvenile yellowtail flounder (*Pleuronectes ferrugineus*). Bull. Aquacult. Assoc. Can. 98, 21–22.

Wieser, W., Medgyesy, N., 1991. Metabolic rate and cost of growth in juvenile pike (*Exox lucius* L.) and perch (*Perca fluviatilis* L.): the use of energy budgets as indicators of environmental change. Oecologia 87, 500–505.

Wiesmann, D., Scheid, H., Pfeffer, E., 1988. Water pollution with phosphorus of dietary origin by intensively fed rainbow trout (*Salmo gairdnieri* Rich.). Aquaculture 69, 263–270.

Wilson, R.P., 1994. Utilization of dietary carbohydrate by fish. Aquaculture 124, 67–80.

Wilson, R.P., 2002. Amino acids and proteins. In: Hardy, R.W., Halver, J. (Eds.), Fish Nutrition. Academic Press, Amsterdam, pp. 143–179.

Wilson, R.P., Halver, J.E., 1986. Protein and amino acid requirements of fishes. Ann. Rev. Nutr. 6, 225–244.

Wilson, R.P., Halver, J.E., 2002. Protein and amino acid requirements of fishes. Ann. Rev. Nutr. 6, 225–244.

Wilson, R.P., Poe, W.E., 1985. Relationship of whole body and egg essential amino acid pattern to amino acid requirement in channel catfish (*Ictalurus punctatus*). Comp. Biochem. Physiol. 80B, 385–388.

Wilson, R.P., Poe, W.E., 1987. Apparent inability of channel catfish to utilize dietary mono- and disaccharides as energy sources. J. Nutr. 117, 280–285.

Wilson, R.P., Poe, W.E., 1988. Choline nutrition of fingerling channel catfish. Aquaculture 68, 65–71.

Wilson, R.P., Naggar, G.E., 1992. Potassium requirement of fingerling channel catfish, *Ictalurus punctatus*. Aquaculture 108, 169–175.

Wilson, R.P., Harding, D.E., Garling Jr., D.L., 1977. Effect of dietary pH on amino acid utilization and the lysine requirement of fingerling channel catfish. J. Nutr. 107, 166–170.

Wilson, R.P., Allen Jr., O.W., Robinson, E.H., Poe, W.E., 1978. Tryptophan and threonine requirements of fingerling channel catfish. J. Nutr. 108, 1595–1599.

Wilson, R.P., Poe, W.E., Robinson, E.H., 1980. Leucine, isoleucine, valine and histidine requirements of fingerling *channel catfish*. J. Nutr. 110, 627–633.

Wilson, R.P., Robinson, E.H., Gatlin III, D.M., Poe, W.E., 1982. Dietary phosphorus requirement of channel catfish. J. Nutr. 112, 1197–1202.

Wilson, R.P., Bowser, P.R., Poe, W.E., 1983. Dietary pantothenic acid requirement of fingerling channel catfish. J. Nutr. 113, 2124–2134.

Wilson, R.P., Bowser, P.R., Poe, W.E., 1984. Dietary vitamin E requirement of fingerling channel catfish. J. Nutr. 114, 2053–2058.

Winfree, R.A., Stickney, R.R., 1981. Effect of dietary protein and energy on growth, feed conversion efficiency and body composition of *Tilapia aurea*. J. Nutr. 111, 1001–1012.

Winfree, R.A., Stickney, R.R., 1984. Starter diets for channel catfish: effects of dietary protcin on growth and carcass composition. Prog. Fish Cult. 46, 79–83.

Wirth, M., Steffens, W., Meinelt, T., Steinberg, C., 1997. Significance of docosahexaenoic acid for rainbow trout (*Oncorhynchus mykiss*) larvae. Fett/Lipid 99, 251–253.

Woodall, A.N., Ashley, L.M., Halver, J.E., Olcott, H.S., van der Veen, J., 1964. Nutrition of salmonid fishes. XIII. The a-tocopherol requirement of chinook salmon. J. Nutr. 84, 125–135.

Woodward, B., 1990. Dietary vitamin B_6 requirements of young rainbow trout (*Oncorhynchus mykiss*). FASEB J. 4, 3748. (abstract).

Woodward, B., Frigg, M., 1989. Dietary biotin requirements of young trout (*Salmo gairdneri*) determined by weight gain hepatic biotin concentration and maximal biotin-dependent enzyme activities in liver and white muscle. J. Nutr. 119, 54–60.

Wu, D.H., Liao, C.X., Huang, Z.Z., 1990. Studies on the requirement of grass carp for a-tocopherol in early growth [in Chinese]. J. Chinese Acad. Fish. Sci. 3, 42–47.

Wu, F., Jiang, M., Zhao, Z., Liu, A., Liu, W., Wen, H., 2007a. The dietary niacin requirement of juvenile *Ctenopkaryngodon idellus* [in Chinese]. J. Fish. China 32, 65–70.

Wu, F., Wen, H., Jiang, M., Zhao, Z., Liu, A., Liu, W., 2007b. Effects of dietary vitamin B12 on growth, body composition and hemopoiesis of juvenile grass carp (*Ctenopharyngodon idellus*) [in Chinese]. J. Jilin Agric. Univ. 29, 695–699.

Xia, M.-H., Huang, X.-L., Wang, H.-L., Jin, M., Li, M., Zhou, Q.-C., 2014. Dietary niacin levels in practical diets for *Litopenaeus vannamei* to support maximum growth. Aquacult. Nutr. http://dx.doi.org/10.1111/anu.12210

Xiao, L.D., Mai, K.S., Ai, Q.H., Xu, W., Wang, X.J., Zhang, W.B., et al., 2009. Dietary ascorbic acid requirement of co-bia. *Rachycentron canadum* Linneaus. Aquacult. Nutr. http://dx.doi.org/10.1111/j.1365-2095.2009.00695.X

Xie, Z., Niu, C., 2006. Dietary ascorbic acid requirement of juvenile ayu (*Plecoglossus altivelis*). Aquacult. Nutr. 12, 151–156.

Xie, S., Cui, Y., Yang, Y., Liu, J., 1997. Bioenergetics of Nile tilapia, *Oreochromis niloticus*: effects of food ration size on metabolic rate. Asian Fish. Sci. 10, 155–162.

Xie, N.B., Feng, L., Liu, Y., Jiang, J., Jiang, W.D., Hu, K., et al., 2011. Growth, body composition, intestinal enzyme activities and microflora of juvenile Jian carp (*Cyprinus carpio* var. Jian) fed graded levels of dietary phosphorus. Aquacult. Nutr. 17, 645–656.

Xie, F., Zeng, W., Zhou, Q., Wang, H., Wang, T., Zheng, C., et al., 2012. Dietary lysine requirement of juvenile Pacific white shrimp, *Litopenaeus vannamei*. Aquaculture 358–359, 116–121.

Xu, X., Ji, W., Castell, J.D., O'Dor, R., 1993. The nutritional value of dietary n-3 and n-6 fatty acids for the Chinese prawn (*Penaeus chinensis*). Aquaculture 118, 277–285.

Xu, X.L., Ji, W.J., Castell, J.D., O'Dor, R.K., 1994. Essential fatty acid requirement of the Chinese prawn, *Penaeus chinensis*. Aquaculture 127, 29–40.

Xu, Z., Dong, X., Liu, C., 2007. Dietary Zinc requirement of juvenile cobia (*Rachycentron canadum*). Fish. Sci. 26, 138–141.

Xu, Q.Y., Xu, H., Wang, C., Zheng, Q., Sun, D., 2011. Studies on dietary phosphorus requirement of juvenile Siberian sturgeon *Acipenser baerii*. J. Appl. Ichthyol. 27, 709–714.

Yamamoto, H., Satoh, S., Takeuchi, T., Watanabe, T., 1983. Effects on rainbow trout of deletion of manganese or trace elements from fish meal diet. Nippon Suisan Gakkaishi 49, 287–293.

Yang, X., Tabachek, J.L., Dick, T.A., 1994. Effects of dietary n-3 polyunsaturated fatty acids on lipid and fatty acid composition and haematology of juvenile Arctic charr *Salvelinus alpinus* (L.). Fish Physiol. Biochem. 12, 409–420.

Yang, Y., Guo, Q., Han, Y., Fan, Z., 2005. The apparent digestibility of phosphorus in different phosphorus sources on carp [J]. J. Northeast Agric. Univ. 6, 762–766.

Yarzhombek, A.A., Shcherbina, T.V., Shmakov, N.F., Gusseynov, A.G., 1984. Specific dynamic effect of food on fish metabolism. J. Ichthyol. 23, 111–117.

Ye, C.X., Liu, Y.J., Tian, L.X., Mai, K.S., Du, Z.Y., Yang, H.J., et al., 2006. Effect of dietary calcium and phosphorus on growth, feed efficiency, mineral content and body composition of juvenile grouper, *Epinephelus coioides*. Aquaculture 255, 263–271.

Ye, C.X., Liu, Y.J., Mai, K.S., Tian, L.X., Yang, H.J., Niu, J., et al., 2007. Effect of dietary iron supplement on growth, haematology and microelements of juvenile grouper, *Epinephelus coioides*. Aquacult. Nutr. 13, 471–477.

Ye, C.X., Tian, L.X., Yang, H.J., Liang, J.J., Niu, J., Liu, Y.J., 2009. Growth performance and tissue mineral content of juvenile grouper (*Epinephelus coioides*) fed diets supplemented with various levels of manganese. Aquacult. Nutr. 15, 608–614.

Ye, C.X., Tian, L.X., Mai, K.S., Yang, H.J., Niu, J., Liu, Y.J., 2010. Dietary magnesium did not affect calcium and phosphorus content in juvenile grouper, *Epinephelus coioides*. Aquacult. Nutr. 16, 378–384.

Yone, Y., 1978. Essential fatty acids and lipid requirements of marine fish. In: Dietary Lipids in Aquaculture (ed. by the Japanese Society of Scientific Fisheries). Koseisha-Koseik-Abu, Tokyo, Japan, pp. 43–59.

Yossa, R., Sarker, P.K., Mock, D.M., Vandenberg, G.W., 2014. Dietary biotin requirement for growth of juvenile zebrafish *Danio rerio* (Hamilton-Buchanan). Aquacult. Res. 45, 1787–1797.

Yu, T.C., Sinnhuber, R.O., 1979. Effect of dietary o3 and o6 fatty acids on growth and feed conversion efficiencies of coho salmon (*Oncorhyncus kisutch*). Aquaculture 16, 31–38.

Yu, H.-R., Zhang, Q., Xiong, D.-M., Huang, G.-Q., Li, W.-Z., Liu, S.-W., 2013. Dietary available phosphorus requirement of juvenile walking catfish, Clarias leather. Aquacult. Nutr. 19, 483–490.

Yue, Y., Zou, Z., Zhu, J., Li, D., Xiao, W., Han, J., et al., 2014. Dietary threonine requirement of juvenile Nile tilapia, *Oreochromis niloticus*. Aquacult. Int. 22, 1457–1467.

Zarate, D.D., Lovell, R.T., Payne, M., 1999. Effects of feeding frequency and rate of stomach evacuation on utilization of dietary free and protein-bound lysine for growth by channel catfish, *Ictalurus punctatus*. Aquacult. Nutr. 5, 17–22.

Zehra, S., Khan, M.A., 2012. Dietary vitamin C requirement of fingerling, *Cirrhinus mrigala* (Hamilton), based on growth, feed conversion, protein retention, hematological indices, and liver vitamin C concentration. J. World Aquacult. Soc. 43, 648–658.

Zein-Eldin, Z.P., Corliss, J., 1976. The effect of protein levels and source on growth of *Penaeus aztecus*. In: Pillay, T.V.R., Dill, W.A. (Eds.), Avances in aquaculture. FAO Technical Conference on Aquaculture. Kyoto, Japan, 26 May–2 June 1976, pp. 592–596.

Zeitoun, I.H., Ullrey, D.E., Halver, J.E., Tack, P.I., Magee, W.T., 1974. Influence of salinity on protein requirements of coho salmon (*Oncorhynchus kisutch*) smolts. J. Fish. Res. Board Can. 31, 1145–1148.

Zhang, C., Mai, K., Ai, Q., Zhang, W., Duan, Q., Tan, B., et al., 2006. Dietary phosphorus requirement of juvenile Japanese seabass, *Lateolabrax japonicus*. Aquaculture 255, 201–209.

Zhao, Z., Wen, H., Wu, F., Liu, A., Jiang, M., Liu, W., 2008. Dietary folic acid requirement for grass carp fingerling. *Ctenopharyngodon idella* [in Chinese]. J. Shanghai Fish. Univ. 17, 187–192.

Zheng, F., Takeuchi, T., Yoseda, K., Kobayashi, M., Hirokawa, J., Watanabe, T., 1996. Requirement of larval cod for arachidonic acid, eicosapentaenoic acid, and docosahex-aenoic acid using by their enriched *Artemia nauplii*. Nippon Suisan Gakkaishi 62, 669–676.

Zheng, X., Seiliez, I., Hastings, N., Tocher, D.R., Panserat, S., Dickson, C.A., et al., 2004. Characterization and comparison of fatty acyl Δ 6 desaturase cDNAs from freshwater and marine teleost fish species. Comp. Biochem. Physiol. B 139, 269–279.

Zhou, Q., Ding, Y., Zheng, S., Su, S., Zhang, L., 2004a. The effect of dietary vitamin C supplementation on growth and anti-disease ability of shrimp. *Penaeus vannamei*. Acta Hydrobiol. Sin. 28, 592–598.

Zhou, Q.C., Liu, Y.J., Mai, K.S., Tian, L.X., 2004b. Effect of dietary phosphorus levels on growth, body composition, muscle and bone mineral concentrations for orange-spotted grouper *Epinephelus coioides* reared in floating cages. J. World Aquacult. Soc. 35, 427–435.

Zhou, Q.-C., Wu, Z.-H., Tan, B.-P., Chi, S.-Y., Yang, Q.-H., 2006. Optimal dietary methionine requirement for Juvenile Cobia (*Rachycentron canadum*). Aquaculture 258, 551–557.

Zhou, Q.-C., Wu, Z.-H., Chi, S.-Y., Yang, Q.-H., 2007. Dietary lysine requirement of Juvenile Cobia (*Rachycentron canadum*). Aquaculture 258, 634–640.

Zhou, L.B., Wang, A.L., Ma, X.L., Zhang, W., Zhang, H.F., 2009a. Effects of dietary iron on growth and immune response of red drum (*Sciaenops ocellatus*). Oceanologia Et Limnologia Sinica 40, 663–668.

Zhou, L.B., Zhang, W., Wang, A.L., Ma, X.L., Zhang, H.F., Liufu, Y.Z., 2009b. Effects of dietary zinc on growth, immune response and tissue concentration of juvenile Japanese seabass *Lateolabrax japonicus*. Oceanologia Et Limnologia Sinica 01.

Zhou, X., Wang, Y., Gu, Q., Li, W., 2009c. Effects of different dietary selenium sources (selenium nanoparticle and selenomethionine) on growth performance, muscle composition and glutathione peroxidase enzyme activity of crucian carp (*Carassius auratus* gibelio). Aquaculture 291, 78–81.

Zhou, F., Xiong, W., Xiao, J.X., Shao, Q.J., Bergo, O.N., Hua, Y., et al., 2010a. Optimum arginine requirement of juvenile black sea bream. *Sparus macrocephalus*. Aquacult. Res. 41, e418–e430.

Zhou, F., Shao, J., Xu, R., Ma, J., Xu, Z., 2010b. Quantitative L-lysine requirement of juvenile black sea bream (*Sparus macrocephalus*). Aquacult. Nutr. 16, 194–204.

Zhou, Q.-C., Wang, Y.-L., Wang, H.-L., Tan, B.-P., 2012. Dietary arginine requirement of juvenile Pacific white shrimp, *Litopenaeus vannamei*. Aquaculture 364–365, 252–258.

Zhou, Q.-C., Zeng, W.-P., Wang, H.-L., Wang, T., Wang, Y.-L., Wang, H.-L., Xie, F.-J., 2013. Dietary threonine requirements of juvenile Pacific white shrimp, *Litopenaeus vannamei*. Aquaculture 392–395, 142–147.

Zhu, C.-B., Dong, S.-L., Wang, F., Zhang, H.-H., 2006. Effects of seawater potassium concentration on the dietary potassium requirement of *Litopenaeus vannamei*. Aquaculture 258, 543–550.

Zhu, Y., Chen, Y., Liu, Y., Yang, H., Liang, G., Tian, L., 2012. Effect of dietary selenium level on growth performance, body composition and hepatic glutathione peroxidase activities of largemouth bass *Micropterus salmoide*. Aquacult. Res. 43, 1660–1668.

Zwingelstein, G., Bodennec, J., Brichon, G., Abdul-Malak, N., Chapelle, S., El Babili, M., 1998. Formation of phospholipid nitrogenous bases in euryhaline fish and crustaceans. I. Effects of salinity and temperature on synthesis of phosphatidylserine and its decarboxylation. Comp. Biochem. Physiol. B Biochem. Mol. Biol. 120, 467–473.

Functional feed additives in aquaculture feeds

Pedro Encarnação
Biomin Singapore Pte Ltd, Singapore

5.1 Introduction

Aquaculture feeds are formulated with a vast pool of ingredients which, when fed to the animal, are intended to supply its nutritional requirements to perform its normal physiological functions, including maintaining a highly effective natural immune system, growth, and reproduction. To ensure the dietary nutrients are ingested, digested, absorbed, and transported to the cells, an increasing diversity of nonnutritive feed additives are being used in aquatic feeds.

The range of feed additives used in aquatic feeds is very diverse. Certain feed additives target the feed quality, including pellet binders, antioxidants, and feed preservatives (antimold and antimicrobial compounds). Enzymes are used to improve the availability of certain nutrients (proteases, amylases) or to eliminate the presence of certain antinutrients (phytase, nonstarch polysaccharides (NSP) enzymes).

Other additives are used to improve the animals' performance and health. The concept of functional aquafeeds represents an emerging new pattern to develop diets for fish and crustaceans. Feed transformation into biomass gain is a process that starts in the digestive system of the animal. As such, its health status and its functionality correlate directly with the economic results of the farmer. From mammalian research it is well known that the gastrointestinal tract is responsive and sensitive to a wide range of stressors. Some of the more common features are degeneration of the intestinal mucosa and perturbation of its barrier function and uptake mechanisms. Closely connected with the state of health of the gut is a well-balanced intestinal microflora, which helps the digestive and absorptive process and protects the host against invading pathogens.

Several studies have also shown that different feed ingredients and changes in diet composition can affect gut structure and microbiota balance influencing digestive and absorptive functions (Ringo and Olsen, 1999). Alteration of the intestinal microbiota composition and consequent reduction of protective gut microflora may contribute to pathogenesis in the gut (Ringo and Olsen, 1999). Management of the gut flora is therefore an important issue to achieve good feed efficiency, animal growth, and animal health.

S. Nates (Ed): Aquafeed Formulation. DOI: http://dx.doi.org/10.1016/B978-0-12-800873-7.00005-1

Different strategies have been used to combat problems arising from bacterial and viral threats, with chemotherapy being the most commonly used approach, by using large amounts of antibiotics and chemical products. Nowadays, there are more sustainable ways to manage gut microflora and fish performance by supplementing the feeds with nutraceuticals or functional foods to modulate the health and performance of farmed animals. There are several options available to manage and regulate the fish gut environment, which include the use of probiotics, prebiotics, immune-stimulants, phytogenic substances, and organic acids.

5.2   Phytogenics

Phytogenics are a relatively young class of feed additives that are gaining interest within the aquaculture industry. Phytogenic feed additives (PFAs) are plant-derived products which are added to the feed in order to improve animal performance.

PFAs are an extremely heterogeneous group of feed additives originating from leaves, roots, tubers, or fruits of herbs, spices, or other plants. They are available either in a solid, dried, or ground form, as extracts or essential oils. Within PFAs, the content of active substances in products may vary widely, depending upon the plant part used (e.g., seeds, leaf, root, and bark), harvesting season, and geographical origin (Steiner, 2006).

These plant active ingredients (e.g. phenolic and flavonoids) can exert multiple effects on the animal, such as stimulation of appetite, antimicrobial action, direct reduction of gut bacteria, stimulation of gastric juices, enhancement of immune system, and anti-inflammatory and antioxidant properties (Windisch et al., 2008; Chakraborty and Hancz, 2011; Chakraborty et al., 2014; Reverter et al., 2014). Most phytochemicals are redox active molecules possessing antioxidant characteristics that may improve the general physiological condition of fish (Chakraborty and Hancz, 2011). The antimicrobial mode of action is considered to arise mainly from the potential of hydrophobic essential oils to intrude into the bacterial cell membrane, disintegrate membrane structures, and cause ion leakage (Krosmayr, 2007).

Feed additives containing essential oils of aromatic plants are the most common type of phytogenic products used in aquatic feeds. Essential oils are concentrated hydrophobic liquid compounds characterized by a strong odor and are formed by aromatic plants as secondary metabolites. They contain most of the plant's active compounds, a variety of volatile molecules such as terpenoids, phenol-derived aromatic components, and aliphatic components (Chakraborty et al., 2014).

Many essential oil components are generally recognized as safe and have been used for many years in food, cosmetic, and pharmaceutical industries. Among the herbs and spices used in animal nutrition, oregano is probably used most frequently, as it is rich in carcavol and thymol, known to have strong antibacterial and antioxidative properties, and is described as acting synergistically (Burt, 2004; Zhou et al., 2007a; Baser, 2008). The value of garlic (*Allium sativum*) extract for bacterial disease control and immunostimulation has previously been demonstrated by a number of cultured fish (Nya et al., 2010; Militz et al., 2013). The observed antimicrobial and immunostimulant activity of garlic in teleosts is largely explained by the transient phytochemical allicin (diallyl thiosulfinate) and its derivatives (Nya et al., 2010).

Essential oils extracted from rosemary (*Rosmarinus officinalis*) also seem particularly interesting due to their high concentration of components such as carnasol and carnosic acid, which have high antioxidant properties (Abutbul et al., 2004). In addition, peppermint (Talpur, 2014) and cinnamon (Ahmad et al., 2011) also seem to be possible candidates to be used as PFAs in aquatic species to improve growth performance, fish health status and to reduce microbial challenge in the gut.

Most studies on application of essential oils in animal nutrition are conducted in swine and poultry. However, there is increasing evidence that the application of phytogenics can also be beneficial for some aquaculture species, such as fish and shrimp. Application of PFAs in aquaculture feeds varies from as low as 100 g/ton feed (0.01% diet) to 2–3 kg/ton feed (0.2–0.3% diet), depending on the type and concentration of the active compounds in the product.

Zheng et al. (2009) reported that in channel catfish, diets supplemented with a commercial product containing oregano essential oil and its phenolic compounds carvacrol and thymol was shown to act as a growth promoter. After eight weeks into the feeding trial, fish fed the diet supplemented with the commercial product (0.05%) showed significantly higher ($P < 0.05$) weight gain, protein efficiency ratio, and improved FCR compared to fish fed all other diets.

In addition, Peterson et al. (2014) reported an improvement in fillet composition (higher protein and lower lipid) when channel catfish were fed diets supplemented (0.02%) with a commercial product containing oregano, anise, and citrus oils. In a laboratory study it was observed that fish fed the essential oil-based product consumed 30% more feed and gained 44% more weight during a 12-week study (Peterson et al., 2014). However, in two subsequent studies under field pond conditions, fish fed diets supplemented with the essential oil product achieved higher performance but not significantly different ($P > 0.05$) from fish fed the control feed (Peterson et al., 2014).

Positive effects of PFAs were also reported in trout. Giannenas et al. (2012) conducted a study to evaluate the effect of two feed additives, one rich in carvacrol (12 g/kg) and other rich in thymol (6 g/kg) with an application of 1 g/kg feed (0.1% diet). Results showed that dietary phytogenic supplementation with both products improved ($P < 0.05$) feed efficiency compared to the control, body weight gain was unaffected by phytogenic supplementation (Giannenas et al., 2012). Supplementation with the PFAs also increased antioxidant protective capabilities in the trout fillet at storage, expressed by a significant reduction in malondialdehyde formation (indicative of lipid peroxidation) after 5 days of refrigerated storage.

In another trial testing the application of a PFA in trout diets, it was reported that supplementation of a thymol–carvacrol-based product at levels of 0.1%, 0.2%, and 0.3% diet improved fish growth and weight gain compared to control (Ahmadifar et al., 2011). Improvement in FCR was only reported at levels of 0.2% and above.

The effect of phytogenic supplementation in trout health parameters was also observed in both studies. The number of lymphocytes increased when a thymol–carvacrol-based product was used at 0.2% and 0.3% diet (Ahmadifar et al., 2011).

Levels of lysozyme and total complement concentrations as well as catalase activity were higher in the phytogenic supplemented group (Giannenas et al., 2012).

Several other scientific studies focused on the therapeutic effects of essential oils on pathogenic bacteria and parasites. Abutbul et al. (2004) tested the effects of rosemary *R. officinalis* extracts as a treatment for *Streptococcus iniae* in tilapia. A significant reduction in mortality of infected tilapia was then obtained when fish were fed a diet containing ethyl acetate rosemary extract (1:24 w/w) or leaf powder (Abutbul et al., 2004). In white leg shrimp (*Litopenaeus vannamei*), Cardozo et al. (2008) tested the application of an encapsulated combination of two active ingredients of oregano (thymol and carvacrol 1:1 at a concentration of 0.003% diet), under normal and stress challenge conditions. There were no significant differences in growth but the group fed the phytogenic substances showed an improved FCR (1.21 vs. 1.29). Shrimp fed the thymol-carvacrol diet showed a significantly higher survival rate than the control group when shrimp were challenged with *Vibrio harveyii* (10^6 CFU/ml) (Cardozo et al., 2008). The performance and survival benefits of the thymol-carvacrol mixture correlated directly with improvements in the immune parameters with a significant increase in shrimp phagocytosis index and prophenoloxidase activity, indicative of a more resistant immune system against *V. harveyii* (Cardozo et al., 2008).

In European sea bass, dietary supplementation with carvacrol at 0.025% and 0.05% did not affect growth performance, feed intake, or feed conversion ratio, but provided an appreciable resistance to a challenge with *Listonela anguillarum* (Volpati et al., 2014).

Thus, there is enough evidence of the beneficial properties and efficacy of PFAs on the performance and health of cultured aquatic species. However, their efficacy is related to the type of plant, method of extraction, and extract concentrations. As such, it is necessary to quantify and characterize chemically the plant extracts in order to identify active molecules, and facilitate the establishment of a standardized protocol, including extract concentration and adequate dosing.

5.3 Organic acids

Dietary acidification by the addition of organic acids has been widely used in animal nutrition and organic acids, and organic acids have become a promising feed additive to improve gut health and performance. Organic acids, such as acetic, butyric, citric, formic, lactic, propionic, malic, and sorbic acids and their salts have been used as acidifiers in animal feeds (NRC, 2011).

Their different strengths (in terms of their minimal inhibitory concentrations) and properties mean that it is very usual to find liquid blends of different acids. Furthermore, for customer convenience, organic acids are available on the market in a variety of forms, mainly: adsorbates – liquid acids or mixtures of acids adsorbed onto a solid, inert substrate, usually silica or vermiculite; salts – usually solids (except, for example, ammonium propionate and ammonium formate which are liquid), in order to minimize or improve some functional properties in the feed mill, such as corrosion, volatility, and odor (FEFANA, 2014).

Organic acids are applied directly into feedstuffs and compound feed. Liquid acids and blends are sprayed onto the feed, whereas solid acids and acid salts are added directly or via special premixtures. The mode of action of organic acids in the intestinal tract involves two different actions: the pH-decreasing action of organic acids in stomach and small intestine contributes to an improved activity of digestive enzymes creates an impaired environment for pathogens and inhibits the growth of Gram-negative bacteria through the dissociation of the acids and production of anions in the bacterial cells (Lückstädt, 2008).

Available information on the beneficial effects of dietary inclusion of organic acids, their salts or their combination on fish and shrimp performance appears to vary depending on many factors such as fish species, fish size, or age; type and level of organic acids, salts, or their combination; composition and nutrient content of experimental diets; buffering capacity of dietary ingredients; culture and feeding management; and water quality (FEFANA, 2014).

Positive effects of organic acids on protein hydrolysis and diet digestibility have been demonstrated in several species. In rainbow trout dietary supplementation with 1% sodium diformate increased the digestibility of protein, lipids, and amino acids (Morken et al., 2011). Liebert et al. (2010) showed that when 0.3% of sodium diformate was added in the diet of tilapia fingerlings (*Oreochromis niloticus*), the protein efficiency ratio and protein retention significantly improved. In contrast, Ng et al. (2009) reported no significant ($P < 0.05$) effects on growth, feed utilization, and nutrient digestibility when red hybrid tilapia feeds were supplemented with various dietary levels (0%, 0.1%, 0.2%, or 0.3%) of a commercial organic acid blend (six organic acids, 38% of total acids, adsorbed onto a silicic-acid-based carrier) and 0.2% potassium diformate. Nevertheless, there was a trend towards improved results with fish fed the organic-acid-supplemented diets (Ng et al., 2009).

Improved digestibility is also reported in shrimp. Tiger shrimp (*Penaeus monodon*) diets supplemented with 1% of commercial acidifiers with sodium butyrate enhanced the digestibility of dry matter, crude protein, and energy, leading to numerical improvements in weight gain, survival, and feed conversion ratio (Nuez-Ortin, 2011).

Several other studies focused on the effects of organic acids in improving the utilization of phosphorus and other minerals. A study with rainbow trout showed that the apparent digestibility of phosphorus could be significantly increased in fish that were fed a fishmeal-based diet supplemented with 10 ml/kg formic acid (Vielma and Lall, 1997). Furthermore, studies showed that using 0.5–3% citric acid favors nitrogen (N) and phosphorus (P) retention in sea bream (*Pagrus major*) (Hossain et al., 2007; Sarker et al., 2005, 2007). It is advocated that the positive effects of organic acids in phosphorus digestibility are related to the solubilization of bone minerals in fish bone or fish meal as well as a chelating effect that reduces the antagonistic interaction between calcium and phosphorus (Sarker et al., 2005).

Dietary supplementation with organic acids can also have a synergetic effect of the action of phytase enzyme. Baruah et al. (2007) reported a significant increase in phosphorus digestibility due to dietary addition of citric acid (30 g/kg) in diets of Indian carp (*Labeo rohita*), and attributed the growth improvement in fish to the beneficial effect of citric acid in releasing minerals from the phytic acid complex of the soybean-meal-based diet.

In a comprehensive study to evaluate the potential of several organic acid salts (sodium acetate, sodium butyrate, sodium citrate, sodium formate, sodium lactate, and sodium propionate), to be used as feed additives in shrimp diets, it was concluded that sodium propionate had the most potential as a feed supplement in diets of white shrimp (Silva et al., 2013). The application of 2% sodium butyrate in commercial diets increased the diet attractiveness, feed intake, and increased the coefficient of apparent digestibility of energy and phosphorus (Silva et al., 2013).

Regarding the stability of salts in the diets, molecules with a higher carbon number (C), such as butyrate (4C), lactate (3C), and citrate (6C), showed less leaching. Silva et al. (2013) also clearly indicated that salts of organic acids possess inhibitory activity against pathogenic *Vibrio* species in marine shrimp, with propionate, butyrate, and acetate salts exhibiting the highest inhibitory capacity.

Many organic acid feed additives are supplemented to the diets because of their antimicrobial effects. Growth rates of many Gram-negative bacteria are reduced below pH 5. Organic acids are weak acids, which means that a certain proportion of the molecules remain undissociated, depending on the acid's pKa value and the ambient pH level. Moreover, small acids are lipophilic and can diffuse across the cell membrane of Gram-negative bacteria. In the more alkaline cytoplasm they dissociate and cause pH reduction. This reduction alters cell metabolism and enzyme activity thus inhibiting growth of intraluminal microbes, especially pathogens (FEFANA, 2014).

The antimicrobial effect of organic acids has been shown in some challenge test studies where the application of organic acids in the feed reduced mortality when fish are exposed to pathogenic bacteria. Total fecal and adherent gut bacterial count significantly decreased in fish fed an organic acid blend or potassium diformate-containing diets (Ng et al., 2009). Cumulative mortality 15 days after a challenge with *Streptococcus agalactiae* was reduced in fish fed diets supplemented with the organic acid blend or formate (Ng et al., 2009). Another study reported that potassium diformate at dietary levels of 0.2%, 0.3%, or 0.5% significantly improved growth and feed conversion in Nile tilapia. Mortality at 15 days post-challenge with *Vibrio anguillarum* (orally challenged after 10 days of feeding) was lower in the group fed the formate-containing diets (Ramli et al., 2005).

A new emerging strategy to modulate gut microflora and reduce pathogens in the gut is the combination of organic acids with other antimicrobial substances, such as phytochemicals or permeabilizers, in an attempt to use possible synergism to more effectively combat pathogenic bacteria.

Zhou et al. (2007b) studied thymol and carvacrol, combined with acetic acid against *Salmonella typhimurium*. In the presence of acetic acid, the antibacterial activity from the combinations of thymol (100 mg/l) plus acetic acid (0.10%) and of carvacrol (100 µl/l) plus acetic acid (0.10%) achieved those reached with 400 mg/l thymol and 400 µl/l carvacrol alone. This demonstrates a powerful synergistic antimicrobial effect.

Riemensperger and Santos (2011a,b), reported that when a permeabilizer substance was added to an antimicrobial mixture of organic acids (formic, propionic, and acetic) plus a phytochemical (cinnamaldehyde) their inhibition against *Aeromonas veronni* and *Yersinia ruckeri* was greatly enhanced. When this synergistic mixture was supplemented in tilapia feed at a level of 1 g/kg, fish fed the supplemented feed attained significantly higher growth and improved FCR (Riemensperger and Santos, 2011a,b).

In addition, when the organic acid + permeabilizer + phytochemical mixture was supplemented in white shrimp diets at a level of 0.6 g/kg results showed a significant effect in weight gain and FCR (Riemensperger and Santos, 2012). At the end of the growth trial shrimp where challenged with *Vibrio parahaemolyticus* (injected with 0.2 ml at 5×10^7 CFU). Results showed that mortality rate in shrimp fed the synergistic blend was considerably delayed (LT50 of 10 h for the control compared to 96 h for the synergistic blend) and after 124 h there was a significant reduction in mortality, 97% in control vs. 75% in the supplemented diet (Riemensperger and Santos, 2012).

Though there are only a limited number of published studies on the use of acidifiers in growth promotion and disease prevention in aquaculture, results from the studies indicate promising potential and compel aquafeed manufacturers to consider using acidifiers. The use of acidifiers can be an efficient tool to achieve sustainable, economical, and safe fish and shrimp production.

5.4 Yeast products

Yeast and its components have long been used in terrestrial and aquatic animals. The reasons for this extensive use are its excellent nutritional content and availability. Further applications are being developed for yeast as a functional feed additive as probiotic live yeast, yeast fractions (yeast cell walls, yeast extracts), or as a source for more purified products such as β-glucans and nucleotides.

Very few species of yeast are used commercially in the aquaculture industry. The most widely commercialized species is *Saccharomyces cerevisiae*, known as "baker's yeast" which is also used in breweries to make beer and ale (brewer's yeast), distilleries to make spirits, and industrial alcohol (ethanol yeast) and wineries to make wine (Tacon, 2012). Other yeasts, such as Torula yeast, are also increasingly being tested and applied as feed additives.

Baker's yeast comes as a pure and primary culture, grown under strict conditions on a sugar substrate such as molasses. The primary grown culture controlled process also makes a very consistent base for the production of yeast extracts, autolyzed yeast, yeast cell walls, and their derivate, nucleotides and β-glucans. Yeast cell walls produced from baker's yeast usually have a high content of mannans (Tacon, 2012). Brewer's yeast can be identified either as the ferment used in brewery industries (yeast primary production) or the by-product of these industries which is the form mainly used in aquaculture. The nutritional content is similar to the one in baker's yeast, but contains more trace minerals such as selenium and chromium (Tacon, 2012). Brewer's yeast can be used to produce yeast fractions, but the quality is less consistent than in baker's yeast. Ethanol yeast is harvested after having performed alcoholic fermentation and distillation for the conventional production of bioethanol from sugarcane, beet sugar, or grain syrup. Selling prices are normally low, however the quality, and the protein content, are very inconsistent (Tacon, 2012).

It has been shown that yeasts can enhance growth, survival, gut maturation, and improve the immune and antioxidant systems in fish and shrimps. Yeast has immunostimulatory properties mainly due to components such as β-glucan, mannoproteins, chitin (as a minor component), and nucleic acids (Meena et al., 2013). However, the influence of yeast products in the animal organism is complex to access and will depend on the way the product (whole yeast, autolyzed yeast, mechanical or chemical disruption, and live) is given to the animal. Therefore, depending on the techniques used to obtain the yeast product, we will obtain biochemical differences, as well as functionality differences. Consequently, not all yeast products will generate similar results when fed to aquatic animals, depending on the processing method and delivery method, each product should be considered to have unique properties, even if it is the same yeast species (Tacon, 2012).

Inactive dried yeast and autolyzed yeast come from primary grown cultures or brewer's yeast. They are used in aquaculture feeds as a source of protein and nitrogen. Inactive yeast is a yeast that has been deactivated by high-temperature drying (often spray drying). The cells come as a whole and the cell wall is not ruptured, making the access to intracellular material (amino acids, vitamins, glucans, etc.) difficult. A way to access these materials is to partially hydrolyze the yeast cell wall to let the cellular content be partially released from the cell. This can be facilitated by activating the internal autolytic enzymes of the live yeast (autolysis), adding external enzymes (notably proteolysis) or playing on the osmotic pressure to rupture the cell wall (plasmolysis). Different grades of autolyzed yeast can be obtained depending on the level of autolysis (from partial to total). The final product is a mixture of cellular content and yeast cell wall. Furthermore, the autolysis process degrades protein and forms peptides (dipeptides to tetra-peptides) and oligonucleic acids which are readily digestible by the animal. Nevertheless, depending on the original yeast material used, autolyzed and inactive yeast quality can be very different (Tacon, 2012).

The improvement in the nutritional value of brewer's dry yeast for salmonid fishes by disruption of the cell wall was reported by Rumsey et al. (1991). When the yeast cells were fully disrupted, the absorption of nitrogen increased by more than 20% and the metabolizable energy of the yeast by more than 10%. Energy and nitrogen digestibility were further increased after the removal of all wall material and separation of nitrogen into amino acid and nucleic acid fractions (Rumsey et al., 1991).

Several studies have shown a beneficial effect of dietary administered *S. cerevisiae* in fish. Improved growth, feed efficiency, blood biochemistry, survival rate, and nonspecific immune responses were found in different species (Oliva-Teles and Gonçalves, 2001; Li and Gatlin III, 2004; Hisano et al., 2007). The use of *S. cerevisiae* has improved the growth performance and feed efficiency in tilapia cultivation (Lara-Flores et al., 2003; Hisano et al., 2007). Dietary supplementation of yeast also improved the immune system and the resistance against bacterial infections in hybrid striped bass, *Morone chrysops* × *M. saxatilis* (Li and Gatlin III, 2004) and it enhanced the cellular innate immune response of gilthead sea bream (Ortuno and Cuesta, 2002). Rainbow trout (*Oncorhynchus mykiss*) fry fed two yeast strains, *S. cerevisiae* and *Saccharomyces boulardii*, registered improvements in the intestinal microbiota and the brush border enzyme activities.

Cellular yeast components can also provide important nonnutritive compounds that may benefit fish health, including: mannan oligosaccharides (MOS); glucose polymers (β-glucans), chitin, as well as nucleic acids (Rumsey et al., 1992). Yeast β-glucans have been applied in aquaculture as immune-stimulants to modulate the innate immune system of fish and shrimp and improve their survival (Sealey and Gatlin III, 2001; Bricknell I and Ra, 2005; Welker et al., 2012; Meena et al., 2013; Bai et al., 2014).

β-Glucans are insoluble polysaccharides consisting of repeating glucose units that can be joined through β1-3 and β1-6 linkages (Welker et al., 2012). The source and extraction process from which these glucans are obtained can greatly affect their immune-stimulatory capacity (Sealey and Gatlin III, 2001). β-Glucans applied as feed additives will exert their primary effects at the intestinal level through the induction of cytokines, which will affect the systemic immune response in fish. The positive influence of β-glucans has been demonstrated for various fish species (Sahoo and Mukherjee, 2002; Couso et al., 2003; Li and Gatlin III, 2004; Welker et al., 2012). Several studies have shown the immune effects of β-glucans specifically on antibody production, expression of immune system genes, survival, resistance to infectious diseases, and improvement in stress resistance (Meena et al., 2013). However, β-glucans have to be used carefully in aquaculture as some experiments report negative effects in fish when used for prolonged periods at high concentrations. This can be avoided by carefully choosing the source of β-glucan and using them either at high concentration (2 kg/T) for a short period or by pulse feeding, or a low concentration continuously (0.5 g/kg) (Welker et al., 2012).

MOS are cellular yeast components that can promote the growth of bacterial populations associated with a healthy, well-functioning intestine. MOS can also bind to certain Gram-negative bacteria and prevent intestinal cloning, acting as a mechanism for bacteria removal from the gut (Dimitroglou et al., 2010). Enhanced growth performance, feed efficiency, increase in absorption surface of the gut, as well as an up-regulation of the activities of specific digestive enzyme have been documented by several authors (Staykov et al., 2007; Dimitroglou et al., 2010). MOS can also stimulate the immune system; in European sea bass (*Dicentrarchus labrax*) improvements in gut mucus and lysozyme activity were observed (Torrecillas et al., 2011). Good results on an improved immune system were also obtained by other authors (Staykov et al., 2007; Dimitroglou et al., 2010).

Yeast extracts come from the further hydrolysis and purification of autolyzed yeast. Insoluble yeast cell walls are separated from the cellular content by centrifugation. Yeast extracts are very soluble, rich in peptides (up to 65–70% of the product) and free amino acids, such as glutamic acid, and vitamins. They also contain a high level of nucleic acid which can be further purified to increase the level of 5′ nucleotides. They are used in aquaculture in functional feeds and hatcheries as a source of nucleotides complementing the de novo synthesis of cells in multiplication and helping boost immunity and antistress mechanisms (Tacon, 2012).

Nucleotides also have beneficial effects, as they positively affect the immune system, hepatic function, lipid metabolism, disease resistance, development of small intestine, and growth (Burrells et al., 2001a,b; Low et al., 2003; Gatlin and Li, 2007). It is also considered that under conditions of limited nucleotide intake or rapid growth, dietary nucleic acids may have a protein-sparing effect, as it limits the *de novo* synthesis of these molecules from its amino acid precursors. A beneficial effect of dietary nucleotide supplementation on growth rate was also observed in Atlantic salmon under conditions of management and environmental stressors (vaccination, grading, salt water transfer, etc.) (Burrells et al., 2001a; Gatlin and Li, 2007). Immunomodulatory effects were also found, particularly under stress conditions imposed by management or infectious diseases by several authors (Burrells et al., 2001a; Gatlin and Li, 2007). These same authors also found that inclusion of nucleotides at very low levels (0.03%), besides increasing resistance to challenge infections, enhancing the efficacy of vaccination, and lowering mortality after challenge, also improved the growth rate following salt-water transfer.

5.5 Probiotics

Probiotics are live microbial feed supplements which beneficially affect the host animal by improving its intestinal microbial balance. The interaction of probiotics with the digestive tract and its endogenous microflora is a subject that has been studied extensively and its benefits on aquatic animal health have been well documented and scientifically reviewed (Verschuere et al., 2000; Irianto and Austin, 2002; Merrifield et al., 2010; Merrifield and Carnevali, 2014).

Intake of probiotics has been demonstrated to modulate the gut microbiota, and therefore assist in returning a microbiota that has been disturbed to a beneficial composition. Mechanisms by which this is achieved include the production of inhibitory compounds against pathogens (lactoferrin, lysozyme, bacteriocins), competition for essential nutrients and adhesion sites, inhibition of virulence gene expression or disruption of quorum sensing, the supply of essential nutrients and enzymes resulting in enhanced nutrition, and the modulation of interactions with the environment and the development of beneficial immune responses (Balcazar et al., 2006, 2007; Merrifield et al., 2010).

A list of characteristics for potential probiotic bacteria to be used in aquaculture species has been reported by Vine et al. (2006) and extended by Merrifield et al. (2010). Among the essential properties to be a candidate are: being a nonpathogenic microorganism, being free of plasmid-encoded antibiotic resistance genes, and being resistant to bile salts and low pH. Other favorable properties are adequate and rapid growth at host rearing temperature, antagonistic properties against key pathogens, capacity to produce extracellular enzymes that improve feed utilization, viability under normal storage conditions, and acceptable survival under processing conditions.

Since it is unlikely to find a candidate that will fulfill all of these characteristics, efforts should focus on exploring the possibilities of simultaneously using several probiotics or the use of probiotics with prebiotics (termed synbiotics). Through the combined application of multiple favorable probiotic candidates it may be possible to produce greater benefits than the application of individual probionts (Merrifield et al., 2010). Probiotic preparations consisting of several different probiotic strains are described to be more effective than single-strain probiotics (Wang et al., 2008b).

Among the various probiotics, lactic acid bacteria (e.g., *Lactobacillus* spp., *Pediococcus* spp., *Enterococcus* spp.) and *Bacillus* spp. are often the most popular, and have been shown to improve the growth and/or nutrient utilization of various aquaculture animals (Lara-Flores et al., 2003; Wang et al., 2008a; Apún-Molina et al., 2009; Merrifield et al., 2010; Avella et al., 2011) as well as their resistance to pathogenic bacteria (Aly et al., 2008a,b; Ringø et al., 2010; Ai et al., 2011; Burbank et al., 2011).

Prerequisite for any effect directly in the host is the establishment of the probiotic bacteria as part of the indigenous gut microbiota, at least transiently. A significant alteration of the gut microbiota composition in aquatic animals after probiotic supplementation has been exhaustively demonstrated (Merrifield et al., 2010).

Several probiotic strains produce digestive enzymes, thus facilitating feed utilization and digestion. As a result, growth performance is improved. Supplementation of enzyme-producing probiotics accordingly improved the digestive enzyme activity (lipase, proteases, amylase) and feed utilization in gilthead sea bream (*Sparus aurata*, L.) larvae (Cüneyt Szer et al., 2008), the common carp (*Cyprinus carpio*) (Wang, 2006), but also in the white shrimp *L. vannamei* (Wang, 2007).

Probiotics are also used to control the emergence of opportunistic pathogens by competing for nutrients, space, and attachment sites in the gut. Direct antimicrobial effects through excreted molecules have also been described for a variety of different probiotic strains (Merrifield et al., 2010; Merrifield and Carnevali, 2014).

In order to exert their effects, probiotics must be viable at their site of action, meaning that the selected microorganism must survive the stressful conditions of feed processing and storage. One of the main challenges is high temperature used to produce aquafeeds, in particular extruded fishfeeds. This is seen as a major drawback and could explain the limited development of in-feed application of probiotics for use in aquaculture (Castex et al., 2014). Spore-forming bacteria, particularly *Bacillus* species, are often interesting potential probiotics for feed applications: they can be heat-stable, conferring fairly good viability after pelleting, and have high resistance to gastric conditions (Hong et al., 2005). However, their survivability during the higher temperature pelletization or extrusion encountered in fish and shrimp feeds is questionable. Suggestions to include spore-forming bacteria directly in aquaculture feeds before pelletization and/or extrusion may not be ideal (Castex et al., 2014).

Novel approaches such as inducing heat tolerance in the potential microorganism by optimizing the production conditions, or protecting viable probiotic cells, by microencapsulation of microbial cells are being attempted (Kailasapathy, 2002). Despite this, applications to shrimp feeds or extruded fish feeds are still at an early stage, necessitating approaches where feed manufacturers are looking internally for options involving optimization of their processes and/or the equipment used. Post-pelleting applications, by spraying, are generally alternatives that have shown some promise in terms of homogeneity, stability, and conformity at both laboratory and industrial levels (Castex et al., 2014).

5.6   Enzymes

Exogenous enzymes are a well-accepted class of feed additives in diet formulations for poultry and swine, to overcome the negative effects of antinutritional factors, and to improve the digestion of dietary components and animal performance. A number of microbial/fungal enzyme products are available in the market. However, almost all of these products are oriented to the livestock market and later promoted for use in aquaculture feeds.

Most reported studies conducted on the application of enzymes in aquaculture species are on phytase application. The application and benefits of phytase in animal species are well documented. Up to 80% of phosphorus in plant seeds is in the form of phytate. The digestibility and availability of phytate phosphorus for fish is very low. Further, phytate forms chelates with a large number of mineral cations (K, Mg, Ca, Zn, Fe, and Cu) and complexes with proteins and amino acids, thereby reducing the bioavailability of other minerals and digestibility of proteins (Debnath et al., 2004).

The in vivo efficacy of phytase enzymes is largely determined by their intrinsic properties, such as activity at different pH conditions, resistance against proteolytic degradation, and thermal stability (Sinha et al., 2011).

Rodehutscord and Pfeffer (1995) conducted a study with trout where they were fed two vegetable diets (mainly based on soybean meal) with or without 1,000 U/kg phytase supplementation. At a water temperature of 15°C, both digestibility and utilization of phosphorus increased in the control vs. treatment group from 25% to 57% and from 17% to 49% respectively. Conversely, Cheng and Hardy (2002) showed that microbial phytase supplementation of diets at 500 U/kg diet containing barley, canola meal, wheat, and wheat middlings improved availability of energy and phosphorus in rainbow trout. Debnath et al. (2005) showed that phytase at 500 U/kg feed gave higher weight gain, apparent net protein utilization and energy retention value in *Pangasius pangasius* fingerlings. In a related study, Debnath et al. (2004) showed that the whole-body content of Ca, P, Zn, Fe, Cu, and Co was elevated in fish fed diets supplemented with phytase levels more than 250 U/kg. Liebert and Portz (2005) reported better growth, feed conversion ratio, protein efficiency ratio, and mineral deposition in juvenile Nile tilapia, *O. niloticus*, fed diets supplemented with microbial phytase.

Trials with carp (Schafer et al., 1995) also showed improved phytate phosphorus utilization as well as a reduction in phosphorus excretion compared with monocalcium–phosphate addition. Five hundred U/kg of phytase increased phosphorus utilization by 20% and reduced phosphorus excretion by 28% compared with the unsupplemented (negative control) group (Schafer et al., 1995).

There are also some studies on improving carbohydrate and protein digestion using microbial enzymes. Stone et al. (2003) reported that starch digestibility was improved in juvenile silver perch by a commercial α-amylase. They found that the effect of the commercial enzyme was greater on diets containing raw wheat starch than on diets containing gelatinized wheat starch. Conversely, Kumar et al. (2006) also found that dietary supplementation of a corn-based feed with α-amylase (*Aspergillus* origin) at 50 mg/kg enhanced growth and protein sparing in rohu fingerlings. However, when gelatinized corn replaced raw corn in the diet, the enzyme supplementation provided no advantage.

Protease supplementation in aquafeeds has been tested mainly in salmonids. Addition of 250 g/ton of a commercial protease to a coextruded canola–pea-based diet resulted in significant improvement in feed efficiency and apparent digestibility of crude protein, energy, lipid, and dry matter ($P < 0.05$) in rainbow trout (Drew et al., 2005). However, when the same protease was added to a cold-extruded flax–pea-based diet there was no effect on digestibility (Drew et al., 2005). In another in vivo study conducted with three species of salmonids (coho salmon, Atlantic salmon, and rainbow trout), both protein and carbohydrate digestibility were improved significantly in fish fed treatment diets containing a commercial protease, when compared with those fed the control diets (Chowdhury, 2014).

High levels of NSPs such as cellulose, xylans, and mannans reduce the nutritive value of many plant ingredients. Intestinal enzymes to digest these carbohydrates are not produced by most aquatic animals (Sinha et al., 2011). The improvement in the digestibility of NSPs is achieved by supplementation of NSP-degrading enzymes in the diet (Sinha et al., 2011).

The application of NSP enzymes in fish diets was tested in several species. Ng and Chong (2002) reported that dry matter and energy digestibility coefficients of palm kernel meal in red hybrid tilapia improved when supplemented with pure mannanase and a commercial feed enzyme mix that included protease, cellulase, glucanase, and pectinase.

Li et al. (2009) tested the application of 1 g/kg of NSP-degrading enzyme (50 fungal β-glucanase units/g) in tilapia feed and reported increased activity of amylase in the hepatopancreas and intestine by 11.4% and 49.5%, respectively. The application of an enzyme blend (containing β-glucanase and β-xylanase) at concentrations of 75, 150, and 300 l/kg in diets containing 30% wheat or dehulled lupin had no remarkable effect on dry matter, energy, or protein digestibility when fed to silver perch (Stone et al., 2003). The supplementation of two NSP-degrading enzymes, 400 mg "VP" (contains mainly glucanase, pentosanase, and cellulase, each at 50 IU/g) and 800 mg "WX" (contains mainly xylanase, 1,000 IU/g) in basal feed of Japanese sea bass (*Lateolabrax japonicus*) significantly enhanced the specific growth rate, feed efficiency ratio, nitrogen retention, and reduced ammonia excretion (Ai et al., 2007).

While considering the effectiveness of enzyme applications, one must also take into account findings that have shown no impact of enzyme application on fish or shrimp performance (Carter et al., 1992, 1994; Divakaran and Velasco, 1999; Stone et al., 2003). Other than the likelihood that species-related differences exist in the effectiveness of enzymes, it is also possible that ingredients used in diets determine the effectiveness of an enzyme. According to Drew et al. (2005), benefits of protease addition should be assessed for individual ingredients.

Most feed enzymes are applied in powder or granular form prior to feed conditioning. Feed conditioning (e.g. pelleting, expansion, extrusion) may adversely affect the efficacy of phytase since these enzymes are rapidly inactivated at temperatures above 50–60°C. Post-pellet spray application with liquid phytase represents an alternative to the incorporation of solid enzymes prior to conditioning. However, such systems require additional efforts in terms of labor and investments. Therefore, it has been a major target of research to improve the thermal tolerance of enzymes.

At present, the application of enzymes in aquaculture feeds is more limited and is directly related to some constrains regarding the effectiveness of enzyme applications in aquafeeds. These concerns are mainly related to: (i) The effect of temperature on the stability of enzymes applied in the feed prior to pelleting or extrusion; (ii) leaching loss of the enzyme in water, when enzymes are coated on the feed after thermal processing; and (iii) the effectiveness of microbial enzymes that have 37°C optimum temperature, when applied in cold water aquatic animals that have low body temperatures. These concerns will have to be addressed for the aquafeed industry to widely embrace enzyme application as a way to improve feed performance.

5.7   Mycotoxin binders

Given the trend and the economical need to replace expensive marine-derived proteins such as fish meal with less expensive and more available plant protein sources, the impact of mycotoxin contamination in aquaculture feeds will have the tendency to increase due to the higher susceptibility for mycotoxin contamination in ingredients of plant origin. Mycotoxins are secondary metabolites produced by different species of fungi (Binder, 2007) that have the potential to reduce the growth and health status of aquatic organisms consuming contaminated feed (Santacroce et al., 2008; van Hoof et al., 2012; García-Pérez et al., 2013).

Although mycotoxin contamination of feed and feed ingredients represents an increased threat to aquaculture operations, there are a number of options available to feed manufacturers to prevent or reduce the risk of mycotoxicosis associated with mycotoxin contamination. These range from careful selection of raw materials, maintaining good storage conditions for feeds and raw materials, and using feed additives to combat the widest possible range of different mycotoxins that may be present.

The most commonly used strategy of reducing the exposure to mycotoxins is to decrease their availability by the inclusion of various mycotoxin-binding agents or adsorbents, which leads to a reduction in mycotoxin uptake and distribution to the blood and target organs (Binder, 2007). Various substance groups have been tested and used for this purpose, with aluminum silicates, in particular clay and zeolitic materials, as the most commonly applied groups (Binder, 2007). These are added to the feed during feed formulation at levels that can range from 1 to 10 g/kg depending on the type of product used.

Critical parameters concerning the use of adsorbents for binding mycotoxins are their efficacy, specificity, as well as the mechanism of the adsorption process (chemisorption/physisorption) (Vekiru et al., 2007). The stability of the sorbent–toxin bond needs to be strong to prevent desorption of the toxin, and they need to be effective in a broad pH level since a product needs to work throughout the gastrointestinal tract (Binder, 2007).

A comprehensive study on the ability of various adsorbents to bind aflatoxin B1 (AfB1), concluded that different binding materials have different efficiencies to adsorb AfB1 (Vekiru et al., 2007). According to the evaluated chemisorption index, AfB1 was in general strongly bound to bentonites indicating an adsorption process due to chemisorption. Also, results of the potential adsorption of selected vitamins confirmed that bentonite binders neither adsorb vitamin B5 nor vitamin H, having a clear preference to adsorb AfB1 (Vekiru et al., 2007).

The different impacts of the binders on AfB1 and vitamin adsorption, as well as the significant reduction of AfB1 adsorption of some binders, when adsorption tests were performed in real gastric juice instead of buffer solution, highlighted the differences in the selectivity of adsorption.

The different capacities for different binder products to effectively reduce the negative impacts of AfB1 in shrimp feeds was reported by García-Pérez et al. (2013). When testing the supplementation of four different binders in shrimp feeds contaminated with 75 ppb AfB1, it was observed that only one binder was completely effective in reducing the negative effects of AfB1 in shrimp (García-Pérez et al., 2013).

In fish, an in vivo trial using pangasius catfish (*Pangasionodon hypoththalmus*) as a model reported that AFB1 contamination in catfish diets at a level of 50 μg AFB1/kg and above can affect fish performance and disease resistance. This reduction in performance however, was not observed ($P > 0.05$) in fish fed diets containing 500 μg AFB1/kg plus a bentonite-based aflatoxin binder at a level of 1.5 g/kg feed, attesting that the adsorption capacity of the binder was effective (Tu, 2010). Conversely, a study testing two types of clay adsorbents (bentonite and montmorillonite) in tilapia feed showed that when incorporated at a rate of 5 g/kg feed, both products were able to improve all the tested parameters of tilapia when fish were fed a diet with 1.5 ppm AfB1 (Hassan et al., 2010).

Unfortunately, different mycotoxin groups are completely different in their chemical structure, and therefore it is impossible to equally deactivate all mycotoxins by using only a single strategy. Adsorption works perfectly for aflatoxin, but less or nonabsorbable mycotoxins (such as ochratoxins, zearalenone, and the whole group of trichothecenes) have to be deactivated by using a different approach. Biotransformation is defined as detoxification of mycotoxins using microorganisms or enzymes which specifically degrade the toxic structures to nontoxic metabolites. Different microorganisms, live bacteria, and yeast strains expressing specific mycotoxin-degrading enzymes have been identified (Schatzmayr et al., 2006). Their incorporation in animal feeds has been shown to counteract mycotoxin-induced performance deterioration in piglet and broilers (Binder, 2007); no results of this novel approach are available for aquatic species yet.

References

Abutbul, S., Golan-Goldhirsh, A., Barazani, O., Zilberg, D., 2004. Use of *Rosamarinus officinalis* as a treatment against *Streptococcus iniae* in tilapia (*Oreochromis* sp.). Aquaculture 238, 97–106.

Ahmad, M., El Mesallamy, A., Samir, F., Zahran, F., 2011. Effect of Cinnamon (*Cinnamomum zeylanicum*) on growth performance, feed utilization, whole-body composition, and resistance to *Aeromonas hydrophila* in Nile Tilapia. J. Appl. Aquacult. 23, 289–298.

Ahmadifar, E., Falahatkar, B., Akrami, R., 2011. Effects of dietary thymol–cravacrol on growth performance, hematological parameters and tissue composition of juvenile rainbow trout, *Oncorhynchus mykiss*. J. Appl. Ichtyol. 27, 1057–1060.

Ai, Q., Mai, K., Zhang, W., Xu, W., Tan, B., Zhang, C., 2007. Effects of exogenous enzymes (phytase, non-starch polysaccharide enzyme) in diets on growth, feed utilisation, nitrogen and phosphorus excretion of Japanese seabass, *Lateolabrax japonicus*. Comp. Biochem. Physiol. 147A, 502–508.

Ai, Q., Xu, H., Mai, K., Xu, W., Wang, J., Zhang, W., 2011. Effects of dietary supplementation of *Bacillus subtilis* and fructooligosaccharide on growth performance, survival, non-specific immune response, and disease resistance of juvenile large yellow croaker, *Larimichthys crocea*. Aquaculture 317, 155–161.

Aly, S.M., Ahmed, Y.A.G., Ghareeb, A.A.A., Mohamed, M.F., 2008a. Studies on *Bacillus subtilis* and *Lactobacillus acidophilus*, as potential probiotics, on the immune response and resistance of *Tilapia nilotica* (*Oreochromis niloticus*) to challenge infections. Fish Shellfish Immunol. 25, 128–136.

Aly, S.M., Mohamed, M.F., John, G., 2008b. Effect of probiotics on the survival, growth and challenge infection in *Tilapia nilotica* (*Oreochromis niloticus*). Aquacult. Res. 39, 647–656.

Apún-Molina, J.P., Santamaría-Miranda, A., Luna-González, A., Martínez-Diaz, S.F., Rojas-Contreras, M., 2009. Effect of potential probiotic bacteria on growth and survival of tilapia *Oreochromis niloticus* L., cultured in the laboratory under high density and suboptimum temperature. Aquacult. Res. 40, 887–894.

Avella, M.A., Olivotto, I., Silvi, S., Ribecco, C., Cresci, A., Palermo, F., et al., 2011. Use of *Enterococcus faecium* to improve common sole (*Solea solea*) larviculture. Aquaculture 31, 384–393.

Bai, N., Gu, N., Zhang, W., Xu, W., Mai, K., 2014. Effects of β-glucan derivatives on the immunity of white shrimp *Litopenaeus vannamei* and its resistance against white spot syndrome virus infection. Aquaculture 426–427, 66–73.

Balcazar, J.L., Decamp, O., Vendrell, D., De Blas, I., Ruiz-Zarzuela, I., 2006. Health and nutritional properties of probiotics in fish and shellfish. Microbiol. Ecol. Health Dis. 18, 65–70.

Balcázar, J.L., de Blas, I., Ruiz-Zarzuela, I., Vendrell, D., Calvo, A.C., Márquez, I., et al., 2007. Changes in intestinal microbiota and humoral immune response following probiotic administration in brown trout (*Salmo trutta*). Br. J. Nutr. 97, 522–527.

Baser, K.H.C., 2008. Biological and pharmacological activities of carvacrol and carvacrol bearing essential oils. Curr. Pharm. Des. 14, 3106–3119.

Baruah, K., Sahu, N.P., Pal, A.K., Jain, K.K., Debnath, D., Mukherjee, S.C., 2007. Dietary microbial phytase and citric acid synergistically enhances nutrient digestibility and growth performance of *Labeo rohita* (Hamilton) juveniles at sub-optimal protein level. Aquacult. Res. 38, 109–120.

Binder, E.M., 2007. Managing the risk of mycotoxins in modern feed production. Anim. Feed Sci. Technol. 133, 149–166.

Bricknell, I., Ra, D., 2005. The use of immunostimulants in fish larval aquaculture. Fish Shellfish Immunol. 19 (5), 457–472.

Burbank, D.R., Shah, D.H., LaPatra, S.E., Fornshell, G., Cain, K.D., 2011. Enhanced resistance to coldwater disease following feeding of probiotic bacterial strains to rainbow trout. Aquaculture 321, 185–190.

Burrells, C., Williams, P.D., Forno, P.F., 2001a. Dietary nucleotides: a novel supplement in fish feeds 1. Effects on resistance to disease in salmonids. Aquaculture 199, 159–169.

Burrells, C., Williams, P.D., Southgate, P.J., Wadsworth, S.L., 2001b. Dietary nucleotides: a novel supplement in fish feeds 2. Effects on vaccination, salt water transfer, growth rates and physiology of Atlantic salmon (*Salmo salar* L.). Aquaculture 199, 171–184.

Burt, S., 2004. Essential oils: their antibacterial properties and potential applications in food – a review. Int. J. Food Microbiol. 94, 223–253.

Cardozo, P., Kamel, C., Greathead, H.M.R., Jintasataporn O., 2008. Encapsulated plant extracts as dietary enhancers of growth, feeding efficiency and immunity in white shrimp (*Litopenaeus vannamei*) under normal and stress conditions. Aqua 2008. X Congresso Ecuatoriano de Acuicultura & Aquaexpo. October 6–9. Guayaquil. Ecuador (Abstract).

Carter, C.G., Houlihan, D.F., McCarthy, I.D., 1992. Feed utilization efficiencies of Atlantic salmon (*Salmo salar* (L.)) parr: effect of a supplementary enzyme. J. Comp. Biochem. Physiol. 101 (2), 369–374.

Carter, C.G., Houlihan, D.F., Buchanan, B., Mitchell, A.I., 1994. Growth and feed utilization efficiencies of seawater Atlantic salmon, *Salmo salar* L., fed a diet containing supplementary enzymes. Aquacult. Fish. Manage. 25, 37–46.

Castex, M., Durand, H., Okeke, B., 2014. Issues with industrial probiotic scale-up. In: Merrifield, D., Ringo, E. (Eds.), Aquaculture Nutrition, Gut Health, Probiotics and Prebiotics John Wiley & Sons, Chichester, West Sussex, pp. 347–359.

Chakraborty, S.B., Hancz, C., 2011. Application of phytochemicals as immunostimulant, antipathogenic and antistress agents in finfish culture. Rev. Aquacult. 3, 103–119.

Chakraborty, S.B., Horn, P., Hancz, C., 2014. Application of phytochemicals as growth-promoters and endocrine modulators in fish culture. Rev. Aquacult. 6, 1–19.

Cheng, Z.J., Hardy, R.W., 2002. Apparent digestibility coefficients of nutrients and nutritional value of poultry by-product meals for rainbow trout *Oncorhynchus mykiss* measured *in vivo* using settlement. J. World Aquacult. Soc. 33, 458–465.

Chowdhury, M.A.K., 2014. Use of a heat-stable protease in salmonid feeds – Experiences from Canada and Chile. Int. Aquafeed, 30–32.

Couso, N., Castro, R., Magarinos, R., Obach, A., Lamas, J., 2003. Effect of oral administration of glucans on the resistance of gilthead seabream to pasteurellosis. Aquaculture 219, 99–109.

Debnath, D., Sahu, N.P., Pal, A.K., Jain, K.K., Yengkokpam, S., Mukherjee, S.C., 2004. Mineral status of *Pangasius pangasius* (Hamilton) fingerlings in relation to supplemental phytase: absorption, whole body and bone mineral content. Aquacult. Res. 36, 326–335.

Debnath, D., Pal, A.K., Narottam, P.S., Jain, K.K., Yengkokpam, S., Mukherjee, S.C., 2005. Effect of dietary microbial phytase supplementation on growth and nutrient digestibility of *Pangasius pangasius* (Hamilton) fingerlings. Aquacult. Res. 36, 180–187.

Dimitroglou, A., Merrifield, D.L., Spring, P., Sweetman, J., Moate, R., Davies, S.J., 2010. Effects of dietary mannan oligosaccharides (MOS) and soybean meal on growth performance, feed utilisation, intestinal histology and gut microbiota of gilthead sea bream (*Sparus aurata*). Aquaculture 300, 182–188.

Divakaran, S., Velasco, M., 1999. Effect of proteolytic enzyme addition to a practical feed on growth of the Pacific white shrimp, *Litopenaeus vannamei* (Boone). J. Aquacult. Res. 30 (5), 335–339.

Drew, M.D., Racz, V.J., Gauthier, R., Thiessen, D.L., 2005. Effect of adding protease to co-extruded flax:pea or canola:pea products on nutrient digestibility and growth performance of rainbow trout (*Oncorhynchus mykiss*). Anim. Feed Sci. Technol. 119, 117–128.

FEFANA, 2014. Organic Acids in Animal Nutrition. Fefana Publication, Brussels, 97 pp.

García-Pérez, O., Tapia-Salazar, M., Nieto-López, M., Cavazos, D., Cruz-Suárez, E., Ricque-Marie, D., 2013. Effectiveness of aluminosilicate-based products for detoxification of aflatoxin contaminated diets for juvenile Pacific white shrimp, *Litopenaeus vannamei*. Ciencias Marinas 39 (1), 1–13.

Gatlin III, D.M., Li, P., 2007. Nucleotides. In: Nakagawa, H., Sato, M., Gatlin III, D.M. (Eds.), Dietary Supplements for the Health and Quality of Cultured Fishes CABI, Oxon, UK, pp. 193–209.

Giannenas, I., Triantafillou, E., Stavrakakis, S., Margaroni, M., Mavridis, S., Steiner, T., et al., 2012. Assessment of dietary supplementation with carvacrol or thymol containing feed additives on performance, intestinal microbiota and antioxidant status of rainbow trout (*Oncorhynchus mykiss*). Aquaculture 350–353, 26–32.

Hassan, A.M., Kenaway, A.M., Abbas, W.T., Abdel-Wahhab, M.A., 2010. Prevention of cytogenetic, histochemical and biochemical alterations in *Oreochromis niloticus* by dietary supplement of sorbent materials. Ecotoxicol. Environ. Saf. 73, 1890–1895.

Hisano, H., Barros, M.M., Pezzato, L.E., 2007. Levedura e zinco como pró-nutrientes em rações para tilápia do Nilo (*Oreochromis niloticus*): aspectos hematológicos. Bol. Inst. Pesca 33 (1), 35–42.

Hong, H.A., le Duc, H., Cutting, S.M., 2005. The use of bacterial spore formers as probiotics. FEMS Microbiol. Rev. 29, 813–835.

Hossain, M.A., Pandey, A., Satoh, S., 2007. Effects of organic acids on growth and phosphorus in red sea bream *Pagrus major*. Fish. Sci. 73, 1309–1317.

Irianto, A., Austin, B., 2002. Probiotics in aquaculture. J. Fish Dis. 25, 633–642.

Kailasapathy, K., 2002. Microencapsulation of probiotic bacteria: technology and potential applications. Curr. Issues Intest. Microbiol. 3, 39–48.

Krosmayr, A., 2007. Experimental studies of the gastrointestinal effects of essential oils in comparison to avilamycin in weaned piglets (PhD dissertation). Universität für Bodenkultur Wien. 91 pp.

Kumar, S., Sahu, N.P., Pal, A.K., Choudhury, D., Mukherjee, S.C., 2006. Studies on digestibility and digestive enzyme activities in *Labeo rohita* (Hamilton) juveniles: effect of microbial a-amylase supplementation in non-gelatinised or gelatinised corn-based diet at two protein levels. Fish Physiol. Biochem. 32, 209–220.

Lara-Flores, M., Olvera-Novoa, M.A., Guzmán-Méndez, B.E., López-Madrid, W., 2003. Use of the bacteria *Streptococcus faecium* and *Lactobacillus acidophilus* and the yeast *Saccharomyces cerevisiae* as growth promoters in Nile tilapia (*Oreochromis niloticus*). Aquaculture 216, 193–201.

Li, P., Gatlin III, D.M., 2004. Dietary brewer's yeast and the prebiotic Grobiotic AE influence growth performance, immune responses and resistance of hybrid striped bass (*Morone chrysops* x *M. saxatilis*) to *Streptococcus iniae* infection. Aquaculture 231, 445–456.

Li, J.S., Li, J.L., Wu, T.T., 2009. Effects of non-starch polysaccharides enzyme, phytase and citric acid on activities of endogenous digestive enzymes of tilapia (*Oreochromis niloticus* x *Oreochromis aureus*). Aquacult. Nutr. 15, 415–420.

Liebert, F., Portz, L., 2005. Nutrient utilization of Nile tilapia *Oreochromis niloticus* fed plant based low phosphorus diets supplemented with graded levels of different sources of microbial phytase. Aquaculture 248, 111–119.

Liebert, F., Mohamed, K., Lückstädt, C., 2010. Effects of diformates on growth and feed utilization of all male Nile Tilapia fingerlings (*Oreochromis niloticus*) reared in tank culture. In: XIV International Symposium on Fish Nutrition and Feeding, Qingdao, China, Book of Abstracts, 190 pp.

Low, C., Wadsworth, S., Burrells, C., Secombes, C.J., 2003. Expression of immune genes in turbot (*Scophthalmus maximus*) fed a nucleotide-supplemented diet. Aquaculture 221, 23–40.

Lückstädt, C., 2008. The use of acidifiers in fish nutrition. CAB Rev. Perspect. Agri. Vet. Sci. Nutr. Nat. Resour. 3 (44), 1–8.

Meena, D.K., Das, P., Kumar, S., Mandal, S.C., Prusty, A.K., Singh, S.K., et al., 2013. Beta-glucan: an ideal immunostimulant in aquaculture (a review). Fish Physiol. Biochem. 39 (3), 431–457.

Merrifield, D.L., Carnevali, O., 2014. Probiotic modulation of the gut microbiota of fish. In: Merrifield, D., Ringo, E. (Eds.), Aquaculture Nutrition, Gut Health, Probiotics and Prebiotics John Wiley & Sons, Chichester, West Sussex, pp. 185–222.

Merrifield, D.L., Dimitroglou, A., Foey, A., Davies, S.J., Baker, R.T.M., Bøgwald, J., et al., 2010. The current status and future focus of probiotic and prebiotic applications for salmonids. Aquaculture 302, 1–18.

Militz, T.A., Southgate, P.C., Carton, A.G., Hutson, K.S., 2013. Dietary supplementation of garlic (*Allium sativum*) to prevent monogenean infection in aquaculture. Aquaculture 408-409, 95–99.

Morken, T., Kraugerud, O.F., Barrows, F.T., Sørensen, M., Storebakken, T., Øverland, M., 2011. Sodium diformate and extrusion temperature affect nutrient digestibility and physical quality of diets with fish meal and barley protein concentrate for rainbow trout (*Oncorhynchus mykiss*). Aquaculture 317, 138–145.

Ng, W.K., Chong, K.K., 2002. The nutritive value of palm kernel and the effect of enzyme supplementation in practical diets for red hybrid tilapia (*Oreochromis* sp). Asian Fish Sci. 15, 167–176.

Ng, W.-K., Chik-Boon, K., Kumar, S., Siti-Zahrah, A., 2009. Effects of dietary organic acids on growth, nutrient digestibility and gut microflora of red hybrid tilapia, *Oreochromis* sp., and subsequent survival during a challenge test with *Streptococcus agalactiae*. Aquacult. Res. 40, 1490–1500.

Nuez-Ortin, W.G., 2011. Gustor-Aqua: an effective solution to optimize health status and nutrient utilization. International Aquafeed. May–June, 18–20.

Nya, E.J., Dawood, Z., Austin, B., 2010. The garlic component, allicin, prevents disease caused by *Aeromonas hydrophila* in rainbow trout, *Oncorhynchus mykiss* (Walbaum). J. Fish Dis. 33, 293–300.

NRC (National Research Council), 2011. Nutrient Requirements of Fish. National Academy Press, Washington, DC, 376 pp.

Oliva-Teles, A., Gonçalves, P., 2001. Partial replacement of fishmeal by Brewers yeast *Saccaromyces cerevisiae*/in diets for sea bass *Dicentrarchus labrax* juveniles. Aquaculture, 202 Ž2001, 269–278.

Ortuno, J., Cuesta, A., 2002. Oral administration of yeast, *Saccharomyces cerevisiae*, enhances the cellular innate immune response of gilthead seabream (*Sparus aurata* L.). Vet. Immunol. Immunopathol. 85 (1–2), 41–50.

Peterson, B.C., Bosworth, B.G., Li, M.H., Beltran, R., Santos, G., 2014. Assessment of a Phytogenic Feed Additive (Digestarom P.E.P. MGE) on growth performance, processing yield, fillet composition, and survival of channel catfish. J. World Aquacult. Soc. 5 (2), 206–212.

Ramli, N., Heindl, U., Sunanto, S., 2005. Effect of potassium-diformate on growth performance of tilapia challenged with *Vibrio anguillarum*. World Aquaculture 2005, Bali, Indonesia (Abstract).

Reverter, M., Bontemps, N., Lecchini, D., Banaigs, B., Sasal, P., 2014. Use of plant extracts in fish aquaculture as an alternative to chemotherapy: current status and future perspectives. Aquaculture 433, 50–61.

Riemensperger, A., Santos, G., 2011a. Breakthrough in natural growth promotion. International Aquafeed. July–August. 18–20.

Riemensperger, A., Santos, G., 2011b. Maintaining health in shrimp culture. Aqua Culture Asia Pacific. March–April. 28–30.

Ringo, E., Olsen, R.E., 1999. The effect of diet on aerobic bacterial flora associated with intestine of Arctic charr (*Salvelinus alpinus* L.). J. Appl. Microbiol. 86, 22–28.

Ringø, E., Lovmo, L., Kristiansen, M., Bakken, Y., Salinas, I., Myklebust, R., et al., 2010. Lactic acid bacteria vs. pathogens in the gastrointestinal tract of fish: a review. Aquacult. Res. 41, 451–467.

Rodehutscord, M., Pfeffer, E., 1995. Effects of supplemental microbial phytase on phosphorus digestibility and utilization in rainbow trout (*Oncorhynchus mykiss*). Water Sci. Technol. 31 (10), 141–147.

Rumsey, G.L., Hughes, S.G., Smith, R.R., Kinsella, J.E., Shetty, K.J., 1991. Digestibility and energy values of intact, disrupted and extracts from brewer's dried yeast fed to rainbow trout *Oncorhynchus mykiss*. Anim. Feed Sci. Technol. 33, 185–193.

Rumsey, G.L., Winfree, R.A., Hughes, S.G., 1992. Nutritional values of dietary nucleic acids and purine bases to rainbow trout. Aquaculture 108, 97–110.

Sahoo, P.K., Mukherjee, S.C., 2002. The effect of dietary immunomodulation upon *Edwardsiella tarda* vaccination in healthy and immunocompromised Indian major carp (*Labeo rohita*). Fish Shellfish Immunol. 12, 1–16.

Santacroce, M.P., Conversano, M.C., Casalino, E., Lai, O., Zizzadoro, C., Centoducati, G., et al., 2008. Aflatoxins in aquatic species: metabolism, toxicity and perspectives. Rev. Fish. Biol. Fish. 18, 99–130.

Sarker, S.A., Satoh, S., Kiron, V., 2005. Supplementation of citric acid and amino acid chelated trace element to develop environment-friendly feed for red sea bream, *Pagrus major*. Aquaculture 248, 3–11.

Sarker, S.A., Satoh, S., Kiron, V., 2007. Inclusion of citric acid and/or acid-chelated trace elements in alternate plant protein source diets affects growth and excretion of nitrogen and phosphorus in red sea bream *Pagrus major*. Aquaculture 262, 436–443.

Schafer, A., Koppe, W.M., Meyer–Burgdorff, K.H., Gunther, K.D., 1995. Effects of microbial phytase on utilization of native phosphorus by carp in diet based on soybean meal. Water Sci. Technol. 31, 149–155.

Schatzmayr, G., Zehner, F., Schatzmayr, D., Taubel, M., Binder, E.M., 2006. Microbials for deactivating mycotoxins in contaminated feed. Mol. Nutr. Food Res. 50 (6), 543–551.

Sealey, W.M., Gatlin III, D.M., 2001. Overview of nutritional strategies affecting the health of marine fish. In: Lim, C., Webster, C.D. (Eds.), Nutrition and Fish Health The Haworth Press Inc., Binghamton, NY, pp. 103–118.

Silva, B.C., Vieira, F.N., Mouriño, J.L., Ferreira, G.S., Seiffert, W.Q., 2013. Salts of organic acids selection by multiple characteristics for marine shrimp nutrition. Aquaculture 384–387, 104–110.

Sinha, A.K., Kumar, V., Makkar, H.P.S., De Boeck, G., Becker, K., 2011. Non-starch polysaccharides and their role in fish nutrition – a review. Food Chem. 127, 1409–1426.

Stone, D.A.J., Allan, G.L., Anderson, A.J., 2003. Carbohydrate utilisation by juvenile silver perch, *Bidyanus bidyanus* (Mitchell). IV. Can dietary enzymes increase digestible energy from wheat starch, wheat and dehulled lupin? Aquacult. Res. 34, 135–147.

Staykov, Y., Spring, P., Denev, S., Sweetman, J., 2007. Effect of mannan oligosaccharide on the growth performance and immune status of rainbow trout (*Oncorhynchus mykiss*). Aquacult. Int. 15, 153–161.

Steiner, T., 2006. Managing Gut health. Natural Growth Promoters as a Key to Animal Performance. Nottingham University Press, 98 pp.

Szer, C.D., Kaaci, O.H., Saka, S., Firat, K., Otgucuoglu, O., Kucuksari, H., 2008. *Lactobacillus* spp. bacteria as probiotics in gilthead sea bream (*Sparus aurata*, L.) larvae: effects on growth performance and digestive enzyme activities. Aquaculture 280, 140–145.

Tacon, P., 2012. Yeast in Aquaculture. International AquaFeed. November–December. 14–18.

Talpur, A.D., 2014. *Mentha piperina* (Peppermint) as feed additive enhanced growth performance, survival, immune response and disease resistance of *Asian seabass, Lates calcarifer* (Bloch) against *Vibrio harveyi* infection. Aquaculture 420–421, 71–78.

Torrecillas, S., Makol, A., Caballero, M.J., Montero, D., Gines, R., Sweetman, J., et al., 2011. Improved feed utilization, intestinal mucus production and immune parameters in sea bass (*Dicentrarchus labrax*) fed mannan oligosaccharides (MOS). Aquacult. Nutr. 17 (2), 223–233.

Tu, D.C., 2010. Aflatoxin B1 reduces growth performance, physiological response, histological changes and disease resistance in Tra catfish (*Pangasianodon hypophthalmus*) (In Submission of MSc Thesis). Nong Lam University. Ho Chi Min City. Vietnam.

van Hoof, L., van Leeuwen, J., van Tatenhove, J., 2012. All at sea; regionalisation and integration of marine policy in Europe. Marit. Stud. 11, 9.

Vekiru, E., Fruhauf, S., Sahin, M., Ottner, F., Schatzmayr, G., Krska, R., 2007. Investigation of various adsorbents for their ability to bind Aflatoxin B1. Mycotoxin Res. 23, 27–33.

Verschuere, L., Rombaut, G., Sorgeloos, P., Verstraete, W., 2000. Probiotic bacteria as biological control agents in aquaculture. Microbiol. Mol. Biol. Rev. 64, 655–671.

Vine, N.G., Leukes, W.D., Kaiser, H., 2006. Probiotics in marine larviculture. FEMS Microbiol. Rev. 30, 404–427.

Vielma, J., Lall, S.P., 1997. Dietary formic acid enhances apparent digestibility of minerals in rainbow trout, *Oncorhynchus mykiss* (Walbaum). Aquacult. Nutr. 3, 265–268.

Volpati, D., Bulfon, C., Tulli, F., Galeotti, M., 2014. Growth parameters, innate immune response and resistance to *Listonella* (vibrio) anguillarum of *Dicentrarchus labrax* fed carvacrol supplemented diets. Aquacult. Res. 45, 31–44.

Wang, Y.B., 2006. Effect of probiotics for common carp (*Cyprinus carpio*) based on growth performance and digestive enzyme activities. Anim. Feed Sci. Technol. 127, 283–292.

Wang, Y.B., 2007. Effects of probiotics on growth performance and digestive enzyme activity of the shrimp *Penaeus vannamei*. Aquaculture 269, 259–264.

Wang, Y.-B., Tian, Z.-Q., Yao, J.-T., Li, W.-E., 2008a. Effect of probiotics, *Enteroccus faecium*, on tilapia (*Oreochromis niloticus*) growth performance and immune response. Aquaculture 277, 203–207.

Wang, Y.-B., Li, J.-R., Lin, J., 2008b. Probiotics in aquaculture: challenges and outlook. Aquaculture 281, 1–4.

Welker, T.L., Lim, C., Yildirim-Aksoy, M., Klesius, P.H., 2012. Effect of short-term feeding duration of diets containing commercial whole-cell yeast or yeast subcomponents on immune function and disease resistance in channel catfish, *Ictalurus punctatus*. J. Anim. Physiol. Anim. Nutr. 96 (2), 159–171.

Windisch, W., Schedle, K., Plitzner, C., Kroismayr, A., 2008. Use of phytogenic products as feed additives for swine and poultry. J. Anim. Sci. 86, E140–E148.

Zheng, Z.L., Justin, Y.W., Tan, H.Y., Liu, X.H., Zhou, X.X., Wang, K.Y., 2009. Evaluation of oregano essential oil (*Origanum heracleoticum* L.) on growth, antioxidant effect and resistance against *Aeromonas hydrophila* in channel catfish (*Ictalurus punctatus*). Aquaculture 292, 214–218.

Zhou, F., Ji, B.P., Zhang, H., Yang, Z., Li, J., Yan, W., 2007a. The antibacterial effect of cinnamaldehyde, thymol, carvacrol and their combination against the foodborne pathogen *Salmonela typhimurium*. J. Food Saf. 27, 124–133.

Zhou, F., Ji, B., Zhang, H., Jiang, H., Yang, Z., Li, J., et al., 2007b. Synergistic effect of thymol and carvacrol combined with chelators and organic acids against *Salmonella typhimurium*. J. Food Prot. 70 (7), 1704–1709.

Optimizing nutritional quality of aquafeeds

Karthik Masagounder[1], Sheila Ramos[2], Ingolf Reimann[1] and Girish Channarayapatna[2]

[1]Evonik Industries AG, Hanau-Wolfgang, Germany [2]Evonik (SEA) Pte. Ltd, Singapore

6.1 Introduction

Aquaculture is one of the fastest growing industries with an annual growth rate of 8% (FAO, 2011). The feed industry has to grow in parallel to support the growth of the farming industry. Maintaining quality standards is a key driver for the success of every industry and quality control is an integral part of feed mills. In order to ensure quality standards, feed producers declare if they are certified (e.g., ISO 9000 series; Good Manufacturing Practice; Best Aquaculture Practice) by any accredited certification body. Feed bags are usually labeled declaring the minimum content of critical nutrients which should comply with the legislation of a certified organization or the country where the feed is manufactured and sold. Feed labels often include proximate nutrients along with major feed ingredients used. There is no doubt that proximate nutrients give a broad overview of the nutritional quality of feed; however, growth of fish is dependent on various other essential nutrients. For example, fish don't have requirements for crude protein (CP) but for amino acids that comprises the protein (NRC, 2011). The committee of National Research Council (NRC, 2011) has listed more than 40 nutrients that are needed for the normal growth of fish and shrimp, and some of those (e.g., methionine, lysine, essential fatty acids) are commonly limited in fish and shrimp feed. Feed quality therefore means beyond proximate nutrients, and is directly related to how best the feed meets needs for all the nutrients of the target species. Additionally, for a commercial feed producer, the challenge is to produce feed with the same level of nutrients in every batch and at a competitive price. Feed quality overall is influenced by various factors including: (i) knowledge of nutrient requirements of the target species under a given growing condition, (ii) raw material quality and dietary composition, and (iii) feed manufacturing process and pellet quality. While the feed producers can have better control over an optimal feed manufacturing process, the former two factors need special attention. Over the past few years, prices and quality of most feed raw materials have fluctuated extensively. Selecting raw materials based on not only nutrient content and price, but also their variation, play a key role in controlling feed quality and reducing feed cost. The objectives of this chapter are to provide an overview of different sources of the nutrient database for raw materials, describe nutrient variability, and discuss common measures to control nutrient variability and optimize feed quality.

S. Nates (Ed): Aquafeed Formulation. DOI: http://dx.doi.org/10.1016/B978-0-12-800873-7.00006-3

6.2   Sources of nutrient database

Nutritionists obtain information about the nutrient content of raw materials from external sources, such as published books and reports, web sites (e.g., www.feedipeida.org), or from internal laboratory analysis (e.g., wet chemistry analysis, near infrared reflectance spectroscopy (NIRS)). Among the nutrients, protein is the most expensive qualitative nutrient which is dictated by the amino acid profile of the ingredients. It is critical that the diet meets the animal's needs for amino acids in order to avoid performance reduction. At the same time, it is equally important not to waste amino acids with over-formulation which otherwise will simply increase feed cost. Thus, it is challenging for the nutritionist to optimally formulate, with no excess or deficiency, which requires updated and accurate information on the amino acid profile of raw materials used for manufacturing feed.

6.2.1   Book values

Published book values are the first reference for proximate nutrients, amino acid profile, and other nutrients of a raw material. Commonly referred sources include FAO reports, NRC (2011), Evonik publications (e.g., AMINODat 4.0, 2010; AMINODat Aqua 1.0, 2013; annual crop reports), Feedipedia (online database), and other regional publications. Although it is easy to use book values, they become quickly outdated, and often can be very far from the actual nutrient content of raw material. Variation of nutrients for some ingredients, especially for animal protein sources, is often huge. Therefore, using one book value every time can lead to over- or under-formulation of diets, causing nutrient wastage or a reduction in performance. Unfortunately not all the book sources provide information on nutrient variation and therefore, nutritionists cannot predict the nutrient variation of formulated feed when relying on those sources. Factors such as the number of samples analyzed, analytical technique used, proficiency of the laboratory, origin of samples, date analyzed, etc. determine the robustness of data but those information are usually not provided. Thus, caution must be exercised while using such book values.

6.2.2   Regression equation

Prediction equations have been published to determine amino acid levels based on analyzed proximate nutrients. The prediction accuracy of a regression equation is indicated by R^2 value ranging from 0 to 1, with 0 indicating no relationship between the two variables and 1 indicating 100% relationship, that is, 100% variation can be explained by independent variables. A value of 0.50 is considered to be the minimum R^2 requirement for accepting the values from the prediction equation (Redshaw, 2011). AMINODat 4.0 (Evonik, 2010) database provides a regression equation developed for certain ingredients for individual essential amino acids (EAA), based on crude protein levels. Values obtained from this method are expected to be better than book values, as the amino acid profile is matched close to the analyzed crude

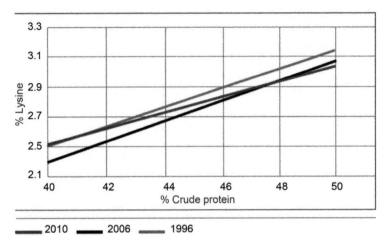

Figure 6.1 Relationship between crude protein and % Lys of soybean meal (SBM) samples based on regression equations published in 1996, 2006, and 2010 (AMINODat editions, Evonik).

protein levels. However, a regression equation cannot produce better values than a wet chemistry method, and the robustness of this method depends on the data behind the prediction equation. Additionally, a regression equation could be quickly outdated because of changing genetics and processing methods of raw material. For example, Figure 6.1 compares the regression lines for the Lys content in the soybean meal (SBM) from 1996, 2006, and 2010 (Evonik AMINODat editions). This shows clearly that levels of this amino acid had decreased from 1996 to 2006 for the same level of CP. In 2010, SBM with lower CP (<46% CP) had higher Lys content compared with SBM in 2006, while SBM with higher CP (>48%) had lower levels of Lys content. This demonstrates the importance of updating regression equations to take account of changing raw material values over time. Feed producers who analyze a large number of samples every year can develop their own regression equations.

6.2.3 Wet chemistry

Most feed mills have an analytical laboratory as a part of their quality control or assurance program and have the capacity to analyze proximate nutrients. Laboratories typically follow standard procedures, such as those described by Association of the Official Analytical Chemists, International (AOAC, International), American Oil Chemist Society (AOCS), and American Association of Cereal Chemists (AACC) for different nutrients and antinutritional factors. While this can be a golden standard for a feed mill, the quality of those data depends on the method adopted in analysis and the quality assurance program adopted by the laboratory. Quality assurance procedures to be practiced by the feed analysis laboratory can be referred from FAO manual (FAO, 2011) which is based on ISO 17025:2005.

As a part of a quality control program, a feed mill should strictly follow procedures for validating the data for their accuracy and precision. Accuracy refers to the closeness of a measured value to a true or standard value, whereas precision indicates how close a group of analyzed values are to each other. Precision therefore refers to repeatability or reproducibility of a set of analyzed values and is determined by calculating the standard deviation or coefficient of variation. For testing accuracy, labs typically analyze pure standards for which the value is already known, and the accuracy is calculated based on % recovery from the actual value. Additionally, ring test (interlaboratory test) is also performed in order to cross-check the own values with those produced in a standard laboratory for the same sample. For example, external laboratories such as AAFCO (the Association of American Feed Control Officials, USA), BIPEA (Bureau Interprofessionnel d'Etudes Analytiques, France), and VDLUFA (Association of German Agricultural Analytic and Research Institutes, Germany) can be used for performing ring tests.

The downside of wet chemistry is that it is the most time-consuming and expensive method to obtain information about the nutrient profile of raw material and feed. Additionally, it requires highly trained laboratory technicians. Although feed mills have the capacity to analyze proximate nutrients based on wet chemistry, only very few are able to analyze amino acids. The other limiting factor is analytical laboratories can handle only a limited number of samples per day, which is a serious setback for a big feed mill that requires hundreds of samples to be processed every day.

6.2.4 Near infrared reflectance spectroscopy

NIRS is now established worldwide as a rapid and reliable method to predict the concentrations of various nutrients in feed ingredients and sometimes, even in finished feed. Traditionally, NIRS analysis of raw materials has been used to rapidly determine proximate nutrients such as moisture, protein, starch, fat, and fiber content, and thereby enable flexibility of decision making. Since the mid-1990s, NIRS has been used to also determine the amino acid content of raw material. The advantages of NIRS are: (i) it does not require any chemical reagents, (ii) results can be obtained within minutes, and (iii) the results are not limited to one or two nutrients but could be several depending on the available calibrations. Overall, NIRS analysis is often very economical and fast, as well as more precise compared with the conventional wet chemistry methods. For routine proximate analysis, the cost by NIRS is only about one-third of the cost by wet chemistry.

6.2.4.1 Principles of NIRS

The basis behind NIRS is that different chemical bonds (O–H, C–H, N–H) within the various organic components will absorb and reflect light to different degrees. For amino acid predictions in raw materials we refer to a NIR spectral region of 1,100–2,500 nm. In order to use the pattern of NIR absorbance to measure the concentration of dietary components, multiple regression equations must be developed from laboratory analysis (wet chemistry) of a representative set of samples. This mathematical

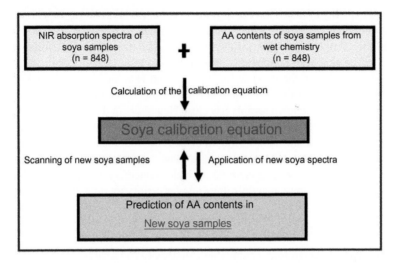

Figure 6.2 Development and application of the soya calibration equation.

process is called chemometrics wherein mathematical equations developed are referred to as "prediction models" or "calibrations." As an example, Figure 6.2 shows the procedure that was used in the Evonik laboratory to establish the calibration for soya (SBM and full fat) analysis. In this example, more than 800 unique samples were used to establish the regression equation which could then be used to predict, using NIRS, the amino acid content of new soya samples. Thus, effective NIRS predictions depend heavily on the accuracy of the calibration equation.

6.2.4.2 How to get an accurate calibration equation

The accuracy of NIRS predictions depends heavily on the accuracy of the calibration equation. Therefore, the performance of a calibration can never be better than the performance of the reference data (analyzed by wet chemical analysis) on which it is produced. To build a solid calibration for a raw material, the reference samples should cover a wide range of nutrient levels, obtained from diverse genetic origins of different locations, grown under various conditions and having undergone various storage and processing methods. The more the samples are used for calibration development, the stronger is the prediction accuracy and robustness. Ideally, a minimum of 100 unique samples should be used for calibrations. The reference lab data are the wet chemistry data upon which the calibrations are created. It is important to remember that the reference data should be a fair and representative subset of the population for which the data calibration will be used. If the reference data are more homogeneous than the entire population or if it is substantially more heterogeneous it is likely that the quality of the calibrations will suffer. All spectra and reference figures must be checked, and the final calibration should be checked by independent samples. These steps must be repeated on a regular basis to keep the calibrations updated.

Once the quality of the calibration is established, the Global-H and Neighborhood-H statistics must be monitored when analyzing samples. These two statistics indicate whether the analyzed samples fall within the calibration data. A Global-H value below 3 indicates that the predicted value of a given sample is within or is similar to the samples existing in the population. If the Global-H value is more than 3, then the sample is an outlier. On the other hand, a Neighborhood-H value <0.60 indicates that the spectrum of the unknown sample has close Neighbors in the spectral population or is indicative that the prediction is very robust. In instances wherein GH is below 3 but NH is more than 0.60, the sample should be sent for wet chemical analysis so that it can become part of future calibration updates. Raw material quality changes due to differences in genetics, processing methods, climate change, among others that influence the nutrient content in raw materials, thus regular calibration updates are necessary.

6.2.4.3   Benefits

The advent of this technology enabled nutritionists to run a large number of samples every day as a part of the quality control program and use the information to update their nutrient matrices almost in real time. By formulating to the actual nutrient content of raw material, nutritionists can increase the confidence in closely meeting nutrient requirements of the target species, while minimizing or removing safety margin. Additionally, immediate analysis of incoming raw material is highly critical for a purchasing department, as CP is not always proportional to the level of amino acids such as Met and Lys, and therefore paying for CP is not always economical. This capability is most valuable to feed millers when dealing with ingredients that routinely vary in their nutrient concentrations.

Among the four sources described, wet chemistry is the best method to obtain the actual nutrient profile of a raw material. However, because of the practical difficulties associated with the wet chemistry method, NIRS is the most practical approach to obtain information about proximate nutrients and amino acids.

6.3   Nutrient levels and variability in commonly used raw materials

Nutrient levels of raw material can vary over time and also by region or supplier. Animal protein sources are generally more variable than plant protein sources. Variation of nutrients in raw materials if not taken into account in the feed formulation can result in either an excess or deficiency of dietary nutrients. Nutrient variation in raw materials ultimately increases the variation of nutrients in the final feed. Increased CV for dietary protein resulting in reduced body weight gain and feed conversion efficiency were documented in broiler chicken (Duncan, 1988). Therefore, it is critical to understand and manage nutrient variation in raw materials. Variation of nutrients in a given raw material is caused by differences in various factors which can be classified into three main categories including: (i) factors involved prior to harvesting or manufacturing of raw materials (e.g., genetics, season or environmental condition under which the plant or animal is grown, soil fertility (in case of plants), food, or feed consumed (in case of animals)), (ii) factors involved during the manufacturing process of raw materials (e.g., temperature, pressure, differences in proportions of different parts used for producing raw materials), and (iii) factors involved post manufacturing of raw materials (e.g., sampling procedures, analytical method adopted for estimating nutrients). Apart from these, variation can also come from unknown factors (e.g., human error in sampling or analyzing) which cannot be avoided (St-Pierre and Weiss, 2008). Understanding potential sources of variation in raw material helps determine which raw material to use and how much to include. In this section, nutrient profile and variations of two commonly used animal sources (fish meal and MBM) and two plant protein sources (SBM and rapeseed meal, RSM) are discussed. Samples of these raw materials were collected during 2010, 2012, and 2014 from the Asia south region but originated from different countries. Samples were analyzed for CP and amino acid levels largely by NIRS (>98%), but also the wet chemistry method. Proximate nutrient levels were reported only for 2014, and were analyzed by NIRS. All the values are reported at 91% dry matter basis for animal protein sources and at 88% dry matter basis for plant protein sources.

6.3.1 Fish meal

Fish meal has excellent nutritional properties, with a high content of digestible protein and balanced EAA. Moreover, fish meal is also rich in other nutrients such as essential fatty acids, phospholipids, cholesterol, minerals, and certain vitamins (Tacon et al., 2009). Table 6.1 provides the CP and amino acid levels of fish meal used in the Asia south region over a number of years. More than 80% of samples originated from the India, Indonesia, Philippines, Thailand, and Vietnam with crude protein levels ranging roughly from 40% to 75%. Fish meal processed in recent years (2012 and 2014) contained higher levels of CP and EAA compared with those produced in 2010. Protein and amino acid content of fish meal varied widely, with CV recorded to be 7–10% for CP and >10% for amino acids.

Variations in fish meal are largely attributed to species and seasonal differences with fish meal production. Differences in the amino acid profile of fish meal owing to differences in fish species are shown in Table 6.2. Here, although herring fish meal contained 3.90% higher CP relative to menhaden, its Lys level is 0.31% lower, indicating that CP level does not indicate the same amino acid profile for fish meal of different species. Sometimes, seafood waste is also used in the production of fish meal, which further exacerbates the variation in the nutrient levels of fish meal.

The relationship between crude protein and Met, Met + Cys, or Lys level (expressed in % CP) of fish meal samples analyzed in 2014 is illustrated in Figure 6.3. Amino acid content is often assumed to be directly proportional to the content of CP which ideally should give a zero slope between CP and any amino acid expressed in % CP. A positive slope ($R^2 = 0.31–0.34$) for Met, Met + Cys, and Lys, however, indicates that as the CP content increased, the content of these amino acids increased at a higher rate. Therefore, in this example, if fish meal is priced merely based on CP, fish meal with high CP would be more valuable than those with low CP. For example,

Table 6.1 Crude protein/EAA content of fish meal and their variations over years

Year	n	CP	MET	Met + Cys	LYS	THR	TRP	ARG	ILE	LEU	VAL	HIS	PHE
Mean %													
2014	3,906	59.70	1.51	2.01	4.10	2.34	0.58	3.38	2.26	4.02	2.70	1.50	2.24
2012	2,429	59.53	1.52	2.00	4.11	2.34	0.57	3.38	2.22	4.08	2.65	1.46	2.14
2010	1,571	58.55	1.46	1.97	4.00	2.30	0.57	3.29	2.21	3.91	2.65	1.50	2.17
CV %													
2014	3,906	7.76	14.52	13.61	16.49	11.00	20.04	8.54	15.08	13.27	12.95	28.84	12.23
2012	2,429	8.78	14.27	14.56	15.61	11.26	21.86	9.63	15.78	13.48	14.77	31.51	15.25
2010	1,571	9.94	18.70	16.52	19.74	14.19	21.89	13.98	16.76	15.93	14.81	38.12	14.70

n = number of samples.

Table 6.2 Crude protein/EAA content of different species fish meal

Fish species	n	CP	MET	Met + Cys	LYS	THR	TRP	ARG	ILE	LEU	VAL	HIS	PHE
Mean %													
Herring	2	69.80	1.85	2.42	4.79	2.83	0.68	4.57	2.70	4.74	3.27	1.62	2.48
Menhaden	2	65.90	1.84	2.42	5.10	2.75	0.77	3.74	2.89	4.89	3.34	1.84	2.70
Sardine	1	65.00	1.70	2.10	4.60	2.50	0.73	3.70	2.72	4.54	3.15	1.99	2.43

Source: AMINODat Aqua.
n = number of samples.

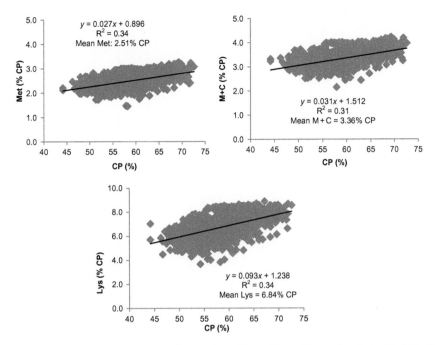

Figure 6.3 Relationship between crude protein and Met, Met + Cys, or Lys level (% CP) of fish meal samples analyzed in 2014.

mean Met content was estimated to be 2.51% CP which indicates 1.38% Met for 55% CP fish meal and 1.63% Met for 65% CP fish meal. However, the actual value was found to be 1.30% Met for 55% fish meal and 1.73% Met for 65% CP fish meal. Data overall suggest that fish meal quality is highly variable and CP is not a reliable indicator of amino acid profile in fish meal.

6.3.2 Meat and bone meal

Among animal protein sources, MBM is a commonly used rendered animal protein source in fish feed because of its competitive price relative to fish meal. Its protein levels are generally >40% but can go up to 70%, and it is a good source of amino acids. However, certain amino acids such as Lys and Met are generally lower than those found in fish meal, with the Met to Lys ratio found to be lower in MBM than in fish meal (27 vs. 37) (AMINODat 1.0 Aqua). Additionally, it has a very high ash content, and low digestibility values for proteins and amino acids limits their inclusion levels (Li et al., 2008). Analytical data show that MBM received during 2014 contain higher levels of amino acids despite being slightly low in CP level compared with MBM received in 2012 (Table 6.3). MBM samples received in 2010 contained slightly higher CP and amino acid levels relative to those received in the later years. Similar to fish meal, MBM contained high variation in the amino acid profile with variation levels recorded to be >10%. However, over the years, the variation has

Table 6.3 Crude protein/EAA content of MBM and their variations over years

Year	n	CP	MET	Met + Cys	LYS	THR	TRP	ARG	ILE	LEU	VAL	HIS	PHE
Mean %													
2014	11,184	48.29	0.64	1.10	2.46	1.50	0.29	3.33	1.32	2.88	2.02	0.90	1.60
2012	2,540	48.82	0.62	1.06	2.41	1.49	0.28	3.32	1.30	2.81	2.01	0.84	1.59
2010	3,378	50.08	0.67	1.12	2.46	1.54	0.30	3.34	1.39	2.89	2.12	0.89	1.69
CV %													
2014	11,184	6.78	13.77	16.57	12.11	12.57	22.19	6.65	14.44	13.79	13.35	20.01	13.80
2012	2,540	7.41	18.30	19.43	13.76	14.67	25.42	7.98	17.51	14.99	13.97	21.92	14.42
2010	3,378	9.08	22.08	22.95	16.15	18.13	33.35	7.80	22.43	17.75	16.65	23.66	16.74

n = number of samples.

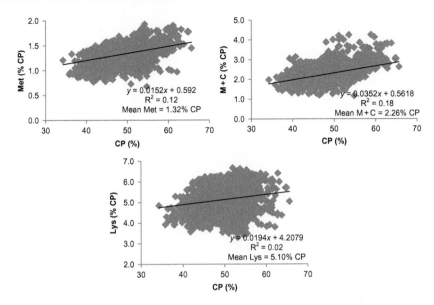

Figure 6.4 Relationship between CP and EAA (Met, Met + Cys, Lys) contents of MBM samples analyzed in 2014.

declined for all the amino acids, suggesting improvements in the consistency of MBM production. Variation in the nutrient profile of MBM is quite common and was also observed by other authors (Parsons et al., 1997; Hendriks et al., 2002). High nutrient variation in MBM is largely due to differences in the proportion of bones and meat, and the type of mammalian species (e.g., cattle, pig, and poultry) used.

The relationship between CP and EAA (expressed as % CP) of MBM samples analyzed during 2014 showed a positive slope for Met, Met + Cys, and Lys (Figure 6.4), indicating that at high CP levels these amino acid levels are proportionately higher than those found at a low CP level. However, the variation is still high. For example, although the average Lys level was found to be 5.10% CP, it varied from 3.30% to 6.75% CP. Therefore, CP is not a good indicator of MBM quality, and assuming one book value for MBM involves high risk.

6.3.3 Soybean meal

Soybean meal is the most commonly used alternative protein source to fish meal in fish and shrimp diets, because of its highly digestible protein and amino acid profiles. However, SBM is low in Met and high inclusion of SBM in the diet requires Met supplementation. The price of SBM has been considerably lower than that of fish meal which makes SBM based diet more economical despite the added cost from supplemental amino acids. Compared to animal protein meals, SBM contains more consistent protein and amino acid levels with their CV values recorded to be <5% over the years (Table 6.4).

Table 6.4 Crude protein/AA content of SBM and their variations over years

Year	n	CP	MET	Met + Cys	LYS	THR	TRP	ARG	ILE	LEU	VAL	HIS	PHE
Mean %													
2014	28,155	47.28	0.64	1.34	2.90	1.84	0.63	3.46	2.12	3.57	2.22	1.26	2.38
2012	20,382	47.92	0.64	1.33	2.93	1.85	0.63	3.52	2.14	3.61	2.24	1.27	2.42
2010	8,707	47.87	0.63	1.32	2.93	1.84	0.63	3.55	2.16	3.63	2.25	1.27	2.44
CV %													
2014	28,155	3.33	4.21	4.84	3.72	3.21	3.72	4.06	3.61	3.35	3.29	3.20	3.66
2012	20,382	3.53	4.53	4.93	4.12	3.61	4.11	4.00	3.84	3.78	3.71	3.73	3.95
2010	8,707	3.16	3.96	5.00	3.33	3.05	3.38	3.63	3.30	3.21	3.18	2.99	3.36

n = number of samples.

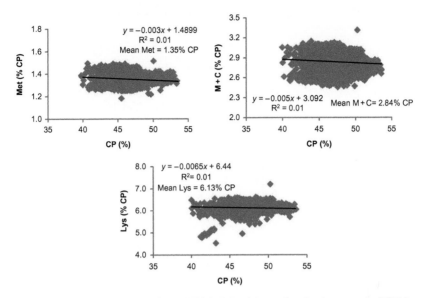

Figure 6.5 Relationship between CP and EAA (Met, Met + Cys, Lys) contents of SBM analyzed during 2014.

The relationship between CP and EAA such as Met, M+C or Lys (% CP) for the year 2014 showed a slightly negative slope with R^2 values at 0.01 (Figure 6.5). Zero R^2 indicates no relationship between the two variables, and a constant value can be assumed. However, the variation in SBM cannot be still ignored. For example, the Met level was found to be 1.35% of the CP level, but it varied widely from 1.25% to 1.45% CP. This means that the Met level in SBM, for example, with 48% CP can vary from 0.60% to 0.70% with its mean value being 0.65%. This variation is partly caused by mixing SBM samples of different origin. Therefore, feed producers still need to pay attention to the variation in SBM although the CV value is <5%.

6.3.4 Rapeseed meal

After SBM, RSM is the second most commonly used plant protein source in aquafeeds. Its protein varies generally from 25% to 40%. It is a good source of amino acids and contains higher levels of Met and Met + Cys relative to SBM or MBM. Additionally, its price is often half that of SBM, and much lower than that of fish meal. Analyzing the data over the years shows that the mean CP level is consistently about 37% and the amino acid profile did not vary much over the years (Table 6.5). It's CV was detected to be <5% for CP and majority of the EAA. However, this CV cannot be ignored and still gives a wide range of amino acid level as illustrated in Figure 6.6. The relationship of Met, Met + Cys, or Lys versus CP (Figure 6.6) showed a negative slope with R^2 to be around 0.30%, indicating that the concentration of

Table 6.5 Crude protein/EAA content of RSM and their variations over years

Year	n	CP	MET	Met + Cys	LYS	THR	TRP	ARG	ILE	LEU	VAL	HIS	PHE
Mean %													
2014	6,974	36.49	0.69	1.61	1.81	1.48	0.50	2.25	1.40	2.45	1.77	1.00	1.43
2012	2,179	36.74	0.69	1.61	1.78	1.49	0.50	2.27	1.40	2.47	1.79	1.00	1.45
2010	832	36.81	0.70	1.60	1.80	1.52	0.50	2.22	1.42	2.49	1.82	0.98	1.45
CV %													
2014	6,974	3.88	4.01	4.45	7.63	4.05	4.62	6.31	4.07	3.59	4.17	4.68	4.16
2012	2,179	4.37	4.87	5.77	8.50	4.00	5.42	7.97	4.38	4.05	4.06	6.46	4.29
2010	832	3.21	2.86	4.46	4.48	2.57	4.49	6.21	3.41	2.64	2.82	5.10	3.39

n = number of samples.

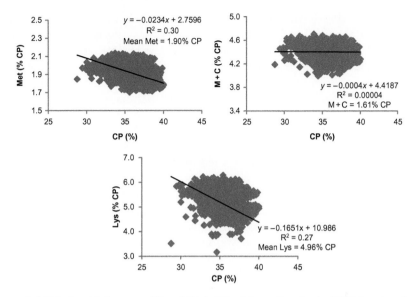

Figure 6.6 Relationship between CP and EAA (Met, Met + Cys, Lys) of RSM analyzed during 2014.

these amino acids is proportionately lower at a higher CP level relative to their low CP. In this case, if the price of RSM is paid in proportion to the CP level, it would be a loss to the purchaser. Therefore, attention has to be paid to the amino acid profile of raw material not only when formulating the diet, but also when making decisions on purchasing.

6.4 Impact of heat damage on the amino acid level and their variability

Heat treatment is commonly applied during processing of protein raw materials. While heat processing undoubtedly has many advantages (e.g., improves oil extraction process; destroys antinutritional factors; kills microbes), excess heat can also damage protein quality. Excess heat treatment destroys certain amino acids (e.g., Lys, Arg, Cys) as well as reduces their availability, as proteins undergo protein-reducing carbohydrate reactions and protein–protein interactions (Papadopoulos, 1989; Glengross et al., 2007). Mostly commonly affected raw materials include SBM, Distillers Dried Grains with Solubles (DDGS), RSM, and sunflower meal. Fontaine et al. (2007) demonstrated that heat treatment reduces the Lys (total and reactive) content of both SBM and corn DDGS. Similarly, Amezcua and Parsons (2007) recorded declines in the Lys content of DDGS owing to different types of thermal treatment (autoclaving, oven drying, autoclaving + oven drying), with autoclaving showing the greatest declines in Lys level (by more than 50%). In another investigation, steam treatment of SBM at 135°C for 20 min reduced their Lys level by 10%, Cys by 15%, and Arg by 6% while extending the heat treatment to 40 min have further reduced the amino acid levels by 8%, 10%, and 5%, respectively (Helmbrecht et al., 2010). Jensen et al. (1995) showed that heat treating RSM at 100°C for 30 and 100 min reduced total Lys content from 5.93 g to 5.72 g and 4.91 g per 16 g N, respectively. Furthermore, the study has also recorded reductions in Cys level (by 12% after 120 min of toasting) as well as in mono-, di-, and oligosaccharide contents of RSM. Similar destruction effects on amino acids due to heat treatment were recorded in other oilseed meals, for example, peanut meal, sunflower meal, and cottonseed meal (Fernandez et al., 1994; Zhang and Parsons, 1994, 1996; Evonik, 2012). Overall, various studies have demonstrated that excess heat treatment would diminish the nutritional quality of various raw materials by reducing the total content of certain amino acids. Furthermore, studies in poultry have shown that although only Lys, Arg, and Cys content decline when examining the total content of amino acids in raw materials, digestibility values decline for all the amino acids (e.g., Fernandez et al., 1994; Zhang and Parsons, 1994, 1996; Helmbrecht et al., 2010). Such studies in fish and shrimp are lacking, and are warranted to determine the effects of excess heat processing on the digestibility of amino acids. Since thermal processing of the same raw material can differ among suppliers, the protein quality of raw material among suppliers can also vary. Therefore, feed producers need to be careful in picking the right supplier who would supply the optimally processed raw material. For RSM analyzed for the years 2012 and 2014, high CV values recorded for Lys (>7%) and Arg (>6%) compared with the majority of other amino acids (<5%) is likely because of the overprocessing of some samples (Table 6.5). Analyzing the amino acid content via NIRS would allow us to capture the differences in the actual content of amino acids among suppliers.

6.5 Proximate nutrients of raw material

Knowledge on proximate nutrients of raw material helps nutritionists to draw a primary decision on the quality of raw material and how they can influence the final feed quality standards that need to be met. Most feed mills have the facility to determine proximate nutrients either by wet chemistry or by NIRS, the later method allowing feed millers to handle large number of raw material samples. As discussed above, proximate nutrients often give a base for the amino acid profile of raw material based on regression equations. Proximate nutrients analyzed via NIRS (AMINOProx®) calibrated for Weende analysis are summarized in Table 6.6. Additionally, phosphorus and gross energy values are also provided. Results indicate that unlike for CP, crude lipid (ether extract) values varied widely for plant protein sources relative to animal protein sources. This is likely because, for the oilseed meals, the oil extraction process varies widely, which increases variation in the oil content and thus, the energy level of those raw materials.

Table 6.6 Proximate nutrients (NIRS-based Weende analysis), phosphorous, gross energy, and their variations in the commonly used raw materials

Raw material	CP (%)	EE (%)	Fiber (%)	Ash (%)	Starch (%)	Total phosphorous (%)	Phytic phosphorous (%)	Gross energy (kcal/kg)
Fish meal (n = 1,399 for total phosphorous; n = 1,405 for others)								
Mean %	59.47	8.85	–	21.75	–	2.88	–	4,582
CV %	7.13	19.02	–	15.03	–	19.68	–	5.12
MBM (n = 5,764 for total phosphorous; n = 5,771 for others)								
Mean	48.57	10.43	–	30.66	–	5.06	–	3,930
CV	6.60	17.11	–	13.49	–	17.35	–	6.95
SBM (n = 131,371 for all)								
Mean	47.40	1.82	4.22	6.67	4.91	0.61	0.37	4,622
CV	3.37	23.93	21.37	4.50	7.58	10.68	10.68	1.30
RSM (n = 2,031 for total and phytic phosphorus; n = 2,033 for others)								
Mean	37.35	2.77	10.23	7.08	–	0.86	0.52	4,446
CV	2.69	28.02	5.49	4.23	–	6.64	6.64	1.30

CP, crude protein; EE, ether extract. Gross energy (kcal/kg) = $(4{,}143 + 56EE\% + 15CP\% - 44Ash\%) \times 0.0041868 \times 239$ (Ewan, 1989). n = number of samples.

6.6   Managing nutrient variation

Variation of nutrients in the incoming raw material need to be properly handled in a feed mill in order to ensure target level of nutrients in the final feed. Some of the commonly practiced measures include:

6.6.1   Safety margin to nutrient specifications

This is the most common measure used by nutritionists when they do not have knowledge about the actual nutrient profiles of raw materials used in the formulation. For example, when using amino acid values of ingredients published in a standard book such as NRC (2011), nutritionists increase the target amino acid levels by 5–10% in order to avoid risks associated with lower than the assumed values for the ingredients. This approach increases diet cost and also leads to nutrient waste if the diet is over-formulated. In Table 6.7 diet cost of USD 446.94/tonne is shown for diet 1 formulated to meet NRC amino acid requirements for Nile tilapia. However, increasing the amino acid recommendations by 5% as a safety margin increased the cost of diet (diet 2) by 8.85% (USD 486.26/tonne). Furthermore, it cannot still be guaranteed, if the assumed safety margin is always enough to meet the requirements of target species. At the same time it is also possible that the diet was over-formulated, if the protein sources contain amino acid levels higher than those expected.

6.6.2   Safety margin to ingredient specifications

The benefit of using an AMINODat-type database over a NRC table is that nutritionists can use amino acid variability of individual ingredients to calculate possible amino acid variability of nutrients in the diet. Once the degree of variability is known, the safety margin needed for an individual amino acid can be determined. For instance, in the example given in Table 6.7, Met in diet 1 contained a standard deviation of 0.068% or a coefficient of variation of 9.7% which is quite high. Adding 1 standard deviation as a safety margin to the Met specifications would change the requirement value from 0.70% to 0.768%. The benefit of this increased requirement value is that at least 83% of the manufactured feed would meet the target specifications of 0.70%. This also indicates that a 5% assumed safety margin would not be enough most of the time because of high nutrient variation.

In an alternative approach, having a closer look at the contribution of individual ingredients (Figure 6.7) on the Met variability reveals that although fish meal is included at only 5% in the diet, 89% of the Met variation is contributed by this ingredient. Also, SBM although included at 42%, contributes only 6% of the total variation. In this case, a safety margin can be included to fish meal by subtracting 1 standard deviation to the amino acid profile of fish meal, or, if possible, highly variable fish meal (CV = 10% for Met in this example) should be avoided in the formulation. Replacing 5% fish meal with SMB level reduces the Met level 0.04% below the specifications which can be overcome by increasing supplemental DL-methionine. This modification minimizes the Met CV to 3.4%, well within the safe limit. Additionally, even adding 1 standard deviation would increase the specification only 0.02%. Therefore, knowing the variation of amino acids for individual ingredients

Table 6.7 **Diets formulated to meet amino acid requirements of Nile tilapia with or without safety margin**

Ingredients	Diet 1 (%)	Diet 2 (%)
SBM, 48% CP	42.36	50.28
Fish meal, 65% CP	5.00	5.00
Cottonseed meal	7.00	7.00
Corn	20.81	28.32
Rice bran, deoiled	22.55	5.27
Monocalcium phosphate	0.52	2.39
Fish oil	1.05	1.00
Vit-Min Premix	0.50	0.50
DL-methionine	0.21	0.24
Price, USD/tonne	446.94	486.26

Nutrients (%)	Requirements (NRC, 2011)	Actual	Requirements with 5% safety margin	Actual
DM		88.41		88.61
GE (kcal/kg)		4,500		4,500
CP		31.63		33.21
CF		6.00		4.85
EE		4.04		4.09
Ash		7.49		8.11
Lys	1.60	1.78	1.68	1.90
Met	0.70	0.70	0.74	0.74
M+C	1.00	1.17	1.05	1.22
Thr	0.95	1.18	1.00	1.24
Trp	0.30	0.39	0.32	0.41
Arg	1.20	2.34	1.26	2.43
Ile	1.07	1.29	1.12	1.38
Leu	1.60	2.36	1.68	2.51
Val	1.26	1.49	1.32	1.54
His	0.83	0.83	0.87	0.87
Phe	1.04	1.54	1.09	1.63

would help increasing the consistency and meeting the nutrient specifications with a reduced cost. However, CV values reported in the book value will not always represent the nutrient variation of raw materials used in a feed mill.

6.6.3 Routine analysis of raw materials

The best approach to control variation of nutrients in feed is to routinely analyze raw materials in order to understand their actual nutrient levels and variations from different suppliers. If a feed mill can analyze only proximate nutrients, they can use a prediction equation to get information about the amino acid profile. This approach can minimize the safety margin for amino acids but does not completely eliminate it as the

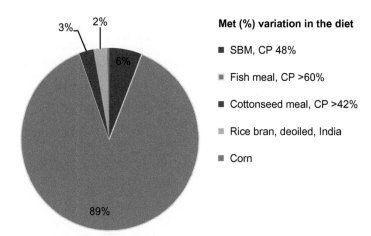

Figure 6.7 Contribution of different ingredients on the methionine variability in the diet.

relationship between proximate nutrients and amino acids is not constant, and varies by time and supplier. A feed mill that has the capacity to analyze amino acids based on NIRS can have better control over raw material quality. This technology enables customers' decisions to be made, through rapid predictions, as to how to best use the raw materials in optimizing feed formulations. However, feed producers have to ensure proper sampling techniques and adequate number of samples when collecting samples for raw material analysis.

6.6.4   Do not mix raw material of different origins or suppliers

Raw material received from different suppliers stored in one common storage bin or silos can be a major source of nutrient variation. MBM samples (global data) analyzed in 2014 were detected to have high variations for amino acid levels (CV >10%) (Table 6.8). When the MBM samples were sorted based on country of origin, the amino acid profile and the variability differed. Samples originating from India contained lower levels of CP and amino acid levels. MBM from Italy contained 3% higher levels of protein, but, lower levels of Met and Lys, compared with MBM from Australia or the United States. This shows that a high CP level can sometimes be deceiving if attention is not paid to the amino acid profile. Between Australia and the United States, both the MBM sources contained similar CP levels, but MBM from Australia showed a higher Lys level than the US MBM (2.60% vs. 2.46%). However, to account for variation, subtracting 1% standard deviation makes Australian MBM no different or slightly inferior to the US MBM with regard to lysine (2.25% vs. 2.26%). Therefore, it is also important to identify a supplier that gives consistent quality with less variation than the one that supplies high content of nutrients but with higher variation. This analysis clearly shows the need of separating raw material based on supplier or quality. Routine analysis of raw material can help minimize nutrient variations and improve feed quality.

In a similar analysis, for SBM samples analyzed in 2014, more than 80% was contributed by Argentina, Brazil, India, and the United States (Table 6.9). Compared

Table 6.8 Crude protein/AA content of MBM and their variations by country of origin analyzed in 2014

Country	n	CP	MET	Met + Cys	LYS	THR	TRP	ARG	ILE	LEU	VAL	HIS	PHE
Mean %													
Global	11,184	48.29	0.64	1.10	2.46	1.50	0.29	3.33	1.32	2.88	2.02	0.90	1.60
Australia	2,402	48.23	0.65	1.08	2.60	1.50	0.30	3.32	1.26	2.97	2.08	0.96	1.64
India	821	45.58	0.52	0.95	2.15	1.28	0.20	3.24	1.12	2.43	1.79	0.68	1.38
Indonesia	889	47.07	0.63	1.09	2.47	1.48	0.29	3.22	1.29	2.88	2.01	0.92	1.60
Italy	731	51.26	0.63	1.12	2.32	1.57	0.29	3.54	1.42	2.95	2.06	0.90	1.68
United States	4,128	47.87	0.64	1.08	2.46	1.49	0.29	3.30	1.32	2.84	1.98	0.88	1.56
CV %													
Global	11,184	6.78	13.77	16.57	12.11	12.57	22.19	6.65	14.44	13.79	13.35	20.01	13.80
Australia	2,402	7.12	12.45	16.02	13.53	13.66	24.34	5.62	11.46	16.91	16.20	23.04	16.70
India	821	8.03	15.61	19.34	12.17	13.55	22.92	8.01	14.56	12.62	12.21	17.77	12.36
Indonesia	889	5.59	11.23	12.33	10.99	10.40	20.88	6.28	10.45	12.70	12.01	19.81	12.64
Italy	731	4.67	10.05	12.84	9.53	8.54	15.28	5.04	10.32	8.95	9.14	8.86	9.03
United States	4,128	4.64	9.91	11.81	8.27	8.57	16.72	5.31	10.24	9.77	9.44	15.62	9.71

n = number of samples.

Table 6.9 Crude protein/EAA content of SBM and their variations by country of origin

Country	n	CP	MET	Met + Cys	LYS	THR	TRP	ARG	ILE	LEU	VAL	HIS	PHE
Mean %													
Global	28,155	47.28	0.64	1.34	2.90	1.84	0.63	3.46	2.12	3.57	2.22	1.26	2.38
Argentina	8,720	46.47	0.64	1.35	2.85	1.83	0.64	3.35	2.08	3.51	2.20	1.24	2.32
Brazil	3,660	48.90	0.66	1.39	2.97	1.89	0.65	3.58	2.20	3.70	2.30	1.30	2.47
India	9,926	47.25	0.62	1.29	2.90	1.82	0.62	3.53	2.10	3.56	2.19	1.26	2.39
United States	1,055	47.70	0.65	1.38	2.93	1.86	0.64	3.47	2.14	3.60	2.25	1.26	2.39
CV %													
Global	28,155	3.33	4.21	4.84	3.72	3.21	3.72	4.06	3.61	3.35	3.29	3.20	3.66
Argentina	8,720	2.21	2.25	2.70	2.90	1.83	2.10	2.84	2.55	2.24	2.27	2.19	2.74
Brazil	3,660	2.53	2.47	3.07	3.08	2.23	2.74	3.15	2.89	2.64	2.66	2.56	3.07
India	9,926	3.58	4.91	4.90	4.05	3.94	4.00	3.48	3.55	3.54	3.39	3.51	3.18
United States	1,055	1.76	1.71	2.54	2.06	1.60	2.11	2.07	2.11	1.91	1.82	1.79	2.32

n = number of samples.

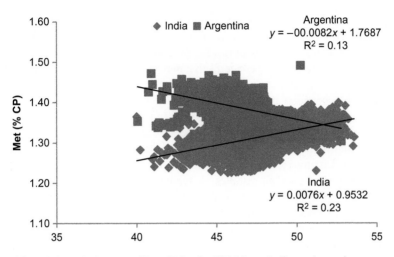

Figure 6.8 Relationship between CP and Met for SBM from India vs. Argentina.

to the global average, SBM originating from Brazil contained higher levels of CP and amino acid levels with lower variation, clearly showing differences to those from other sources. SBM originating from India, although containing a CP level closer to the global average, had levels of many EAA that were lower with higher variability. Plotting Met (% CP) vs. CP (Figure 6.8) for Indian and Argentine SBM revealed that for the same protein level Argentine SBM contains a higher level of Met than Indian SBM indicating that, at the same price, Argentine SBM would be more valuable than that of India. This further bolsters the understanding that the amino acid profile of raw material based on origin of samples would further help decrease variation and increase feed quality.

6.7 Integration of Laboratory Information Management System and formulation

Laboratory Information Management Systems (LIMS) are widely used as the central data backbone of modern analytical laboratories. They are part of the quality control system for feed raw materials and end products built to save time, increase efficiency, and enhance quality and accessibility of analytical results. These are made to provide a central platform to enter results, store the laboratory information, and provide search and data analysis functionality for the benefit of laboratory staff and decision makers.

Vital data for each individual sample should be recorded such as the sample type (e.g., SBM, soybean expeller, or soybean full fat), processing information, country of origin, and supplier information. Supplier information may include the delivery date, the supplier name, and the shipping vessel or delivery truck number. All this information is important when analyzing the accumulated data in the future. Laboratory data are much more than just analytical results. Traceability on who did what, when, and how is important for any future concerns that may arise such as issues on accountability and the like. With all these important pieces of information on a specific sample, the use of LIMS becomes indispensable. From the LIMS, reports and certificates on the analytical values achieved can be produced. In this respect, the system represents and supports quality by structuring lab processes and laboratory work in terms of work flows, and by providing a central data repository on all information connected with a sample. List overviews, like "my open samples" would help the laboratory staff to keep control over all samples. A typical analytical process would include the sampling and the handing over to the lab. In the first step, all samples would be registered in the LIMS. The sample should receive its unique identifier. This may be in the form of a laboratory code consisting of the current year and a continuous number which could be stated on sample bags and bottles in clear text or as a barcode. Calculations based on analytical results might take place in the LIMS automatically, for example, to perform a dry matter standardization of analytical results, taking into consideration the analytically determined dry matter content and the standard dry matter specified for each material (e.g., 88.0% for plant products, 91.0% for products of animal origin). Rules to automatically test and validate analytical results for outliers ("OOS," out of spec) might be included in the LIMS test plans. If a result is found to be untypical or suspicious, based upon reference values or tolerances per material, a retest may be suggested or forced by the LIMS. Certain rules help in receiving reliable and quality-controlled results. It may be defined as the difference between two repetitions of an analysis, such as in instances when there should be <2% relative difference. If it is true that the difference is less, automatically the average of the two repetitions may be used. If the difference is higher, the system may add a third and fourth repetition and may apply additional advanced algorithms for quality control, outlier identification, and averaging as soon as the results for the third and fourth repetitions have been entered. Final control on the integrity of all analytical results and release of results takes place either manually or according to algorithms applied by the system as soon as all analytical results from all requested tests have been entered into the system.

Interpretation of released data could include a statistical summary of results, for example, the minimum value, maximum value, average, standard deviation, and coefficient of variation over a set of samples from one material analyzed during a certain time. The above-mentioned criteria, such as supplier or country of origin, would help grouping and structuring the data while drawing the nutritional value out of the information. From the LIMS system the data can be linked into feed formulation to update a feed matrix manually, by exporting the data to text files and importing them into the feed formulation system, or automatically by invoking the feed formulation system directly. Usually first data are validated, for example, based upon the mentioned criteria min, max, average, and coefficient of variation. After verification analytical results may be transferred to the formulation database for management.

6.8 Summary

Feed quality is influenced largely by raw material quality. Knowledge on the right nutrient profile of raw materials used in a feed mill is very critical for purchasing decisions and feed formulations. Using historical book values and published data of nutrient profile for raw materials can often lead to over- or under-formulations, because of variation of nutrients in the ingredients. Regression equations could be better options than book values, as for example amino acid levels are predicted from the analyzed proximate nutrients. However, they can also become quickly outdated and requires continuous updates. Analyzed values via wet chemistry are the best choice for getting the actual nutrient profile of raw material used, but they are usually expensive and time-consuming. Alternatively, NIRS is the most practical option to obtain information relatively quickly and to handle a large number of samples. Yet, accuracy of NIRS data depends on the accuracy of its calibration equation. As part of a quality assurance program, regular monitoring of data with proper statistical tests and participation in external ring tests are needed to ensure the quality of data.

Raw materials (fish meal, MBM, SBM, RSM) collected from the Asia south region indicate that amino acid levels in the animal protein sources are more variable (CV >10%) than those in the plant protein sources (CV <5%). Analysis indicates that amino acid levels of raw materials are not always directly proportional to their CP level, and therefore, CP is not always a reliable indicator of protein quality. Overprocessing (excess heat treatment) of certain raw materials (e.g., SBM, DDGS, RSM, and sunflower meal) would destroy amino acids (e.g., Lys, Arg, Cys) and thus reduce protein quality and amino acid availability. Mixing heat-damaged raw materials with those of good quality would ultimately increase amino acid variation.

Routinely analyzing the raw materials and implementing proper sampling procedures are the best ways to control raw material variations, as well as to minimize or remove the safety margin in the diet. Variation can be partly controlled by not mixing raw materials of different origins or suppliers. While NIRS can help handle large number of samples, LIMS is useful in managing those data. LIMS, an integral part of modern analytical laboratories, is very useful to store, validate, and transport the data to different end-users (e.g., quality control manager, purchase manager, formulator). Overall, raw material quality needs to be controlled to optimize final feed quality and fish performance.

References

Amezcua, C.M., Parsons, C.M., 2007. Effect of increased heat processing and particle size on phosphorus bioavailability in corn distillers dried grains with solubles. Poult. Sci. 86, 331–337.

Duncan, M.S., 1988. Problems of dealing with raw ingredient variability. Recent Advances in Animal Nutrition. Butterworths, London, UK, pp. 1–11.

Evonik, 2010. AMINODat® 4.0 Evonik Degussa GmbH, p. 566.

Evonik, 2012. Effects of heat treatment of rapeseed meal and sunflower meal on ileal amino acid digestibility in broilers facts and figures. Poultry No. 1599, 1–5.

Ewan, R.C., 1989. Predicting the energy utilization of diets and feed ingredients by pigs. In: van der Honing, Y., Close, W.H. (Eds.), Energy Metabolism, European Association of Animal Production Bulletin No. 43, Pudoc Wageningen, The Netherlands, pp. 271–274.

FAO, 2011. Quality assurance for animal feed analysis laboratories. FAO Animal Production and Health Manual No. 14. Rome, p. 178.

Fernandez, S.R., Zhang, Y., Parsons, C.M., 1994. Effect of overheating on the nutritional quality of cottonseed meal. Poult. Sci. 73, 1563–1571.

Fontaine, J., Zimmer, U., Moughan, P.J., Rutherfurd, S.M., 2007. Effect of heat damage in an autoclave on the reactive lysine contents of soy products and corn distillers dried grains with solubles. Use of the results to check on lysine damage in common qualities of these ingredients. J. Agri. Food Chem. 55, 10737–10743.

Glencross, B.D., Booth, M., Allan, G.L., 2007. A feed is only as good as its ingredients – a review of ingredient evaluation strategies for aquaculture feeds. Aquacult. Nutr. 13, 17–34.

Helmbrecht, A., Redshaw, M.S., Lemme, A., 2010. AMINORED®: an analytical and nutritional concept accounting for heat damage effects in raw materials tested in broilers Aminonews® Special Edition:1–10.

Hendriks, W.H., Butts, C.A., Thomas, D.V., James, K.A.C., More, P.C.A., Verstegen, M.W.A., 2002. Nutritional quality and variation of meat and bone meal. Asian-Aust. J. Anim. Sci. 15 (10), 1507–1516.

Jensen, S.K., Liu, Y.-G., Eggum, B.O., 1995. The effect of heat treatment on glucosinolates and nutritional value of rapeseed meal in rats. Anim. Feed Sci. Technol. 53, 17–28.

Li, M.H., Robinson, E.H., Lim, C.E., 2008. Use of meatpacking by-products in fish diets. In: Lim, C., Webster, C.D., Lee, C. (Eds.), Alternative Protein Sources in Aquaculture Diets. The Haworth Press Taylor and Francis Group, New York, NY, pp. 95–116.

NRC, 2011. Nutrient Requirements of Fish and Shrimp. The National Academies Press, Washington, DC, p. 376.

Papadopoulos, M.C., 1989. Effect of processing on high-protein feedstuffs: a review. Biol. Wastes 29 (2), 123–138.

Parsons, C.M., Castanon, F., Han, Y., 1997. Protein and amino acid quality of meat and bone meal. Poult. Sci. 76, 361–368.

Redshaw, M.S., 2011. Expect the best – the new AMINODat® 4.0 Aminonews® (special issue) 15, 1–4.

St-Pierre, N., Weiss, W.P., 2008. Understanding feed analysis variation and miniming its impact Aminonews® Special Issue 10 (1), 1–12.

Tacon, A.G.J., Metian, M., Hasan, M.R., 2009. Feed ingredients and fertilizers for farmed aquatic animals: sources and composition. FAO Fisheries and Aquaculture Technical Paper. No. 540. Rome, 209 pp.

Zhang, Y., Parsons, C.M., 1994. Effects of overprocessing on the nutritional quality of sunflower meal. Poult. Sci. 73, 436–442.

Zhang, Y., Parsons, C.M., 1996. Effects of overprocessing on the nutritional quality of peanut meal. Poult. Sci. 75, 514–518.

Index

LOA and LNA - pg 65
↳Linolenic
↳Linoleic

Various Notes

) Phospholipids are important for larval + juvenile fish
  ↳ lower Qtys needed for freshwater than marine
  ↳ requirements decrease with age

CPI Antony Rowe
Chippenham, UK
2017-02-22 11:15